Remediation of Plastic and Microplastic Waste

This book provides recent developments and advancements in management of plastic wastes and associated challenges with it. It primarily addresses the issues of plastics that might drastically affect the lifeform in the long run and its prevention by introducing appropriate alternatives and/or finding strategies to mitigate the existing microplastic crisis using suitable approaches. It focusses on efforts on neutralizing and restricting further spread of microplastic pollution, its bioaccumulation and associated human health impacts.

FEATURES:

- Covers technological mitigation of plastics and microplastic wastes along with their remediation in a technical manner.
- Discusses advances in plastic and micro-/nano-plastic pollution and possible pathways of pollution.
- Demonstrates the mitigation measures to minimize such pollution loads, with a special focus on the application of nanotechnology.
- Reviews recycle and value-added products from the waste plastic.
- Focusses on development of alternate clean energy sources.

This book is aimed at researchers and graduate students in environmental and chemical engineering and remediation.

Remediation of Plastic and Microplastic Waste

Edited by
Surajit Mondal, Papita Das, Arnab Mondal,
Subhankar Paul, Jitendra Kumar Pandey and
Tapas K. Das

CRC Press
Taylor & Francis Group
Boca Raton London New York

CRC Press is an imprint of the
Taylor & Francis Group, an **informa** business

Designed cover image: ©Shutterstock Images

First edition published 2024
by CRC Press
2385 NW Executive Center Drive, Suite 320, Boca Raton FL 33431

and by CRC Press
4 Park Square, Milton Park, Abingdon, Oxon, OX14 4RN

CRC Press is an imprint of Taylor & Francis Group, LLC

ISBN: 978-1-032-55559-1 (hbk)
ISBN: 978-1-032-58229-0 (pbk)
ISBN: 978-1-003-44913-3 (ebk)

DOI: 10.1201/9781003449133

Typeset in Times
by Apex CoVantage, LLC

Contents

Chapter 3   Toxic Effect of Food-Borne Microplastics on Human Health 50

*Riashree Mondal, Subarna Bhattacharyya and
Punarbasu Chaudhuri*

Chapter 4   Microplastics in the Atmosphere 60

Moumita Sharma

Chapter 5   Microplastics in the Atmosphere: Identification, Sources
and Transport Pathways.. 71

*Rahul Arya, Jaswant Rathore, Ajay Kumar Mishra and
Arnab Mondal*

Chapter 11

Poushali Chakraborty, Sampad Sarkar, Arkaprava Roy,
Kesang Tamang and Papita Das

Chapter 12

Jesse Joel Thathapudi, Levin Anbu Gomez, Vishruth Vijay,
Vani Chandrapragasam, Ritu Shepherd, Subhankar Paul,
Meng-Jen Lee and Prathap Somu

Uma Sankar Mondal, Anisha Karmakar, Aritri Paul and Subhankar Paul

Editors' Biographies

Dr. Surajit Mondal completed his masters in Energy Systems and Ph.D. on the Renewable Energy domain in 2020. Currently, he is teaching in University of Petroleum and Energy Studies to UG and PG students. He has published more than 32 international research and review articles as an author/co-author. He has published 42 patents and granted eight patents against his name. He has completed two DST- (Govt. of India) funded projects in the field of Energy Systems/Sustainability.

Professor Papita Das earned her Ph.D. in Chemical Engineering from Jadavpur University, India. She is Professor, Department of Chemical Engineering, and Director, School of Advanced Studies in Industrial Pollution Control Engineering, Jadavpur University, Kolkata, India. She is known for her work in water treatment using different novel adsorbents materials. She is also working on biomass-based energy production and polymeric-nanocomposite synthesis and its degradation. She has published more than 175 international journal research articles and reviews and more than 50 book chapters in various SCI and Scopus-indexed journals. She is also ranked in World Ranking of the top 2% Scientists (2020, 2021 and 2022), published by Stanford University, which represents the top 2% the most cited scientists in various discipline. Ranked 432 among 53,348 researchers in the field of Chemical Engineering (2022) based on career-long impact and 171 for the single year 2021 among single year citation of 2021, she ranked 614 among 55,697 researchers in Chemical Engineering (2020) and 534 among 66,189 researchers in the field of Chemical Engineering (2021) based on career-long impact and ranked 217 for single-year 2020 discipline. She has also guided ten Ph.D. students (Completed), two (submitted) and seven (ongoing). She was Editor of two books published by Elsevier and Springer, and she is Editorial Board Member, Editor-in-Chief and Associate Editor of various international journals. She had completed 17 projects funded by govt. agencies and industries.

Mr. Arnab Mondal is currently working as a research project fellow in the field of environmental engineering from CSIR National Environmental Engineering Research Institute (NEERI), Nagpur. He has completed his master's from the Institute of Environment and Sustainable Development (IESD), Banaras Hindu University, Varanasi. He has published several research articles in international journals. He has also been associated with several government-funded projects like Dept. of Water Resources (Ministry of Jal Shakti) and the Ministry of Earth Science.

Professor Subhankar Paul did his Ph.D. in biotechnology from Indian Institute of Technology, Delhi, India. He is basically a chemical engineer and currently teaching in the Department of Biotechnology & Medical Engineering, National Institute of Technology Rourkela, India. He has published more than 70 international research and review articles as an author or co-author. He has collaborated in many foreign laboratories, including University of Sydney (Australia), Indiana University (U.S.A),

Yamagata University and University of Tokyo (Japan). Presently, his research is focused on nanotechnology in environmental remediation. He has already guided many Ph.D. and master's students.

Professor Jitendra Kumar Pandey did his Ph.D. in polymer chemistry at National Chemical Laboratory, Pune. He is known for his exemplary work in the field of water treatment, water purification, alternate energy, biosensors, nano absorbent, etc. He has published more than 120 reviews and research articles, has eight patents and has published four books as well. He has completed more than seven government-funded projects of 3+ Crore INR with national and international collaborations on aligned areas. He has been honoured as a prestigious NASI fellow.

Professor Tapas K. Das holds a B.S. in chemical engineering from Jadavpur University in Kolkata, India, and Ph.D. from Bradford University, U.K. Tapas was a postdoctoral fellow at London's Imperial College of Science and Technology and a visiting scientist at Princeton University. Tapas has been teaching in the School of Engineering at Saint Martin's University since fall 2005. He has wide practical and theoretical experience in various areas, including air toxics and aerosols, advanced industrial wastewater treatment for water reuse, solid waste management and combustion, profitable process pollution prevention, reuse, recycle, redesign, environmental process intensification, circular economy, life cycle process or product design as a tool for innovation, industrial ecology and eco-industrial park, sustainable engineering, and sustainability. Tapas is a registered professional engineer (PE) in the states of Washington and Wisconsin. Tapas is a fellow member of American Institute of Chemical Engineers and Indian Chemical Society. Tapas is the author of two books – *Industrial Environmental Management: Engineering, and Science and Policy* (John Wiley & Sons, 2020, 1 ed.) and *Toward Zero Discharge: Innovative Methodology and Technologies for Process Pollution Prevention* (John Wiley & Sons, 2005) – as well as 16 book chapters and 80 publications.

Contributors

Arya Rahul
Institute of Environmental Studies
Kurukshetra University
Kurukshetra, Haryana, India

Namratha B.
Department of Chemistry
The Yenepoya Institute of Arts,
    Science, Commerce and
    Management
Mangaluru, Karnataka, India

Bandopadhyay Rajib
Microbiology Section, Department of
    Botany
The University of Burdwan
West Bengal, India

Bharti Sri
CSIR-Indian Institute of Toxicology
    Research
Lucknow

Bhattacharjee Jyoti
Department of Chemical Engineering
University of Calcutta
Kolkata, India

Bhattacharyya Subarna
School of Environmental Studies
Jadavpur University
West Bengal, India

Bombaywala Sakina
Environmental Biotechnology and
    Genomics Division
CSIR-National Environmental
    Engineering Research Institute
    (NEERI)
Nagpur, India

Bose Saswata
Department of Chemical Engineering
Jadavpur University
West Bengal, India

Chakraborty Poushali
Department of Chemical Engineering
Jadavpur University
West Bengal, India

Chandrapragasam Vani
Department of Biotechnology, School of
    Agriculture and Biosciences
Karunya Institute of Technology and
    Sciences (Deemed to be University)
Coimbatore, Tamil Nadu, India

Chaudhuri Punarbasu
Department of Environmental Science
University of Calcutta
Calcutta, India

Das Papita
Department of Chemical Engineering
Jadavpur University
West Bengal, India

Dhulap Vinayak P.
Departments of Environmental Science,
    School of Earth Sciences
Punyashlok Ahilyadevi Holkar, Solapur
    University
Solapur (MS), India

Gaonkar Santosh L.
Department of Chemistry
Manipal Institute of Technology,
    Manipal Academy of Higher
    Education
Manipal, Karnataka, India

Ghorui Madhushree
Microbiology Section, Department of
    Botany
The University of Burdwan
West Bengal, India

Ghosh Subhasis
Department of Chemical Engineering
Jadavpur University
West Bengal, India

Gomez Levin Anbu
Department of Biotechnology, School of
    Agriculture and Biosciences
Karunya Institute of Technology and
    Sciences (Deemed to be University)
Coimbatore, Tamil Nadu, India

Karmakar Anisha
Department of Biotechnology and
    Medical Engineering
National Institute of Technology
    Rourkela
Rourkela, Odisha, India

Khan Dibyendu
Microbiology Section, Department of
    Botany
The University of Burdwan
West Bengal, India

Kumar Rachana
CSIR-Indian Institute of Toxicology
    Research
Lucknow

Lee Meng-Jen
Department of Applied Chemistry
Chaoyang University of Technology
Taichung, Taiwan

Mandpe Ashootosh
Department of Civil Engineering
Indian Institute of Technology Indore
Indore, India

Mishra Ajay Kumar
International Rice Research Institute,
    South Asia Regional Centre
Varanasi, Uttar Pradesh, India

Mondal Arnab
Institute of Environment and
    Sustainable Development
Banaras Hindu University
Varanasi, India

Mondal Riashree
School of Environmental Studies
Jadavpur University
West Bengal, India

Mondal Surajit
School of Engineering
University of Petroleum & Energy
    Studies
Dehradun, India

Mondal Uma Shankar
Department of Biotechnology and
    Medical Engineering
National Institute of Technology
    Rourkela
Rourkela, Odisha, India

Mukherjee Avik
Department of Chemical
    Engineering
Jadavpur University
West Bengal, India

Mukherjee Sayan
Department of Chemical Engineering
Jadavpur University
West Bengal, India

Nazir MD
Microbiology Section,
    Department of Botany
The University of Burdwan
West Bengal, India

Ojha Monu Dinesh
Indian Institute of Technology
Delhi, India

Paul Aritri
Delhi Public School, Rourkela,
    Sector 14
Rourkela, Odisha, India

Paul Subhankar
Department of Biotechnology and
    Medical Engineering
National Institute of Technology
    Rourkela
Rourkela, Odisha, India

Pratap Vinay
Environmental Biotechnology and
    Genomics Division
CSIR-National Environmental
    Engineering Research Institute
    (NEERI)
Nagpur, India

Rathore Jaswant
Centre for Atmospheric Sciences
Indian Institute of Technology Delhi
Hauz Khas, Delhi, India

Roy Sanket
Department of Chemical
    Engineering
Jadavpur University
West Bengal, India

Roy Arkaprava
Department of Chemical Engineering
Jadavpur University
West Bengal, India

Roy Subhasis
Department of Chemical Engineering
University of Calcutta
Kolkata, India

Saha Swastika
Department of Chemical Engineering
Jadavpur University
West Bengal, India

Sarkar Sampad
Department of Chemical Engineering
Jadavpur University
West Bengal, India

Sharma Moumita
Department of Chemical Engineering
Jadavpur University
West Bengal, India

Sharma V. P.
CSIR-Indian Institute of Toxicology
    Research
Lucknow

Shaw Rajdeep
Microbiology Section, Department of
    Botany
The University of Burdwan
West Bengal, India

Shepherd Ritu
School of Liberal Arts
Nehru Arts and Science College
    Coimbatore, Shri Gambhrimal
    Bafna nagar
Malumichampatti, Tamil Nadu, India

Skariyachan Sinosh
Department of Microbiology
St. Pius X College
Rajapuram, Kasaragod, Kerala, India

Somu Prathap
Department of Biotechnology,
    Saveetha School of Engineering
Saveetha Institute of Medical and
    Technical Science (SIMATS)
Chennai, India

Sutkar Pankaj R.
Departments of Environmental Science,
    School of Earth Sciences
Punyashlok Ahilyadevi Holkar,
    Solapur University
Solapur (MS), India

Tamang Kesang
Department of Life Science and
    Biotechnology
Jadavpur University
West Bengal, India

Thathapudi Jesse Joel
Department of Biotechnology, School of
    Agriculture and Biosciences
Karunya Institute of Technology and
    Sciences (Deemed to be University),
    Coimbatore
Tamil Nadu, India

Vijay Vishruth
Department of Biotechnology, School of
    Agriculture and Biosciences
Karunya Institute of Technology and
    Sciences (Deemed to be University),
    Coimbatore
Tamil Nadu, India

Preface

Plastics, a marvel of modern chemistry, boast attributes like being lighter, stronger and cheaper. Since their widespread adoption after World War II, plastics have brought about a revolutionary transformation in our world. The advent of plastic packaging has not only reduced transportation costs but also enhanced fuel economy by facilitating the global movement of manufactured goods. Moreover, the innovation in food packaging has significantly extended the shelf life of perishables, leading to reduced food wastage.

The benefits of plastics extend beyond packaging. Improved insulation has yielded substantial energy savings, and in the medical domain, disposable plastic instruments and hermetically sealed packaging have vastly improved sanitation standards. From a consumer standpoint, plastics have brought about greater convenience and reduced product costs. Consequently, traditional materials like wood, stone, horn, bone, leather, metal and glass have experienced a decline in market share as plastic-based products proliferate. Although plastic can be recycled, the current recycling rates remain disappointingly low, with only a fraction being effectively recycled. The mismanagement of plastic waste poses a significant challenge, especially in underdeveloped countries lacking proper waste management infrastructure. Often, plastic waste finds its way into open dumps or even natural environments due to the absence of suitable disposal facilities. Microplastics, defined as plastic particles smaller than 5 mm, present another facet of the global plastic waste problem. These minuscule particles originate from the breakdown of larger plastic items, microbeads in health and beauty products or fibres released during textile washing. Due to their small size, microplastics easily infiltrate ecosystems, particularly the oceans, and eventually enter the food chain. The presence of harmful additives like PCBs and BPA in microplastics raises concerns about potential health effects, especially in pregnant women. Given that many rely on fish as a primary source of protein, this poses significant health risks for the world's population.

Addressing these issues necessitates a comprehensive approach. It involves examining the production, usage and regulation of synthetic plastic polymers, implementing effective strategies for the recovery and remediation of plastic waste, exploring bioplastics and alternative materials and fostering a global perspective on the way forward. Additionally, a case study on the valorisation of plastic waste can offer insights into viable solutions for managing this pressing problem. In conclusion, while plastics have revolutionised various aspects of modern life, their mismanagement and contribution to pollution demand urgent attention. By focusing on sustainable practices, innovative alternatives and global collaboration, we can effectively tackle the challenges posed by plastics and move towards a cleaner and greener future.

1 Efficient Microplastic Remediation through Best Possible Strategies
A Review

Avik Mukherjee, Saswata Bose and Papita Das

1.1 INTRODUCTION

An indispensable and highly utilitarian material of our daily lives is plastic. The discovery of plastics has made lives more easy and "flexible". There has not been, probably, any arena where plastics have not set their foot. Right from the product making to its packaging as well as branding, which includes printing and adhesive bonding, plastics are everywhere. With the advancement of human civilization, the use of plastics has increased in leaps and bounds. Moreover, this use will definitely exaggerate in the coming years. As every coin has two sides, the limitless use of plastics has also become the cause of serious concerns relating to the adverse effects it adds on environmental pollution.

It affects both biotic and abiotic components of the earth by degrading their qualities. The key advantageous feature of them happens to be source of problems associated with plastics. This refers to their robustness and substantial chemical inertness, which account for plastic's prolonged longevity. Plastics take several decades to degrade, and the remnant particles of plastic debris are really too difficult to be dealt with. One such category of small plastic remnants is referred to as "microplastic". By definition, microplastics are small-sized plastics whose dimensions are below 5 mm. The extensive growth of the occurrence of microplastics all across the globe has been one of the recent causes of concern and a great challenge for the researchers. The presence of microplastics has affected both terrestrial and aquatic lives. Many scientific articles and reviews have thrown light on this matter [1–4]. Very recently, it was also reported to have been found in a sample of human blood, which the researchers believe might have been due to the percolation of microplastics via the food chain [5]. Microplastics as well as nanoplastics are believed to have ecotoxicological effects in the environment. Studies have revealed the toxicological effects of microplastics on various living organisms, especially those which ingest MPs mistakenly [6]. MPs are infamously associated with hazardous effects which include internal injuries to various body organs, alimentary canal blockage, dietary intake disorders, circulatory system disorders, etc. [7, 8]. Studies have reported about chronic digestive and epithelial cellular disorders

DOI: 10.1201/9781003449133-1

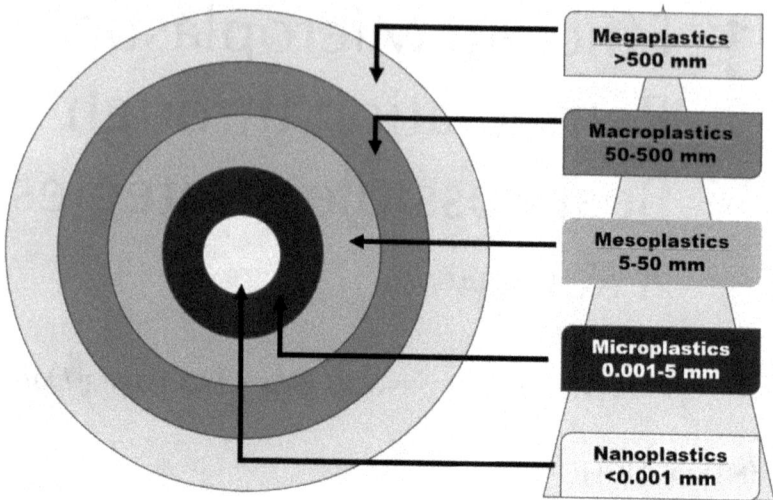

FIGURE 1.1 Classification of plastic debris according to their dimensions.

resulting from HDPE MP ingestion in *Mytilus edulis*, leading to detrimental effects on their tissues and intestinal tracts [9].

The detrimental effects associated with microplastic pollution increase many folds due to the adsorption of co-contaminants, which include plastic additives like stabilizers, antioxidants, dyes, unreacted monomers, remnant catalyst particles and many more which get leached out from the polymer matrix in due course of time [10, 11]. The small, dimensional structure of micro and nanoplastics plays the major negative role in this regard. Microplastics, that are mostly found in the environment may be classified on the basis of their shapes as spheres, fragments and fibres [12, 13]. There have been studies on effects of shape of MPs in their various properties and different methods to assess the quality of MP samples [14–16]. To comprehend and resolve the problem relating to microplastic pollution, one must understand the chemical nature of the concerned plastic, mode of their application/use as well as their behaviour in different climatic and environmental conditions. Based on the environmental conditions, the degradation mechanism followed by different plastics may be categorised as thermal, photo-oxidative, ozone-initiated, catalytic, mechano-chemical and biodegradation [17, 18].

1.2 MICROPLASTICS IN THE ENVIRONMENT

Plastics are introduced to the environment as wastes, a part of which undergoes recycling, and the rest is discarded and dumped mostly as landfills [19]. According to reports, the amount of such plastic wastes accumulating in the natural habitat is significantly high compared to the overall amount of plastics manufactured globally. These discarded plastic wastes are considered to be the potential source of microplastic generation. Microplastics may be broadly classified as primary and secondary based on their source of origin. The primary ones are those which are produced due to direct discharge from personal care products like cosmetics. The secondary types are attributes of environmental weathering of plastic wastes in their macroscopic forms [20]. Moreover, microfibres are generated in municipal waste waters during laundering of synthetic textile goods. The presence of an extensive amount of primary MPs in municipal effluents has been markedly reported in a study by Murphy and his team in 2016 [21]. The presence of MPs has been reported in waterbodies including seas, sea-coasts [22], rivers [23] and lakes [24]. Additionally, land and the atmosphere are also not free from the grasp of MP contamination. Zhou et al. reported their presence in the soil, while the air that we breathe in also contains MPs [25]. Owing to the minute dimensions of MPs, they get readily ingested by a wide range of organisms, which include both aquatic and terrestrial lives, and subsequently, get transported via the food web, spanning from the small-sized zooplanktons to the bigger mammals, including human beings. MP ingestion may lead to various serious health hazards [26–29].

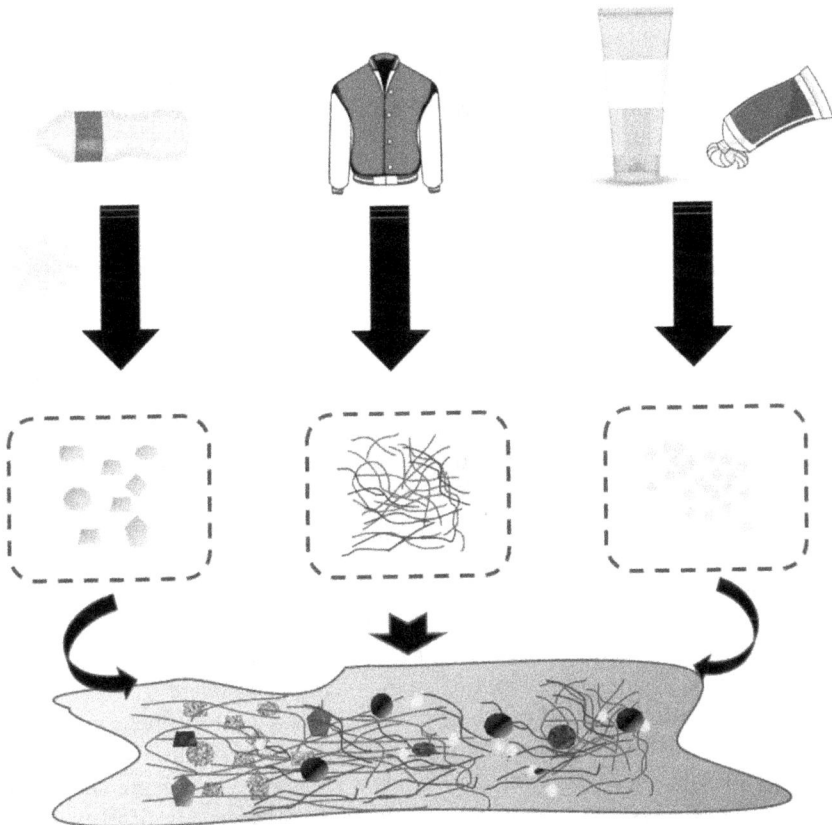

FIGURE 1.2 Various sources introducing plastic into the aquatic environment.

Plastic-derived pollutants are introduced into the marine environment from different sources, which include industries, domiciliary wastes, or even other water bodies [30]. Another source of MP contamination relates to the run-off waters containing microcapsule fertilizers draining out from agricultural fields. Microcapsules are potential sources for secondary MPs [31]. Moreover, activities such as fishing, fish hatching and offshore drilling also act as sources of secondary MP pollution to various water bodies [32]. Owing to the non-availability of competent methods for MP removal in waste water systems, the aquatic ecosystem gets affected by MP pollution [33]. Aerial winds and waves act as vectors by dispersing the light-weight, minute, dimensional MP particles in all directions, thereby exaggerating the problem manifold [34]. Studies have reported that the inadequacies of MP removal methods, especially from the water systems, have been responsible for contributing to environmental pollution on a larger scale [35–37]. Various factors affect the conveyance and distribution of MPs in the atmosphere, which includes weathering, tide currents, wind speed, etc. Along with land, water and air, the plants are also exposed to MP pollution, and they can also function as a medium for transportation of MPs. Research reports have corroborated the fact that microplastics can enter plant bodies through roots [38], while smaller, dimensional microplastics ($<$ 1μm) and nanoplastics have the potential to travel further and get translocated to different, other aerial body parts of a plant [39–43]. It is to be noted that conflicting results have been documented regarding the translocation ability of MPs within a plant body. It has been suggested that the xylem structure has a role to play in the uptake and movement of plastic particles within plant bodies [44].

1.3 HAZARDOUS EFFECTS OF MICROPLASTICS

Since about 71% of the earth's surface is covered by oceans, the majority of MP pollution is concentrated over there. Aquatic plants and animals get severely affected by MPs in different ways and forms. The effect becomes more detrimental as the MP particles adsorb various harmful chemicals onto their surfaces, owing to their small sizes, thereby functioning as a transport vector. Moreover, MPs themselves also leach out harmful chemicals (plastic formulation additives) in the course of time, most of which are highly toxic in nature [10, 11]. According to reports, around 54.5% of MP particles found in the oceanic waters are PE, followed by PP, which comprise about 16.5%; the rest includes PVC, PS, polyester and polyamides. PE and PP, being lighter, are found to float, while the rest, due to their higher densities, sink deep down the oceanic beds, thereby affecting the corals. MPs are known to affect aquatic lives by impacting their activities like feeding, spawning, etc., and ultimately, influencing their growth and existence. The cumulating concentration of MP pollutants lead to their percolation from one trophic level to the other by means of transmission via the food chain, although the extent of such a percolation effect produced by MPs to higher complex food chains could not be quantified [45]. Various reports have confirmed findings about the retention and bioaccumulation of MP granules within various organisms leading to detrimental effects [6]. The overall effect that MPs' injection has on aquatic lives is diminution in their feeding process, growth and also proliferation. The influence of MPs varies from one particular organism to another.

Moreover, the type of plastic and the level of pollutant also have their individual roles to play in affecting any organism.

1.3.1 Effects on Aquatic and Coastal Lives

A study revealed that PS MPs have an unpropitious effect on the reproduction and feeding processes in oysters, resulting from irregular food intake and the subsequent energy distribution within their bodies. PS MPs influence the egg-laying process, thereby reducing their number, which, in turn, affects ovocyte quality and sperm quality. Owing to injection of PS MPs, a retarded sperm speed and sperm count is observed by virtue of which the fertilization process in oysters gets hampered. Reportedly, a 6 μm long PS particle was found to be preferentially ingested by the oysters. The production and growth rate of the progenies (larvae) in such oysters fall by 41% and 18%, respectively [46]. Studies also report about the effect of MPs in gentoo penguins abording in the Antarctic regions, which are considered to be the ideal species for monitoring MP pollution in organisms confined to the region, as they exhibit a sparse movement in the outskirt areas. The evidence of the presence of MPs in the form of fibres, fragments and films have been reported in the droppings of gentoo penguins. Researchers believe that the entry of MPs into penguins must be either due to fallacious feeding or due to intake of contaminated prey [47]. Even the zooplanktons are not are not spared from the effects of MP contamination. A study by Cole and his co-researchers had discussed the effect of MP injection on the food habit, fertility and overall functioning of copepods, a type of zooplankton. It was found that exposure to PS microbeads having a dimension of about 20μm was responsible for retarded growth which was characterised by a reduction in the carbon biomass by about 40%, an energy shortcoming in copepods ultimately leading to death. The report also focussed on the production of small-sized eggs, along with reduced hatchings in copepods caused by prolonged exposure to MPs in the environment [48]. It was even reported by a group of researchers that there occurs a marked increment in the toxicity level when MPs are present along with glyphosate, an herbicide [49]. Studies on the impact of MPs on mussels shows that there is a direct, proportional relationship between the size of mussels and the level of MPs [50]. This may be attributed to the greater volume of water absorption by large-sized mussels. It was also found that the mussels absorbed more fibres than plastics, which may be correlated to the smaller size and less abundant availability of the latter in their habitat. MPs ingestion leads to malnutrition and faecal compaction in mussels, while the extent of damage increases due to the release of hazardous chemicals from various pollutants which remain adsorbed on the MPs.

Quite recently, MPs have been reported, in the UK, to have been introduced into the bodies of demersal sharks. According to Germanov et al., the entry of MPs into sharks is via direct or indirect intake (via food chain) [51]. Researchers believe that such a release of chemicals from microfibres causes damage to their reproductive and immune systems [52]. MP consumption also has reportedly caused damage to marine worms [14, 53]. A study by a group of researchers has shown adverse effects of polyethylene exposure on feeding, growth and mortality rates in *Arenicola marina*. MP pollution has also affected seabirds which are considered as biological markers

FIGURE 1.3 Percolation of MPs through the food chain.

of environmental changes, pointing out at the severity of the situation [47, 54]. Moreover, the transportation of MPs through the food chain has also affected terrestrial organisms, especially human beings. The presence of MPs has also been detected in human bodies. It was first detected in human excreta in 2019 [55] and in human placentas in 2021 [56] and, very recently, in 2022, even in human blood [5].

1.4　REMEDIATION OF MICROPLASTIC POLLUTION AND TRANSFORMATION STRATEGIES

There have been several attempts involving varied, different approaches to remove microplastics from the environment or to at least facilitate their further degradation to the next lower level. The techniques, or methods, that have been commonly employed over the last decade include adsorption [57], coagulation [58], filtration [59], microbial degradation [60], magnetic separation [61], advanced oxidative processes (AOPs) such as Fenton and/or Fenton mechanism [62], supporting ionic liquids [63] and photocatalysis [64]. We shall present a comparative study of the methods. The available methods to remove MPs from the environment can be broadly classified into two categories: the physical trapping mode and the chemical method. The physical mode of trapping MP particles include coagulation, membrane-based filtration and adsorption techniques, whereas the chemical means involve techniques like advanced oxidation processes (AOPS) and photocatalysis. It is quite crucial to assess these available, remedial methods for MP removal.

1.4.1　Physical Methods

Physical strategies of separation involve methods like coagulation, adsorption and sedimentation, membrane-based filtration and others. These methods facilitate removal of MPs from an aqueous medium without affecting the structural and chemical nature of the polymer chains to an extent which leads to the degradation process. Some of the most popularly employed physical methods of MP removal are discussed in the brief account in the following segment.

1.4.1.1 Coagulation Method

Coagulation is a well-known and simple process, adopted in many water treatment plants to produce good-quality drinking water. Commonly used coagulants include sulfates of iron and aluminium. The working action of coagulants involves the coalition of particular pollutants which remain in a suspended state, resulting in the formation of flocks [65, 66]. The flocks containing pollutants can be subsequently eliminated from water stocks via sedimentation processes. In case of MPs, the process of coagulation and flocculation deals with precipitation of them via chemical attachment with the coagulants. The nature of forces operative behind effective interactions between the pollutants and the coagulants decide the fate of MP removal from waste water. Various other factors influencing the coagulation and flocculation of MPs include the chemical nature of the coagulant molecules, along with that of the polluted water system, the pH of the wastewater and also the population of the MPs in the aqueous system. The chemical composition and the size of the MPs, however, play major roles in such cases. The advantage of this process lies in its simplicity, although its application to MP removal, at times, turns out to be difficult, owing to the adsorption of other pollutants, both organic and inorganic, including heavy metals. While the cost involved may become an issue and varies with the grade and amount of the coagulant used, the generation of sludge, which is an integral part of this method, becomes an extra burden, as its disposal is quite challenging. The sludge contaminated with the coagulant-MP assemblies poses a threat of reintroduction of environmental pollution. Among various research studies, Ma and co-workers had used an iron complex as coagulant for removing PE MPs from drinking water. The experimental findings were recorded at a neutral pH condition and under specific laboratory conditions on a small scale [67]. The efficacy of the coagulant was not satisfactory in terms of MP removal, which may be attributed to the entrapment of microplastic granules within flocs possessing weaker interactions with the plastic debris, thereby leading to inefficient coagulation [68]. A successful attempt has been reported concerning PE-MP removal via an electrocoagulation technique applied to synthetically generated wastewater. The process dealt with coagulation and flocculation stimulated by electrical energy applied through electrodes. The electrocoagulation technique is associated with advantages like high efficiency of MP removal, coupled with reduced sludge generation, energy efficiency and being economical. Moreover, the process can also be made automatic [69]. A remarkable improvement, in terms of efficiency, for MP removal via the coagulation mechanism was reported on the addition of a polar phase, PAM. This may be attributed to the ionic interactions between the two oppositely charged phases, which includes the Fe-centric floc phase with a cationic character and the PAM phase having an anionic nature, which are believed to be responsible for the formation of more stable aggregates (flocs), involving both the coagulant and the MPs. Comparative studies report better performance in MP removal of polyacrylamide-based cationic coagulants than the aluminium-based ones in aqueous environments [70]. The interactions occurring between the coagulant molecules and the MP granules depend on the surface properties of the latter, which subsequently affect the efficacy of coagulation involved in the MP-removal process [69]. Inherently, most plastics are substantially inert, which makes MPs granules, as well, much less reactive. But the presence of surface irregularities, developed due to mechanical weathering caused by natural agencies, increases the surface energy of MP granules, thereby imparting a certain degree of attractability between the MPs and the coagulants. Environmental weathering of MPs via various natural agencies may generate polar oxygen-bearing groups and reactive, unsaturated sites, thereby facilitating enhanced interactions with the flocculating agents. Moreover, surface adsorption of various organic compounds may also lead to such an effect. The chemical composition of the plastic plays a vital role in governing the interactions between the MP granules and the flocculating agent. For instance, micro-dimensional debris of polyester has been reported to have responded to the same flocculating agent more effectively, in comparison to that of PE and PS, when subjected to similar conditions. This finding could be accredited to the interactions arising due to abundance of polar carbonyl groups in polyester chains [70]. The efficiency of coagulation and subsequent removal depends on the physical and chemical nature of the MP, varying in between the range of 80% to 99%. As reported, the efficacy of removal of PS-MPs from wastewater has been found to be greater than 99% for inorganic coagulants, PAC (poly-aluminium chloride), $FeCl_3$ and polyamine. Owing to the high removal efficiency, coagulation by using inorganic coagulants has been recommended as the tertiary step associated with a WWTP [70, 71].

Among other metals, aluminium-based coagulants displayed the best performance in terms of PE-MP removal from wastewater [67, 72]. Another report by a group of researchers recommended an effective and unique MP-removal method for cosmetic and textile sectors, involving the adherence of the MP particles via alkoxy-silyl linkages, leading to the formation of large-sized, three-dimensional agglomerates, thereby facilitating pollutant removal by filtration. The choice of the coagulant, along with the amount of addition, have to be very judiciously monitored and controlled in order to obtain best results [73].

1.4.1.2 Membrane-Based Filtration Methods

Waste removal involving filtering membranes has been a very effective technique applied for purifying water. Some of the advantageous features include high efficiency of separation and a precisely smaller dimension of the plant. Membranes may be classified based on various parameters. On the basis of structural morphology, they may be of two categories; namely, symmetric and asymmetric [74]. Asymmetric membranes having a micro- and nano-dimensional, porous bed are known to be highly effective in dealing with the removal of microorganisms. For the purpose of MP removal, UF and RO techniques have been proven to be efficient, to some extent. However, many improvements are yet to be manifested in this regard [75]. The factors influencing efficient MP removal includes membrane quality, the nature of the MP and the environmental conditions. The removal efficiency depends on membrane durability, size, influent flux, and quantity and quality of MPs [76]. Another membrane-based technique that has gained immense popularity, in terms of waste water treatment, is dynamic-membrane technology (DM technology) [77, 78]. This technology involves the presence of a cake layer over a supporting matrix. Thus, there is a dual level of entrapment barriers for the impurities. The intrinsic advantage of this technique is ascribed to the non-involvement of any externally added chemicals which are considered to be secondary sources of contamination. The simplicity of DM filters is associated with their operational flexibility, as their working is merely gravity based, devoid of the necessity of any pump. This is due to the fact that low resistance is encountered during filtration, coupled with the involvement of low trans-membrane pressure. Modern developments involve simultaneous methodologies, combined with membrane-based filtration in order to enhance the efficiency of MP removal from waste water. For instance, membrane bioreactors (MBR) are designed, aiming at a high-end level of treatment of wastewater. The technology involves a membrane system, along with a bioreactor, based on heterogeneous reaction [79]. The pollutants are initially introduced to the bioreactor, where a biologically activated sludge initiates biodegradation of the former. The output of the bioreactor is thereafter allowed to pass through a membrane-filtering system. The combined influence of both the bioreactor and the filtering membrane leads to an overall-efficient wastewater treatment in terms of MP removal. An MBR design fabricated by a team of researchers has been reported to be remarkably effective in MP removal from wastewater in the primary effluent stream in wastewater involving a specifically designed pilot plant [80]. The elimination rate of the pollutants (MPs) was found to be as high as 99.9%, indicating a major success in MP removal. Moreover, the bio reactor has been known to play a major role in the biodegradation of the organic molecules belonging to a different class of chemicals, which remain absorbed onto the MP surfaces. It is noteworthy that the biological reactor permitted the degradation of organic impurities adhered onto MPs, which is beneficial to enhance the accuracy of qualitative analysis of the separated MPs. Studies have reported different ranges of efficiencies of MP removal from an aqueous medium by using different types of filters. Disc filters showed 40–98.5% efficiency, while 97% efficiency was shown by rapid sand-filters; on the other hand, 95% efficiency was found in the case of dissolved air-floatation methods [80]. However, plastics being predominantly based on saturated hydrocarbons exhibit considerable inertness towards biodegradation processes; this renders the MP degradation meagre when applied to MBR processes. Apart from this, the inherent toxicity of the MPs or NPs also intervenes in the process. These makes the effectiveness of the MBR process limited, to a certain extent, when applied for MP remediation [81, 82]. The efficiency of MP removal of membrane bioreactors is dependent on various factors, including the activity of the microorganisms, the population of MP in the aqueous medium to be purified, aeration, temperature and other environmental conditions related to the climate and season and, more importantly, the presence of other substances, including contaminants and nutrients as well [81]. All such factors contribute to the overall complexity of the process mechanism. Hence, the prime shortfall of this process lies in its membrane instability which arises due to clogging and fouling issues leading to poor performance and declining efficiency. Some microorganisms like bacteria, archea, picoeukaryotes, etc., are capable of initiating degradation reactions in MPs in an aqueous medium. Studies reveal that Agios and Souda consortium can bring about degradation in PE-MPs. However, the Souda consortium was reported to out-perform the former [83]. In another study, PE-MPs were successfully removed by the action of *Zalerion maritimum*, a marine fungus. Similarly, among *Bacillus gottheilii* and *cereus* bacterial stains, the former has been reported to be more efficient in MP degradation [84]. Other studies showed about 66% PE-MP removal efficiency by *Tridacna maxima* [85]. However, some processes involving higher efficiency of MP removal were also reported. A sequential batch reactor designed to operate under a combined anaerobic-aerobic conditions exhibited greater than 99% removal efficiency [86]. There have been several attempts to remove MPs from wastewater using MBRs in various ways – using MBRs only or in combination with other methods. When employed singularly, MBRs could remove about 70% MP. When combined with adsorption processes, the efficiency went higher up to about 83%, while a complete removal, 100% efficiency, was achieved when MBR process was combined with an anaerobic method and subsequently followed by an RO filtration. The future prospect of MP degradation as a part of wastewater treatment could be based on MBR technology, along with the use of enzymes. [87–92]. Another study was carried out to remove MPs from wastewater by the use of a four-zone based biofilter. The efficiency of MP removal was found to be 89% [93]. Another report was associated with the use of three different polymer ultrafiltering membranes –PC, CA and PTFE – applied to remove PA and PS-MPs from an aqueous medium. In all the cases, the removal efficiency was found to be around 94% [94]. It is to be noted that, in order to prevent membrane fouling to certain extent, the UF technique was adopted. PE-MPs are probably the most difficult ones to be removed from aqueous medium, as their density nearly matches that of water. A combination of UF and a coagulation process based on iron compositions has been reported to accomplish a 100% PE-MP removal efficiency, along with some resistance to membrane fouling [67, 72]. The removal efficiency depends on membrane durability, size, influent flux, and quantity and quality of MPs. Another filtration technology that has been tested for MP separation from an aqueous medium is gravity-powered filtration, which is a non-pressure, gravity-driven filtration method. The overall process functioned in two different modes: the first one was a filtering step in order to separate the MPs from wastewater, while the second mode, termed the "backflush mode", was rendered to the cleaning of the set-up by removing the collected MPs. For this set-up, a 3D filter operating at a water pressure of 1.68 kPa was found to be best suited [95, 96]. Another gravity-driven filtration technique which has been employed for MP separation is the Dynamic membrane technology. The filter structure is associated with a cake layer which is formed by means of deposition in due course of filtration [97]. Another attribute of this method of filtration is that it can operate at very low pressures; further, lower than that associated with UF and MF techniques [98]. Despite their advantageous facets, membrane-based processes suffer from some key downsides, which include fouling, abrasion and other phenomena which hamper efficiency of water purification, posing a major challenge for researchers for the days to come [81, 82].

1.4.1.3 Adsorption Methods

One of the well-known methods to purify water has been by means of adsorption, wherein the pollutants get absorbed onto an adsorbent, thereby facilitating their removal. The interaction between the pollutants and the adsorbent can be of varied types, the spectrum of which includes ionic interactions, π-π interactions, hydrogen bonds and hydrophobic interactions. Various adsorbents have been developed and tested by many researchers to evaluate the efficacy of adsorption processes for the removal of MPs from water samples.

One such attempt involves the use of 3DRGO – i.e., 3D reduced graphene oxide – for the adsorption of PS microspheres, the average diameter of which is within the micrometre range (5 μm). The presence of the phenyl rings within the polymer chains was responsible for the development of π-π interactions between the PS microspheres and the absorbent 3DRGO. It is generally expected that pollutants, in this case MPs, with abundance of π-bonds would readily interact with RGO, thereby facilitating adsorption.

PS microplastics repossessing such a chemical structure, therefore, get readily adsorbed onto the RGO adsorbent phase. But the MPs belonging to non-polar class of polymers having primarily saturated backbone chains do not exercise such interactions with polar-adsorbing molecules of RGO. The study revealed that the adsorption efficiency of 3D-RGO depended on the pH of the medium, concentration of various ions, adsorption time and temperature [99]. Another study involving a group of researchers, Streb et al., developed a nanoparticle based on iron oxide and microporous silicon dioxide, following coreshell morphology, supported on ionic liquid phases [100]. The iron oxide phase comprised the core, which was responsible for absorbent recycling. The hydrophobic POM-IL phase facilitates a ready interaction with PS particular beads. This unique approach was found to be significantly efficient in terms of MP removal, even when applied to large-scale purification of water. The magnetic character of the novel Fe-based adsorbent also paved the way towards effective separation of the adsorbent phase from the pollutants, which, in comparison to the filtration method, was found to be very time efficient as well. A unique fabrication based on hybrid organic-inorganic silica gel was synthesized to carry out a three-step MP-removal process [101].

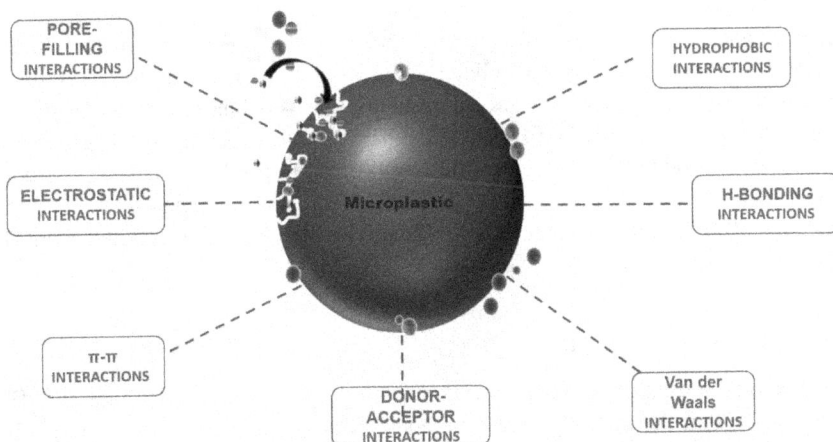

FIGURE 1.4 Different types of interactions on MP surfaces.

Moreover, the challenge stiffens manifolds in dealing with nanoplastics (NPs), as they are more detrimental to organisms, and being, further, minute in size, they get more abundantly spread in the environment, thereby making the removal processes very difficult. A group of researchers, Darbha and coworkers, had been successful in carrying out PS-NP removal from aqueous environment via adsorption process by means of designing a double layer structured Zn-Al hydroxide [102]. However, the absorbent was found be much less effective in media with higher pH, owing to deprotonation in the OH groups. This makes the use of the absorbing system limited when directly applied to real water-purifying systems having various chemical complexities. Many more research outcomes based on adsorption of MPs involve different varieties of absorbing systems, one of which comprised $CNT-Fe_3O_4$, while another was Zr-based MOF foam. These were, reportedly, very efficient in MP removal [103, 104]. According to reports, MPs tend to get strongly adsorbed onto the surfaces of various microalgaes found in different aquatic environments. The factors which are responsible for such interactions include the presence and nature of surface charge and the characteristics of the surface of the microorganisms [105, 106]. A study was conducted on the adsorption characteristics of PS-MPs on seaweed, a marine microalga conferring a sorption efficiency of 94.5%. The adsorptive attachments are attributed to the ionic interactions between the MP surfaces and the surface of the microalgae, which releases polysaccharide alginate possessing anionic groups [107]. The nature of charge on the MPs governs the intensity of interaction with the microorganisms. As per the report, PS-MPs possessing positive surface charges get more strongly adsorbed onto unicellular green algae compared to the MPs with negative surface charges. This finding is attributed to the fact that the chemical composition of microalgaes comprises anionic polysaccharides [108]. Although activated charcoal is a very popular adsorbent, it is significantly ineffective when it comes to MP adsorption, owing to their small size and strong tendency to get desorbed. However, a biochar adsorbent synthesized from biomasses of pine and spruce was found to remove PE-MPs satisfactorily from aqueous solutions. The retention efficiency of the adsorbents was found to be reduced with a decrease in particle size of the MPs. In this regard, the spruce-based adsorbent was found to outperform the pine-based one [109]. Although, the adsorption method plays its part in MP removal, nevertheless, complete eradication of such pollutants is not possible due to the inherent trait of the method, as the sludge collected following an adsorption process is highly contaminated with MPs which are subsequently used as landfills or as fertilizers, either directly or indirectly [110]. Hence, the harmful effect of MPs subtly gets incorporated within the pipeline of agricultural and farming products via sewage sludge treatment processes [111]. Hitherto, despite several developments – the physical strategies of MP removal – complete eradication of the effects associated with MP pollution has not yet been possible.

1.4.2 Chemical Methods

The inherent limitation associated with the physical separation methods of MPs, which deals with the inability of the methods towards complete eradication of MPs from the environment, has prompted the upsurge in the inclination towards chemical degradation strategies. In order to facilitate permanent removal of MPs, the mechanism concerning chemical methods is based on the generation of reactive oxygen species. The involvement and function of catalytic systems have, therefore, been indispensable.

1.4.2.1 Advanced Oxidation Processes

Advanced oxidative processes (AOPs) are considered to be a class of emerging chemical methods which are capable of removing soluble organic molecules from aqueous media by means of oxidative reactions involving hydroxyl radicals ($^{\cdot}$OH). Photodegradation, or photocatalytic degradation, has been considered to be a branch of advanced oxidation process (AOPs). Unlike the physical methods of MP removal, the advanced oxidation processes operate via breakage of covalent bonds, leading to polymeric chain scission. This technique being a substantially highly energy-efficient one has been proved to be very productive for MP degradation in aqueous media. Upon exposure to suitable photo sources, a photocatalytic semiconductor generates electrons and holes via photoexcitation. The combination of these photogenerated holes with H_2O molecules or with OH ions leads to the formation of reactive organic species (ROS), like $O_2^{\cdot -}$ and OH. In detail, the photogenerated holes react chemically with water to generate hydroxyl free radicals, while the photogenerated electrons combine with molecular oxygen to produce super oxide radicals. The combined action of the generated hydroxyl radicals as well as the superoxide ions leads to the degradation of polymer chains. The generated ROS thereafter act upon the MPs, resulting in more significant chain scission within the polymer. Other parallel reactions may involve branching and crosslinking; moreover, complete mineralization processes may also take place, forming CO_2 and H_2O. The photoinitiated degradation of polymeric backbone involves a free-radical mechanism following homolysis. Similar to the free-radical addition polymerization process, the degradation may also be subdivided into three major steps, which are initiation, propagation and termination. Hence, the photoinitiated degradation of polymeric chains may be looked upon as a depolymerization process [96]. The generation of peroxyl radicals is a significant one in the photodegradation processes of a polymer chain, as it marks the onset of the accelerated kinetics. In 2019, Zhu and co-workers suggested that there occurs the formation of environmentally persistent free radicals (EPFR) after a fortnight, following photoirradiation. Thereafter, ROSs are produced which would subsequently act upon the MPs, leading to further polymeric chain scission. Moreover, there occurs a reduction in the molecular weight, resulting in the formation of further, smaller-sized products. The chemical composition and chain structure of the polymer governs the formation and subsequent actions of the EPFRs. Polymers consisting of phenyl rings are more susceptible towards EPFR generation, while the remaining ones are substantially photostable [112]. Thus, the role of photocatalysts becomes immensely significant for the photoinitiated degradation of most polymers.

1.4.2.2 Photocatalysis

Photocatalysis is an eco-friendly, substantially popular and sustainable technique which makes use of the abundantly available sunlight. Owing to the meritorious attributes like higher degradation efficacy, leading to complete mineralization, coupled

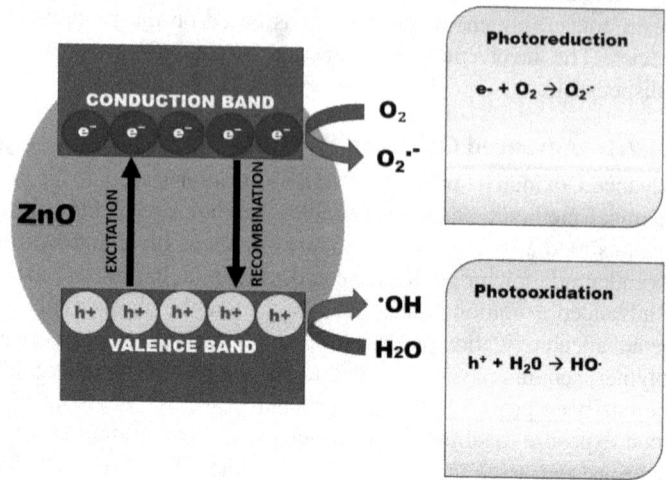

FIGURE 1.5 Photocatalytic action and mechanism.

with non-toxicity of the end products, this method has gained tremendous impetus for the water-purification applications [113]. Photocatalysis is based on the interaction between reactive oxygen species (ROSs), which includes hydroxyl, superoxide free-radicals which are generated on the surfaces of the metal/non-metal oxides (semiconductors) and the non-polar/polar organic matrix. The generated, highly reactive free-radicals play important roles in initiating bond fission within the organic phase, thereby leading to the degradation of the latter. These free-radical-borne degradation processes are generally extremely fast and may lead to complete mineralization of the organic molecules to carbon dioxide and water. The general mechanism of photocatalysis involves the generation of photo-exited holes due to the electronic transition from a lower-energy state to a higher one; in other words, from the valence band to the conduction band. The generated holes subsequently snatch electrons from the relatively inert organic molecules, thereby oxidizing them into carbon dioxide and water [114].

Metal oxide-based semiconductors like ZnO, TiO_2, etc., are the most commonly used and most appropriate candidates for designing photocatalysts. Amongst all, TiO_2 is an immensely popular choice amongst researchers due to its high oxidising efficiency, low cost, low toxicity and higher stability shown in acidic and alkaline environments [115–118]. Even for the purpose of photo degradation of MPs of various kinds, TiO_2 has found its own means of credibility, specially attributed to its high-energy band-gap value (3.2 eV) [119–121]. The capability of a photocatalyst relating redox reactions is dependent on the respective band-gap value in terms of redox potential. Among the most commonly used semiconductors, TiO_2 has got one of the greatest oxidation capabilities. Nevertheless, it is not only the oxidation potential value that plays the major role in photodegradation; light absorption capability is inversely proportional to the band gap. In this regard, ZnS shows inferior photocatalytic activity with respect to TiO_2, the latter being a contender having a proper balance between band-gap value and redox capability. The particle size of TiO_2 plays a decisive role in its photocatalytic activity, as the light absorbing ability is inversely related to the dimension. In nature, there exist three crystalline forms of TiO_2; namely, the anatase grade, rutile grade and brookite [122]. Each of them differs from the other in terms of the structural symmetry of their phases. The basic building block of all the grades is based on an octahedral-shaped TiO_6 unit [123, 124]. For photocatalytic activity, only anatase and rutile grades are commonly known. The brookite grade being structurally unstable, it is not popular for this purpose. The two photolytically active grades of TiO_2 are known to get interconverted to each other on varying temperature. The anatase grade (3.2 eV) is known to possess a higher band-gap value than the rutile grade (3.0 eV), by virtue of which the former functions as a better photocatalyst than the latter, while there some issues regarding the absorptive capacity of the anatase grade. On the other hand, the anatase grade performs better in terms of the recombination rate of photogenerated electrons and holes [125].

1.4.2.2.1 *Photolytic Degradation Mechanism*

The working principle of a photocatalyst involves its irradiation with UV or even sometimes visible light possessing an energy greater than or equal to its bandgap and subsequent excitation of electrons from the lower energy level (the valence band) to the higher energy level (the conduction band), thereby generating an electron-hole pair.

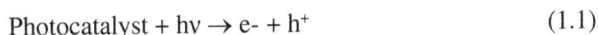

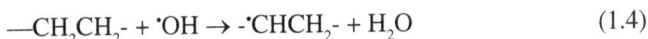

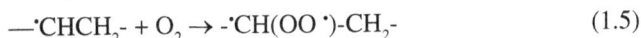

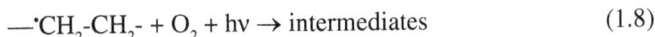

$$\text{Photocatalyst} + h\nu \rightarrow e\text{-} + h^+ \tag{1.1}$$

$$O_2 + e\text{-} \rightarrow O_2^- \tag{1.2}$$

$$H_2O + h^+ \rightarrow {}^{\cdot}OH + H^+ \tag{1.3}$$

$$-CH_2CH_2\text{-} + {}^{\cdot}OH \rightarrow \text{-}{}^{\cdot}CHCH_2\text{-} + H_2O \tag{1.4}$$

$$-{}^{\cdot}CHCH_2\text{-} + O_2 \rightarrow \text{-}{}^{\cdot}CH(OO\,{}^{\cdot})\text{-}CH_2\text{-} \tag{1.5}$$

$$-{}^{\cdot}CH(OO\,{}^{\cdot})\text{-}CH_2\text{-} + h\nu \rightarrow \text{-}CHO\,{}^{\cdot}\text{-}CH_2\text{-} + {}^{\cdot}OH \tag{1.6}$$

$$-CHO\,{}^{\cdot}\text{-}CH_2\text{-} \rightarrow \text{-}CHO + \text{-}{}^{\cdot}CH_2\text{-}CH_2\text{-} \tag{1.7}$$

$$-{}^{\cdot}CH_2\text{-}CH_2\text{-} + O_2 + h\nu \rightarrow \text{intermediates} \tag{1.8}$$

These electrons may be termed "photo-excited electrons" [126]. Some of these electrons return back to the valence band and recombine with the generated holes within a time-span ranging in femtoseconds, while the rest possess a longer lifetime [118]. The phenomenon involving the recombination of electrons and holes hampers the efficacy of the photocatalyst by lowering the availability of photogenerated electrons and holes on the surface of the photocatalyst, which ultimately play the major role in the redox reactions initiating degradation of the targeted substance. Hence, the recombination of the electron-hole pairs has to be prevented. The techniques involving the incorporation of trapping sites and agents which function as scavengers have been a popular and effective practice in this regard [123]. As already mentioned, photocatalytic activity involves a redox reaction in which the photo-excited electrons residing in the conduction band react with aerial oxygen, leading to the formation of superoxide, which subsequently undergo reduction, resulting in the formation of hydrogen peroxides. The production of hydrogen peroxide is a crucial step, as it marks the onset of rapid degradation involving free radicals. The hydrogen peroxide reacts with electrons to produce hydroxides and hydroxyl free radicals. While the foregoing describes the reduction half of the reaction, the oxidative part involves the valence band, where the generated holes react with water present in the aerial moisture to produce hydroxyl free-radicals and protons. The degradation reaction influenced by the hydroxyl free-radicals ultimately leads to mineralization of the organic pollutants, thereby converting the latter into carbon dioxide and water. The overall degradation rate is dependent on the process of interfacial charge transfer that occurs between the semiconductor particles and the organic pollutant such as microplastics [127]. Although the photocatalytic activity of photocatalysts works fine under UV rays, the requirement of the current era seeks the same under mere visible light. UV radiations comprise only about 5% to 8% of the entire solar spectrum. This fact is responsible for the lowering of photocatalytic efficiency of metal-oxide-based photocatalysts. In order to initiate photocatalytic degradation under simply less energetic visible radiations, the basic requirement is to reduce the band gap of the catalyst, thereby facilitating photoexcitation at a lower excitation energy. There are various means to accomplish this goal, amongst which incorporation of dopants is significantly popular. This leads to the generation of structural defects within the crystal of the catalysts which are metal oxide based, most commonly. Khlyustova et al. has discussed various synthetic conditions leading to such defects [128].

1.4.2.2.2 Why Photocatalysis?

While other MP-removal methods are unable to eradicate the pollution concerning plastics completely, photocatalysis and subsequent valorisation have the potential to convert the toxic pollutants to valuable products useful for other purposes. Moreover, the process is considered to be more energy-efficient and eco-friendly. Hence, there has been an immense focus on photocatalytic development, intended for MP remediation, involving varied compositions. Some of such attempts have been discussed in the following section.

1.4.2.2.3 Various Photocatalysts and Doped Photocatalysts Used and Characterization

In this section, we discuss the variety of different photocatalytic materials and dopants used for effective MP degradation and the subsequent characterisation of MPs. As already discussed about TiO$_2$ being the most commonly used photocatalyst, ZnO has also been used effectively for the same purpose by Tofa and his co-workers in the form of nanorods for degradation of LDPE MPs, which confirmed the alterations in the viscoelastic behaviour of the MPs, attributed to the changes in the chemical structure of the chains. The efficacy of the photocatalyst could be proved by establishing the presence of various low-molecular-weighted polar species, including functional groups like peroxides, hydroperoxides, carbonyl groups as well as unsaturated compounds, which are produced following subsequent degradation of the non-polar hydrocarbonaceous HDPE chains [129]. Morphological studies were used to detect surface cracks, wrinkles and cavities to establish the formation of the new compounds through SEM images. The increase in the values of carbonyl and vinyl indices also supported the degradation theory. The same group of researchers extended their work and carried out photodegradation tests on LDPE MPs, using ZnO nanorods doped with plasmonic platinum. In comparison to the weak absorptivity of visible light of ZnO nanorods, Pt-doped ZnO has been reported to be far superior, which is attributed to the surficial plasmon resonance of the Pt nanoparticles. Moreover, the Pt-doped ZnO nanoparticles are successful in preventing the recombination of holes and electrons to a certain extent, resulting in an enhanced photocatalytic activity, as evidenced by various characterization results. Despite exhibiting a much better photocatalytic behaviour than ZnO, Pt-doped ZnO nanorods could not lead to complete degradation of MPs. Micro- and nano-dimensional photocatalytic systems have been designed based on TiO_2 which effectively degrade MPs. The efficiency of the photocatalysts gets enhanced with an increase in their surface area, which also results in a better interaction with the MPs. External factors influencing photocatalysis of MPs include salinity of the aqueous solution and the intensity and energy of the illuminating source [130, 131].

Another group of researchers, Ariza-Tarazona and his co-workers, attempted to achieve complete degradation of HDPE MPs by developing a doping technique of TiO_2 by using non-metals such as carbon and nitrogen. An enhanced photocatalytic activity of the as-synthesized $C,N-TiO_2$ was reported, particularly under low temperature and pH conditions wherein the two conditional factors are believed to lead to a synergistic effect on photocatalytic degradation [132]. The same group had prepared N-doped TiO_2 via two different synthetic routes, one of which exhibited better photocatalytic activity of HDPE MPs under both aqueous and solid ambient, while the other, only in the former environment. The second synthetic method involved the combination of $C,N-TiO_2$ along with SiO_2 [133].

A very innovative and novel method involving the use of smart microrobots were employed by Seyyed Mohsen Beladi-Mousavi et al., microrobots which are capable

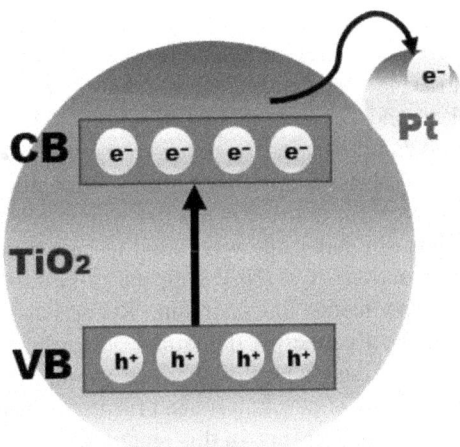

FIGURE 1.6 Role of dopants in enhancing efficiency of photocatalysts by prevention of electron-hole recombination via electron scavenging.

of simultaneous collection and subsequent photodegradation of MPs. The micro-robots are equipped with $BiVO_4$ photocatalyst and Fe_3O_4 having magnetic prop-erties [134]. Apart from PE, PP samples are also photodegraded by researchers. Nanoparticular suspensions of TiO_2 and ZrO_2 in THF respectively were used by Bandara and his group. It was found that ZrO_2 suspension in THF was very effec-tive in photodegradation of both PE and PP, as compared to the TiO_2 suspensions in THF [135]. This difference in the photocatalytic activity could be attributed to the structural morphology of ZrO_2, which is reported to be mesoporous, along with a greater band-gap value, which allows more energetic photon absorption. Moreover, it is more capable, in stabilizing oxygen vacancy, than TiO_2 [136]. A novel approach involving the preparation of hydroxyl group dominated BiOCl ultrathin catalyst was used by Lu and his team for photo-initiated degradation of HDPE microspheres. The report concluded that the hydroxyl free-radicals were primarily responsible for the photocatalytic degradation [137]. Photocatalysis degradation processes, in most cases, lead to the generation of micro-cracks and grooves on the surface of the MPs, thereby facilitating diminution in their sizes. But complete mineralization of the compounds is what the current need demands. This makes photo-reforming a better method to deal with the MP pollution. Photo-reforming processes are associated with harnessing the redox capability of a photocatalyst following photo activation, resulting in the oxidation of the organic pollutant, coupled with the reduction of H^+ ions to H_2 gas under ambient conditions of temperature and pressure. The unique advantage of the method concerns the production of fuel-cell-grade H_2. Thus, the concept of converting "trash to treasure" could be, to some extent, implemented by this method. Back in the year 1981, researchers Kawai and Sakata had first applied the photo-reforming method by using Pt-deposited TiO_2 to degrade PE and PVC samples in NaOH solutions [138]. The platinum phase played the role of a co-catalyst, which is believed to accept the photoexcited electrons from TiO_2, the photocatalyst, thereby leading to the reduction of H_2O or H^+ to H_2. The disadvanta-geous features of the foregoing photocatalytic composition are associated with the involvement of high cost owing to the use of platinum as co-catalyst, along with the higher band-gap value which makes the assembly operative only under energetic UV radiations. In order to avoid such disadvantages, other photocatalytic combinations were developed. Uekert and co-workers worked with CdS/CdO_x quantum dots in aqueous-phase alkaline conditions for the photo-reforming of PUR, PET and PLA [139]. Another attempt involved nontoxic assemblies containing carbon nitrides and nickel phosphide (CN_x/Ni_2P). It is to be noted that, among Pt-TiO_2, CN_x/Ni_2P and CdS/CdO_x quantum dots, the latter one was found to be the most effective one in terms of photocatalytic activity. CN_x/Ni_2P happens to be non-toxic and environ-ment friendly, unlike heavy metals like cadmium. Reports also suggest that photo-reforming processes under alkaline pH domains lead to the production of higher amount of H_2. Nevertheless, the efficacy of photo-reforming, conversion rate, selec-tivity, applicability, etc., of the photocatalyst need further improvement. Researches must try to focus on such developments. In this regard, direct gasification processes find themselves superior. Production of fuel-graded H_2 is not the only pathway avail-able; carbon-based fuels can also be obtained following photocatalysis of MPs under specific conditions. In 2020, a group of researchers headed by Jiao had been successful to synthesize single-unit-celled Nb_2O_5 layers. The catalyst was reported to accomplish complete photodegradation of PE, PP and PVC and subsequent conversion to CH_3COOH in the absence of sacrificial agents. The chemistry behind the mineralization has also been discussed which could be described by two broadly classified steps. The first step of the redox reaction involved an oxidative fission of C-C bond forming CO_2, along with simultaneous reduction of O_2 into $O_2^{\cdot-}$, H_2O_2 and H_2O. While the second redox step was associated with the photoinduced C-C coupling process of carboxyl reactive free-radical ($\cdot COOH$) intermediates resulting in the reduction of CO_2 to CH_3COOH, the oxidation half reaction involved the conversion of H_2O to form O_2. Although the photo-reforming method has been proved to be more effective that mere photocatalysis, the yield of the obtained carbon-based fuel, however, remains a big challenge. Jiao and team, thus, recommended the use of two-component photocatalytic systems for obtaining optimum results. It is to be noted that plastics from various waste materials, which involves single-use bags, disposable containers, food-graded film wraps, etc., can be effectively dealt with to produce CH_3COOH via photo-reforming methods **[140]**.

It is more challenging to deal with the photodegradation of MPs as their presence are mostly populated in aqueous medium, for which solid-phase photocatalysts become non-functional.

The limitations associated with the applicability of photocatalytic assemblies to natural aquatic environments deal with the poorer penetration power of the sunlight which renders the photodegradation process ineffective for the MPs present at depths below the water's surface. The use of an artificial radiation-source poses the risk of detrimental effects on aquatic lives. Moreover, the use of an artificial light-source adds on an additional expense. Additionally, the difficulty increases manifolds owing to the coexistence of various other pollutants in wastewater systems. On the basis of cost, safety, flexibility, sustainability, efficacy and future opportunity, one may compare and evaluate the effectiveness of these methods **[141]**. The chemical degradation methods have an edge over the physical ones, as the latter is associated with a certain level of risk of returning back the trapped MPs into the environment, along with release of organic additives, while the former ensures complete mineralization. Compared to physical trapping methods, photocatalysis offers low-cost involvement and better eco-friendly attributes, ultimately converting the MPs into simpler and non-toxic molecules: water and carbon dioxide **[142]**. Owing to greater flexibility, membrane-based filtration has a better prospect than coagulation and adsorption techniques in terms of efficiency among physical methods. Hence, the chemical methods relating to MPs removal/degradation provide us with a more promising and better alternative over the physical methods. However, the development of a combined physio-chemical pathway could turn out to be more effective than the individual methods by means of providing a synergistic effect. Thus, hybrid materials involving organic and inorganic phases could be the centre of focus in the field of research on this score in the future. There has not been much focus on a comprehensive, summative study/review on MP remediation involving membrane filtration and photocatalysis.

TABLE 1.1

Comparative Study of Different Methodologies Generally Employed for MP Remediation

Methodology/ Strategy	Advantages	Disadvantages	References
Coagulation and Flocculation	• Simple • Separation efficiency ~ 90% • Has an average range of applicability • Relatively cheap	• Poses risk of reincorporation of pollutants in the ecosystem • Poor sustainability in terms of recycling/reuse of coagulants • Possibility of leakage of coagulants • Conversion efficiency = 0, as it is not applicable	[70]
Adsorption	• Fair separation efficiency • Relatively cheap process • Poses substantial opportunity and scope of future research exploration	• Poses risk of reincorporation of pollutants in the ecosystem • Poor sustainability in terms of recycling/reuse of adsorbents • Possibility of leakage of adsorbents • Conversion efficiency = 0, as it is not applicable	[143]
Membrane-based filtration	• Highly reliable • More efficient than coagulation and adsorption • Wide range of applicability • High separation efficiency (Separation efficiency ~ 90% For MBR)	• Membrane fouling • Membrane abrasion • Conversion efficiency = 0, as it is not applicable	[144]
A.O.P.s	• Highly safe in terms of reintroduction of toxic pollutants in the environment • Satisfactory conversion efficiency • Intermediate range of applicability • Average sustainability	• Removal efficiency poorer than membrane-based processes • Cost involved is relatively on the higher side	[145]
Photocatalysis	• Highly safe in terms of reintroduction of toxic pollutants in the environment • 100% conversion efficiency • High range of applicability • High sustainability in terms of stability • Poses high opportunity and scope of future research exploration	• Catalyst poisoning • Mass transfer between photocatalyst and the pollutants • Relatively expensive	[146, 147]

TABLE 1.2

Comparative Study of Different Photocatalytic Systems Employed for Different Types of MP Degradation

Photocatalytic System		Types of MPs Employed	Efficiency (%)	References
TiO$_2$ film		PS	99.99	[148]
TiO$_2$ powder		PS	44.7	
TiO$_2$-PE composite	1 wt%	PE	42	[149]
	0.1 wt %		74.8	
	0.02 wt %		46.9	
TiO$_2$-PE film	0.1 wt%, 50 nm particle size	LDPE	18.1	[150]
	0.1 wt%, 200 nm particle size		7.5	
TiO$_2$-PE nanocomposite film		LDPE	68	[151]
Mesoporous N-TiO$_2$		HDPE	4.7	[152]
Bioinspired N-TiO$_2$ powder		HDPE	6.4	[132]
Bioinspired C,N-TiO$_2$ powder		HDPE	71.8	[133]
PAM-g-TiO$_2$		LDPE	39.9	[153]
ZnO nanorods		LDPE	N./A.	[129]
ZnO-Pt		LDPE	N./A.	[154]
Hydroxy-rich ultrathin BiOCl		PE	N./A.	[155]

1.5 RESEARCH GAPS AND CHALLENGES

The threat to mankind and other forms of life that is posed by microplastics is much greater than that posed by the macro-dimensional ones. Being very small in dimensions, MPs have the potential of being ingested by organisms. Moreover, reports claim there is an abundance of MPs in human food, drink and even in purified edible water. So, the challenge is really difficult. Among other remediation methods, photocatalysis stands as comparatively the more effective one, as discussed previously. Among different types of photocatalysts, the solid-phase polymer-based composites possess capabilities of functioning under mere visible light, but their efficiencies are very low. Moreover, they become ineffective while dealing with MPs in aqueous conditions. The method of dispersion of photocatalysts in an aqueous medium could serve as effective measures in the removal of MPs from various natural waterbodies, but their subsequent conversion into CO_2 and H_2O may lead to another environmental pollution by inflicting a greenhouse effect. However, if the produced CO_2 could be directly made fuel-grade, then it would be really a boon. Hence, photo-reforming processes are preferred over mere photocatalysis, as the carbon content contained by the MPs – better if it be irrespective of the chemical composition – could directly be converted to carbon-based fuels, coupled with the generation of fuel graded H_2. H_2 production is believed to be facilitated by the electron-donation, or hole-scavenging, action of various sacrificial agents produced via photodegradation of plastic materials [156]. There remains a strong requirement for an effective photocatalytic system which would be successful in executing complete valorization of MPs in aqueous environments. In other words, the carbon and hydrogen contents of MPs must be simultaneously converted into fuels or any other energy sources. Researchers must focus more on the conversion of MPs to fuels. Photolytic valorization of MPs requires more investigation, as it still remains very unexplored compared to other, more conventional photocatalytic applications, which include CO_2 reduction, water-splitting reactions, etc. Researchers must also attempt to win over various current drawbacks relating to the efficacy, durability, selectivity of photocatalysis; moreover, a gap needs to be bridged to make laboratory-scaled innovations practically more applicable. More and more, new types of photocatalytic systems must be developed, aiming at MPs valorization. The role of cocatalysts is also crucial in a catalytic system. In order to obtain high photocatalytic efficiency and valorization of MPs, development of an active co-catalyst is indispensable. This is associated with some major challenges, including inhibition of catalytic photo-corrosion [157], effective prevention of electron-hole recombination and facilitation of adsorption and subsequent activation of MPs.

Another research gap relates to the requirement of alkaline conditions as a sample pre-treatment necessity, which is believed to affect the aquatic habitat. In addition, the alkaline conditions lead to non-selective, random chain-scission in MPs, resulting in the formation of chains with varied lengths, which, in turn, affects the selectivity of the products generated via photocatalysis. An alkaline environment also becomes detrimental for metal-oxide-based photocatalytic systems and leads to catalyst poisoning. Hence, the development of a versatile catalytic system capable of working in any environmental conditions is desirable. Reality also demands the practical implementation of the developed photocatalytic system in waste water treatment plants, for which the housing of the photocatalyst is necessitated within a reasonably flexible and porous matrix, preferably a membrane. Such assemblies could eliminate sedimentation issues and, at the same time, contribute to the recycling of the catalyst. On the hind side, this might affect the mass transfer that occurs between the catalyst and the pollutant [64, 158]. Continuous research must go on in order to achieve an optimum balance of catalytic composition, working conditions and the corresponding reactor design resulting in efficient photocatalytic valorization of MPs. In the past, various photocatalysts have been developed based on transition metal oxides like TiO_2, ZnO, etc., along with the doped ones, which are most responsive to UV radiations and are expensive as well, while some quantum dots like CdS/CdO_x have been reported to operate under visible light but are associated with toxicity issues [139]. Hence, it is challenging to develop a photocatalyst which could effectively convert MPs to essential green fuel, even under the more abundantly available visible light. Moreover, the scarcity of UV radiations in a natural aquatic environment is attributed to the screening effect of the earth's atmosphere, which allows only about 5% to 8% to reach the sea level, and the extent gets further decreased as the depth of the water level increases within natural water bodies or water treatment plants. Apart from this, another major challenge is to develop a photocatalytic system functional in an actual natural environment rather than making a prototype operative on the laboratory scale. There is scope to modify the structural attributes of photocatalytic systems in order to tune the band gap and surface properties. Future research must also attempt to increase the versatility of the photocatalysts by making them sensitive to radiations of wide range of wavelengths. The ideal fabrication could include the entire light spectrum. Preventive measures must be adopted in terms of photocatalytic development, which ensure a high lifetime of photogenerated electrons and holes. This could be achieved by restricting their recombination.

The fabrication of advanced and novel photocatalysts demands the proposition of modern theoretical models and characterization techniques. Despite the current scenario relating to the lacunas involving the practical applicability of photocatalytic systems, there is, indeed, an immense potentiality and future prospect which could be achieved through sustained optimization in the fabrication of photocatalysts. In addition, the expansion in the domain of knowledge regarding the photocatalytic mechanism and related studies, along with the development of the technology involving the relevant characterization techniques, are highly recommended.

1.6 CONCLUSION

An inherent shortfall of the WWTPs, in general, relates to the inefficiency of a tertiary treatment step which allows the reintroduction of MPs from the secondary treatment step into water bodies. In such cases, the use of filtering membranes is mostly adopted, which, in turn, is challenging, owing to certain issues such as clogging, disposition, etc. Among various membrane technologies, RO stands out in terms of efficacy of MP removal from an aqueous medium, while it is also susceptible to issues like membrane fouling, hampering performance. On the other hand, the MP-removal strategy involving adsorption has a downside associated with the disposal of the used adsorbing samples containing the pollutants, while the involved cost curbs the exploration of the green AOPs. Nevertheless, there lies an immense potential for the development of efficient, cost-effective strategies for MP removal, particularly from aqueous systems. Strategies like AOPs and photocatalysis will play crucial roles in the near future. The synergistic effect associated with the combined strategies involving both physical and chemical methods holds the key for such future developments. Photocatalytic membranes, advanced reactor design and novel concepts like biomimetics (and many more) would certainly lead to the breakthrough. Moreover, in today's era of scientific research, economics plays a decisive role. The concept of converting waste to raw materials is highly desirable. Valorization of MPs should be the ultimate goal, which not only ensures complete eradication of the pollutants but provides an alternate source to the production of fuel graded products and other valuable resources. Noteworthily, the tradition of using biodegradable plastics may not actually turn out to be a fruitful one unless the design and disposal of such materials are carefully monitored. Owing to their inherent properties, there is always a great risk of producing a greater amount of MPs; that, too, at a brisk rate.

The future of photocatalyst development intended for MP degradation, especially in the aqueous environment, is very bright, creating numerous opportunities, as this field still remained relatively unexplored by researchers all across the globe. This definitely leaves a lot of scope to focus on. Catalysts possessing high activity, coupled with selectivity, need to be developed which could effectively photodegrade MPs of various chemical natures and sizes and convert them into valuable, non-toxic substances. The quality of the photodegradation products needs to be evaluated; highly pure substances are desirable in order to enhance the efficiency of the conversion and subsequent valorization. The selectivity of the photodegradation of the MPs plays an important in this regard. Techniques like Positron annihilation spectroscopy, DRFTIR spectroscopy, HPLC-HRMS, etc., can be used to describe the mechanism of selective degradation of microplastics. Plastic compounding involves various additives, most of which are toxic; hence, leaching out of such chemicals during the degradation process as well as the adsorption of different chemicals onto MPs make them a source of toxic contaminants. Toxicological tests are required to be carried out on the intermediate products produced during photodegradation of MPs.

REFERENCES

1. Bergmann, M., Gutow, L., and Klages, M. (Eds.), *Marine Anthropogenic Litter*, 2015. doi:10.1007/978-3-319-16510-3.
2. Campanale, C., et al., A detailed review study on potential effects of microplastics and additives of concern on human health, *Int. J. Environ. Res. Public Health.*, 17(4):1212, 2020. doi:10.3390/ijerph17041212.
3. Bhuyan, M.S., Effects of microplastics on fish and in human health, *Front. Environ. Sci.*, 10:827289, 2022. doi:10.3389/fenvs.2022.827289.
4. Dissanayake, P.D. et al., Effects of microplastics on the terrestrial environment: A critical review, *Environ Res.*, 209:112734, 2022. doi:10.1016/j.envres.2022.112734.
5. Leslie, H.A. et al., Discovery and quantification of plastic particle pollution in human blood, *Environ Int.*, 163:107199, 2022. doi:10.1016/j.envint.2022.107199.
6. van Raamsdonk, L.W.D. et al., Current insights into monitoring, bioaccumulation, and potential health effects of microplastics present in the food chain, *Foods.*, 9(1), 2020. doi:10.3390/foods9010072.
7. Murray, F., and Cowie, P.R., Plastic contamination in the decapod crustacean Nephrops norvegicus (Linnaeus, 1758), *Mar. Pollut. Bull.*, 62(6):1207–1217, 2011. doi:10.1016/j.marpolbul.2011.03.032.
8. Browne, M.A. et al., Ingested microscopic plastic translocates to the circulatory system of the mussel, Mytilus edulis (L), *Environ. Sci. Technol.*, 42(13):5026–5031, 2008. doi:10.1021/es800249a.
9. von Moos, N., Burkhardt-Holm, P., and Köhler, A., Uptake and effects of microplastics on cells and tissue of the blue mussel Mytilus edulis L. after an experimental exposure, *Environ. Sci. Technol.*, 46(20):11327–11335, 2012. doi:10.1021/es302332w.
10. Catrouillet, C. et al., Metals in microplastics: Determining which are additive, adsorbed, and bioavailable, *Environ. Sci. Process Impacts.*, 23(4):553–558, 2021. doi:10.1039/d1em00017a.

11. Huang, W. et al., Microplastics and associated contaminants in the aquatic environment: A review on their ecotoxicological effects, trophic transfer, and potential impacts to human health, *J. Hazard. Mater.*, 405, 2021. doi:10.1016/j.jhazmat.2020.124187.

12. Rochman, C.M. et al., Rethinking microplastics as a diverse contaminant suite, *Environ. Toxicol. Chem.*, 38(4):703–711, 2019. doi:10.1002/etc.4371.

13. Gray, A.D., and Weinstein, J.E., Size- and shape-dependent effects of microplastic particles on adult daggerblade grass shrimp (Palaemonetes pugio), *Environ. Toxicol. Chem.*, 36(11):3074–3080, 2017. doi:10.1002/etc.3881.

14. Hermsen, E. et al., Quality criteria for the analysis of microplastic in biota samples: A critical review, *Environ. Sci. Technol.*, 52(18):10230–10240, 2018. doi:10.1021/acs.est.8b01611.

15. Choi, J.S. et al., Toxicological effects of irregularly shaped and spherical microplastics in a marine teleost, the sheepshead minnow (Cyprinodon variegatus), *Mar. Pollut. Bull.*, 129(1):231–240, 2018. doi:10.1016/j.marpolbul.2018.02.039.

16. Botterell, Z.L.R. et al., Bioavailability of microplastics to Marine Zooplankton: Effect of shape and infochemicals, *Environ. Sci. Technol.*, 54(19):12024–12033, 2020. doi:10.1021/acs.est.0c02715.

17. Zeenat, E.A. et al., Plastics degradation by microbes: A sustainable approach, *J. King Saud. Univ. Sci.*, 33(6):101538, 2021. doi:10.1016/j.jksus.101538.

18. Webb, H.K. et al., Plastic degradation and its environmental implications with special reference to poly(ethylene terephthalate), *Polymers*, 5:1–18, 2013. doi:10.3390/polym5010001.

19. Lange, J.P., Managing plastic waste—sorting, recycling, disposal, and product redesign, *ACS Sustain. Chem. Eng.* (47):15722–15738, 2021. doi:10.1021/acssuschemeng.1c05013.

20. Thakuria, T., Concept and transmission of microplastics in human diet, *Int. J. Life Sci. Agric. Res.*, 2(04):35–39, 2023. doi:10.55677/ijlsar/V02I04Y2023–03.

21. Murphy, F. et al., Wastewater treatment works (WwTW) as a source of microplastics in the aquatic environment, *Environ. Sci. Technol.*, 50(11):5800–5808, 2016. doi:10.1021/acs.est.5b05416.

22. Zhou, Q. et al., Separation of microplastics from a coastal soil and their surface microscopic features, *Chinese Sci. Bull.*, 61:1604–1611, 2016. doi:10.1360/N972015-01098.

23. Sanchez, W., Bender, C., and Porcher, J.M., Wild gudgeons (Gobio gobio) from French rivers are contaminated by microplastics: Preliminary study and first evidence, *Environ. Res.*, 128:98–100, 2014. doi:10.1016/j.envres.2013.11.004.

24. Wang, W. et al., Microplastics pollution in inland freshwaters of China: A case study in urban surface waters of Wuhan, China, *Sci. Total Environ.*, 575:1369–1374, 2017. doi:10.1016/j.scitotenv.2016.09.213.

25. Sajjad, A. et al., Development of novel magnetic solid-phase extraction sorbent based on Fe_3O_4/carbon nanosphere/polypyrrole composite and their application to the enrichment of polycyclic aromatic hydrocarbons from water samples prior to GC–FID analysis, *J. Iran. Chem. Soc.(Print)*, 15(1):153–161, 2018. doi:101007/s13738-017-1218-6.

26. Law, K.L., and Thompson, R.C., Microplastics in the seas, *Science.*, 345:144–145, 2014. doi:10.1126/science.1254065.

27. Hodson, M.E. et al., Plastic bag derived-microplastics as a vector for metal exposure in terrestrial invertebrates, *Environ. Sci. Technol.*, 51(8):4714–4721, 2017. doi:10.1021/acs.est.7b00635.

28. He, D. et al., Microplastics in soils: Analytical methods, pollution characteristics and ecological risks, *Trends Anal. Chem.*, 2018. doi:10.1016/j.trac.2018.10.006.

29. Yuan, Z., Nag, R., and Cummins, E., Human health concerns regarding microplastics in the aquatic environment—From marine to food systems, *Sci. Total Environ.*, 823, 2022. doi:10.1016/j.scitotenv.2022.153730.

30. Nizzetto, L. et al., A theoretical assessment of microplastic transport in river catchments and their retention by soils and river sediments, *Environ. Sci. Process Impacts.*, 18(8):1050–1059, 2016. doi:10.1039/c6em00206d.

31. Katsumi, N. et al., The role of coated fertilizer used in paddy fields as a source of microplastics in the marine environment, *Mar. Pollut. Bull.*, 161:111727, 2020. doi:10.1016/j.marpolbul.2020.1117.

32. Barboza, L.G.A. et al., Marine microplastic debris: An emerging issue for food security, food safety and human health, *Mar. Pollut. Bull.*, 133:336–348, 2018. doi:10.1016/j.marpolbul.2018.05.047.

33. Free, C.M. et al., High-levels of microplastic pollution in a large, remote, mountain lake, *Mar. Pollut. Bull.*, 85(1):156–163, 2014. doi:10.1016/j.marpolbul.2014.06.001.

34. Shahul Hamid, F. et al., Worldwide distribution and abundance of microplastic: How dire is the situation? *Waste Manag. Res.*, 0734242X1878573, 2018. doi:10.1177/0734242x18785730.

35. Li, X. et al., Microplastics in sewage sludge from the wastewater treatment plants in China, *Water Res.*, 2018. doi:10.1016/j.watres.2018.05.034.

36. Akarsu, C. et al., Microplastics composition and load from three wastewater treatment plants discharging into Mersin Bay, north eastern Mediterranean Sea, *Mar. Pollut. Bull.*, 110776, 2019. doi:10.1016/j.marpolbul.2019.110776.

37. Naji, A. et al., Microplastics in wastewater outlets of Bandar Abbas city (Iran): A potential point source of microplastics into the Persian Gulf, *Chemosphere*, 128039, 2020. doi:10.1016/j.chemosphere.2020.128039.

38. Giorgetti, L. et al., Exploring the interaction between polystyrene nanoplastics and Allium cepa during germination: Internalization in root cells, induction of toxicity and oxidative stress, *Plant Physiol. Biochem.*, 2020. doi:10.1016/j.plaphy.2020.02.014.

39. Li, Z. et al., The distribution and impact of polystyrene nanoplastics on cucumber plants, *Environ. Sci. Pollut. Res.*, 2020. doi:10.1007/s11356-020-11702-2.

40. Li, L. et al., Effective uptake of submicrometre plastics by crop plants via a crack-entry mode, *Nat. Sustain.*, 2020. doi:10.1038/s41893-020-0567-9.

41. Liu, Y. et al., Uptake and translocation of nano/microplastics by rice seedlings: Evidence from a hydroponic experiment, *J. Hazard. Mater.*, 421:126700, 2022. doi:10.1016/j.jhazmat.2021.126700.

42. Lian, J. et al., Impact of polystyrene nanoplastics (PSNPs) on seed germination and seedling growth of wheat (Triticum aestivum L.), *J. Hazard. Mater.*, 2019. doi:10.1016/j.jhazmat.2019.121620.

43. Dong, Y. et al., Uptake of microplastics by carrots in presence of As (III): Combined toxic effects, *J. Hazard. Mater.*, 411:125055, 2021. doi:10.1016/j.jhazmat.2021.125055.

44. Shukla, P.K., Misra, P., and Kole, C., Uptake, translocation, accumulation, transformation, and generational transmission of nanoparticles in plants, *Plant Nanotechnol.*, 183–218, 2016. doi:10.1007/978-3-319-42154-4_8.

45. Granek, E.F., Brander, S.M., and Holland, E.B., Microplastics in aquatic organisms: Improving understanding and identifying research directions for the next decade, *Limnol. Oceanogr. Lettr.*, 5(1):1–4, 2020. doi:10.1002/lol2.10145.

46. Sussarellu, R. et al., Oyster reproduction is affected by exposure to polystyrene microplastics, *Proc. Natl. Acad. Sci. U.S.A.*, 113(9):2430–2435, 2016. doi:10.1073/pnas.1519019113.

47. Bessa, F. et al., Microplastics in gentoo penguins from the Antarctic region. *Sci. Rep.*, 9:14191, 2019. doi:10.1038/s41598-019-50621-2.

48. Cole, M. et al., The impact of polystyrene microplastics on feeding, function and fecundity in the marine copepod Calanus helgolandicus. *Environ. Sci. Technol.*, 49(2):1130–1137, 2015. doi:10.1021/es504525u.

49. Yu, H. et al., Effects of microplastics and glyphosate on growth rate, morphological plasticity, photosynthesis, and oxidative stress in the aquatic species Salvinia cucullate, *Environ. Pollut.*, 279:116900, 2021. doi:10.1016/j.envpol.2021.116900.

50. Marques, F. et al., Major characteristics of microplastics in mussels from the Portuguese coast, *Environ. Res.*, 197:110993, 2021. doi:10.1016/j.envres.2021.110993.

51. Janardhanam, M. et al., Microplastics in demersal sharks from the Southeast Indian coastal region, *Front. Mar. Sci.*, 9:914391, 2022. doi:10.3389/fmars.2022.914391.

52. Parton, K.J. et al., Investigating the presence of microplastics in demersal sharks of the North-East Atlantic, *Sci. Rep.*, 10(1), 2020. doi:10.1038/s41598-020-68680-1.

53. Wang, L. et al., Birds and plastic pollution: Recent advances, *Avian Res.*, 12(1):59, 2021. doi:10.1186/s40657-021-00293-2.

54. Fackelmann, G. et al., Current levels of microplastic pollution impact wild seabird gut microbiomes, *Nat. Ecol. Evol.*, 7(5):698–706, 2023. doi:10.1038/s41559-023-02013-z.

55. Schwabl, P. et al., Detection of various microplastics in human stool, *Ann. Intern. Med.*, 171(7):453, 2019. doi:10.7326/m19-0618.

56. Ragusa, A. et al., Deeply in plasticenta: Presence of microplastics in the intracellular compartment of human placentas, *Int. J. Environ. Res. Public Health.*, 19(18):11593, 2022. doi:10.3390/ijerph191811593.

57. Padervand, M. et al., Removal of microplastics from the environment. A review, *Environ. Chem. Lett.*, 2020. doi:10.1007/s10311-020-00983-1.

58. Tang, W. et al., The removal of microplastics from water by coagulation: A comprehensive review, *Sci. Total Environ.*, 851(Pt 1):158224, 2022. doi:10.1016/j.scitotenv.2022.158224.

59. Conesa, J.A., and Ortuño, N., Reuse of water contaminated by microplastics, the effectiveness of filtration processes: A review, *Energies.*, 15:2432, 2022. doi:10.3390/en15072432.

60. Othman, A.R. et al., Microbial degradation of microplastics by enzymatic processes: A review, *Environ. Chem. Lett.*, 19(4):3057–3073, 2021. doi:10.1007/s10311-021-01197-9.

61. Martin, L.M.A. et al., Testing an iron oxide nanoparticle-based method for magnetic separation of nanoplastics and microplastics from water, *Nanomaterials (Basel).*, 12(14):2348, 2022. doi:10.3390/nano12142348.

62. Elmobarak, W.F. et al., A review on the treatment of petroleum refinery wastewater using advanced oxidation processes, *Catalysts.*, 11:782, 2021. doi:10.3390/catal11070782.

63. Gupta, R. et al., Magnetically supported ionic liquids: A sustainable catalytic route for organic transformations, *Mater. Horiz.*, 7(12):3097–3130, 2020. doi:10.1039/d0mh01088j.

64. Golmohammadi, M. et al., Molecular mechanisms of microplastics degradation: A review, *Sep. Purif. Technol.*, 122906, 2022.

65. Sun, J. et al., Microplastics in wastewater treatment plants: Detection, occurrence and removal, *Water Res.*, 2019. doi:10.1016/j.watres.2018.12.050.

66. Iyare, P.U., Ouki, S.K., and Bond, T., Microplastics removal in wastewater treatment plants: A critical review, *Environ. Sci. Water Res. Technol.*, 6(10):2664–2675, 2020. doi:10.1039/d0ew00397b.

67. Ma, B. et al., Removal characteristics of microplastics by Fe-based coagulants during drinking water treatment, *J. Environ. Sci.*, 2018. doi:10.1016/j.jes.2018.10.006.

68. Carr, S.A., Liu, J., and Tesoro, A.G., Transport and fate of microplastic particles in wastewater treatment plants, *Water Res.*, 91:174–182, 2016. doi:10.1016/j.watres.2016.01.002.

69. Perren, W., Wojtasik, A., and Cai, Q., Removal of microbeads from wastewater using electrocoagulation, *ACS Omega*, 3(3), 3357–3364, 2018. doi:10.1021/acsomega.7b02037.

70. Lapointe, M. et al., Understanding and improving microplastics removal during water treatment: Impact of coagulation and flocculation, *Environ. Sci. Technol.*, 2020. doi:10.1021/acs.est.0c00712.

71. Rajala, K. et al., Removal of microplastics from secondary wastewater treatment plant effluent by coagulation/flocculation with iron, aluminum and polyamine-based chemicals, *Water Res.*, 2020. doi:10.1016/j.watres.2020.116045.

72. Ma, B. et al., Characteristics of microplastic removal via coagulation and ultrafiltration during drinking water treatment, *Chem. Eng. J.*, 359:159–167, 2019. doi:10.1016/j.cej.2018.11.155.

73. Herbort, A.F. et al., Alkoxy-silyl induced agglomeration: A new approach for the sustainable removal of microplastic from aquatic systems, *J. Polym. Environ.*, 2018. doi:10.1007/s10924-018-1287-3.

74. Xu, X. et al., Cost-effective polymer-based membranes for drinking water purification. *Giant*, 100099, 2022. doi:10.1016/j.giant.2022.100099.

75. Ziajahromi, S. et al., Wastewater treatment plants as a pathway for microplastics: Development of a new approach to sample wastewater-based microplastics, *Water Res.*, 112:93–99, 2017. doi:10.1016/j.watres.2017.01.042.

76. Lares, M. et al., Occurrence, identification and removal of microplastic particles and fibers in conventional activated sludge process and advanced MBR technology, *Water Res.*, 133:236–246, 2018. doi:10.1016/j.watres.2018.01.049.

77. Salerno, C. et al., Influence of air scouring on the performance of a self forming dynamic membrane bioreactor (SFD MBR) for municipal wastewater treatment, *Bioresour. Technol.*, 2016. doi:10.1016/j.biortech.2016.10.054.

78. Ersahin, M.E. et al., Applicability of dynamic membrane technology in anaerobic membrane bioreactors. *Water Res.*, 48:420–429, 2014. doi:10.1016/j.watres.2013.09.054.

79. Gurung, K. et al., Incorporating submerged MBR in conventional activated sludge process for municipal wastewater treatment: A feasibility and performance assessment, *J. Membra Sci. Technol.*, 6:158, 2016. doi:10.4172/2155-9589.1000158.

80. Talvitie, J. et al., Solutions to microplastic pollution—Removal of microplastics from wastewater effluent with advanced wastewater treatment technologies, *Water Res.*, 123:401–407, 2017. doi:10.1016/j.watres.2017.07.005.

81. Hou, L. et al., Conversion and removal strategies for microplastics in wastewater treatment plants and landfills, *Chem. Eng. J.*, 126715, 2020. doi:10.1016/j.cej.2020.126715.

82. Enfrin, M., Dumée, L.F., and Lee, J., Nano/microplastics in water and wastewater treatment processes—origin, impact and potential solutions, *Water Res.*, 2019. doi:10.1016/j.watres.2019.06.049.

83. Tsiota, P. et al., Microbial degradation of HDPE secondary microplastics: Preliminary results, *Proceedings of the International Conference on Microplastic Pollution in the Mediterranean Sea*, 181–188, 2017. doi:10.1007/978-3-319-71279-6_24.

84. Auta, H.S., Emenike, C.U., and Fauziah, S.H., Screening of Bacillus strains isolated from mangrove ecosystems in Peninsular Malaysia for microplastic degradation, *Environ. Pollut.*, 231:1552–1559, 2017. doi:10.1016/j.envpol.2017.09.043.

85. Arossa, S. et al., Microplastic removal by red sea giant clam (Tridacna maxima), *Environ. Pollut.*, 2019. doi:10.1016/j.envpol.2019.05.149.

86. Lee, H., and Kim, Y., Treatment characteristics of microplastics at biological sewage treatment facilities in Korea, *Mar. Pollut. Bull.*, 137:1–8, 2018. doi:10.1016/j.marpolbul.2018.09.050.

87. Camacho-Muñoz, D. et al., Effectiveness of three configurations of membrane bioreactors on the removal of priority and emergent organic compounds from wastewater: Comparison with conventional wastewater treatments, *J. Environ. Monitor.*, 14(5):1428, 2012. doi:10.1039/c2em00007e.

88. Balabanič, D. et al., Comparison of different wastewater treatments for removal of selected endocrine-disruptors from paper mill wastewaters, *J. Environ. Sci. Health A*, 47(10):1350–1363, 2012. doi:10.1080/10934529.2012.672301.

89. Yang, Y., Yang, J., and Jiang, L., Comment on "a bacterium that degrades and assimilates poly(ethylene terephthalate).", *Science*, 353(6301):759–759, 2016. doi:10.1126/science.aaf8305.

90. Dawson, A.L. et al., Turning microplastics into nanoplastics through digestive fragmentation by Antarctic krill, *Nat. Commun.*, 9(1), 2018. doi:10.1038/s41467-018-03465-9.

91. Malankowska, M., Echaide-Gorriz, C., and Coronas, J., Microplastics in marine environment—sources, classification, and potential remediation by membrane technology—A review, *Environ. Sci. Water Res. Technol.*, 2020. doi:10.1039/d0ew00802h.

92. Pizzichetti, A.R.P. et al., Evaluation of membranes performance for microplastic removal in a simple and low-cost filtration system, *Case Stud. Chem. Environ. Eng.*, 3:100075, 2021. doi:10.1016/j.cscee.2020.100075.

93. Liu, F. et al., Microplastics removal from treated wastewater by a biofilter, *Water*, 12(4):1085, 2020. doi:10.3390/w12041085.

94. Beljanski, A., Efficiency and effectiveness of a low-cost, self-cleaning microplastic filtering system for wastewater treatment plants, *Proceedings of The National Conference On Undergraduate Research (NCUR)*, 2016.

95. United States Patent, Marine microplastic removal tool, February 3, 2015, US 8944253 B2.

96. Gewert, B., Plassmann, M.M., and MacLeod, M., Pathways for degradation of plastic polymers floating in the marine environment, *Environ. Sci. Process Impacts.*, 17(9):1513–1521, 2015. doi:10.1039/c5em00207a.

97. Anantharaman, A. et al., Pre-deposited dynamic membrane filtration—A review, *Water Res.*, 115558, 2020. doi:10.1016/j.watres.2020.115558.

98. Li, L. et al., Dynamic membrane for micro-particle removal in wastewater treatment: Performance and influencing factors, *Sci. Total Environ.*, 627:332–340, 2018. doi:10.1016/j.scitotenv.2018.01.239.

99. Yuan, F. et al., Study on the adsorption of polystyrene microplastics by three-dimensional reduced graphene oxide, *Water Sci. Technol.*, 2020. doi:10.2166/wst.2020.269.

100. Misra, A. et al., Water purification and microplastics removal using magnetic polyoxometalate-supported ionic liquid phases (magPOM-SILPs), *Angew. Chem. Int. Ed. Engl.*, 59(4):1601–1605, 2019. doi:10.1002/anie.201912111.

101. Herbort, A.F., and Schuhen, K., A concept for the removal of microplastics from the marine environment with innovative host-guest relationships, *Environ. Sci. Pollut. Res.*, 24(12):11061–11065, 2016. doi:10.1007/s11356-016-7216-x.

102. Tiwari, E. et al., Application of Zn/Al layered double hydroxides for the removal of nano-scale plastic debris from aqueous systems, *J. Hazard. Mater.*, 397:122769, 2020. doi:10.1016/j.jhazmat.2020.122769.

103. Tang, Y. et al., Removal of microplastics from aqueous solutions by magnetic carbon nanotubes, *Chem. Eng. J.*, 126804, 2020. doi:10.1016/j.cej.2020.126804.

104. Lan, Y.-Q. et al., Metal-organic framework based foams for efficient microplastic removal, *J. Mater. Chem. A*, 2020. doi:10.1039/d0ta04891g.

105. Bhattacharya, P. et al., Physical adsorption of charged plastic nanoparticles affects algal photosynthesis, *J. Phys. Chem. C*, 114(39):16556–16561, 2010. doi:10.1021/jp1054759.

106. Nolte, T.M. et al., The toxicity of plastic nanoparticles to green algae as influenced by surface modification, medium hardness and cellular adsorption, *Aquat. Toxicol.*, 183:11–20, 2017. doi:10.1016/j.aquatox.2016.12.005.

107. Martins, M.J.F., Mota, C.F., and Pearson, G.A., Sex-biased gene expression in the brown alga Fucus vesiculosus, *BMC Genom.*, 14(1):294, 2013. doi:10.1186/1471-2164-14-294.

108. Gorokhova, E., Ek, K., and Reichelt, S., Algal growth at environmentally relevant concentrations of suspended solids: Implications for microplastic hazard assessment, *Front. Environ. Sci.*, 8, 2020. doi:10.3389/fenvs.2020.551075.

109. Siipola, V. et al., Low-cost biochar adsorbents for water purification including microplastics removal, *Appl. Sci.*, 10(3):788, 2020. doi:10.3390/app10030788.

110. Zhang, Z., and Chen, Y., Effects of microplastics on wastewater and sewage sludge treatment and their removal: A review, *Chem. Eng. J.*, 122955, 2019. doi:10.1016/j.cej.2019.122955.

111. Mahon, A.M. et al., Microplastics in Sewage sludge: Effects of treatment, *Environ. Sci. Technol.*, 51(2):810–818, 2016. doi:10.1021/acs.est.6b04048.

112. Zhu, K. et al., Formation of environmentally persistent free radicals on microplastics under light irradiation, *Environ. Sci. Technol.*, 2019. doi:10.1021/acs.est.9b01474.

113. Duoerkun, G. et al., Construction of n-TiO_2/p-Ag_2O junction on carbon fiber cloth with Vis–NIR photoresponse as a filter-membrane-shaped photocatalyst, *Adv. Fiber Mater.*, 2(1):13–23, 2020. doi:10.1007/s42765-019-00025-8.

114. Karimi Estahbanati, M.R., Kong, X.Y., Eslami, A., and Soo, H.S., Current developments in the chemical upcycling of waste plastics using alternative energy sources, *ChemSusChem*, 2021. doi:10.1002/cssc.202100874.

115. Lee, S.-Y., and Park, S.-J., TiO_2 photocatalyst for water treatment applications, *J. Ind. Eng. Chem.*, 19(6):1761–1769, 2013. doi:10.1016/j.jiec.2013.07.012.

116. Rajaambal, S., Sivaranjani, K., and Gopinath, C.S., Recent developments in solar H_2 generation from water splitting, *Int. J. Chem. Sci.*, 127(1):33–47, 2015. doi:10.1007/s12039-014-0747-0.

117. Nakata, K., and Fujishima, A., TiO_2 photocatalysis: Design and applications, *J. Photochem. Photobiol. C: Photochem. Rev.*, 13(3):169–189, 2012. doi:10.1016/j.jphotchemrev.2012.

118. Schneider, J. et al., Understanding TiO_2 photocatalysis: Mechanisms and materials, *Chem. Rev.*, 114(19):9919–9986, 2014. doi:10.1021/cr5001892.

119. Ishchenko, O. et al., TiO_2, ZnO, and SnO_2-based metal oxides for photocatalytic applications: Principles and development, *Comptes Rendus. Chimie*, 24(1):103–124, 2021. doi:10.5802/crchim.64.

120. Stephen, L., Titanium dioxide versatile solid crystalline: An overview, *Assorted Dimensional Reconfigurable Materials*, 2020. doi:10.5772/intechopen.92056.

121. Kang, X. et al., Titanium dioxide: From engineering to applications, *Catalysts*, 9(2):191, 2019. doi:10.3390/catal9020191.

122. Manske Nunes, S. et al., Different crystalline forms of titanium dioxide nanomaterial (rutile and anatase) can influence the toxicity of cooper in golden mussel Limnoperna fortunei? *Aquat. Toxicol*, 2018. doi:10.1016/j.aquatox.2018.10.009.

123. Zhang, J. et al., Photocatalysis, *Lect. Notes Chem.*, 2018. doi:10.1007/978-981-13-2113-9.

124. Bickley, R. et al., A structural investigation of titanium dioxide photocatalysts, *J. Solid State Chem.*, 92:178–190, 1991. doi:10.1016/0022-4596(91)90255-G.

125. Zhang, J. et al., New understanding of the difference of photocatalytic activity among anatase, rutile and brookite TiO_2, *Phys. Chem. Chem. Phys.*, 16(38):20382–20386, 2014. doi:10.1039/c4cp02201g.

126. Nishimura, A., et al., Effect of preparation condition of TiO_2 film and experimental condition on CO_2 reduction performance of TiO_2 photocatalyst membrane reactor, *Int. J. Photoenergy*, 1–14, 2011. doi:10.1155/2011/305650.

127. Park, H., et al., Surface modification of TiO_2 photocatalyst for environmental applications, *J. Photochem. Photobiol. C: Photochem. Rev.*, 15:1–20, 2013. doi:10.1016/j.jphotchemrev.2012.

128. Khlyustova, A., et al., Doped TiO_2: Effect of doping element on the photocatalytic activity, *Adv. Mater.*, 2020. doi:10.1039/d0ma00171f.

129. Tofa, T.S. et al., Visible light photocatalytic degradation of microplastic residues with zinc oxide nanorods, *Environ. Chem. Lett.*, 2019. doi:10.1007/s10311-019-00859-z.

130. Wang, L. et al., Photocatalytic TiO_2 micromotors for removal of microplastics and suspended matter, *ACS Appl. Mater. Interfaces.*, 2019. doi:10.1021/acsami.9b06128.

131. Sekino, T., Takahashi, S., and Takamasu, K., Fundamental study on nanoremoval processing method for microplastic structures using photocatalyzed oxidation, *Key Eng.*, 523–524:610–614, 2012. doi:10.4028/www.scientific.net/kem.523-524.610.

132. Ariza-Tarazona, M.C. et al., New strategy for microplastic degradation: Green photocatalysis using a protein-based porous N-TiO_2 semiconductor, *Ceram. Int.*, 2018. doi:10.1016/j.ceramint.2018.10.208.

133. Ariza-Tarazona, M.C. et al., Microplastic pollution reduction by a carbon and nitrogen-doped TiO_2: Effect of pH and temperature in the photocatalytic degradation process, *J. Hazard. Mater.*, 122632, 2020. doi:10.1016/j.jhazmat.2020.122632.

134. Beladi-Mousavi, S.M. et al., A maze in plastic wastes: Autonomous motile photocatalytic microrobots against microplastics, *ACS Appl. Mater. Interfaces.*, 13(21):25102–25110, 2021. doi:10.1021/acsami.1c04559.

135. Bandara, W.R.L.N. et al., Is nano ZrO_2 a better photocatalyst than nano TiO_2 for degradation of plastics? *RSC Adv.*, 7(73):46155–46163, 2017. doi:10.1039/c7ra08324f.

136. Tosoni, S. et al., TiO_2 and ZrO_2 in biomass conversion: Why catalyst reduction helps, *Phil. Trans. R. Soc. A*, 376:20170056, 2017. doi:10.1098/rsta.2017.0056.

137. Jiang, R. et al., Microplastic degradation by hydroxy-rich bismuth oxychloride, *J. Hazard. Mater.*, 124247, 2020. doi:10.1016/j.jhazmat.2020.124247.

138. Kawai, T., and Sakata, T., Photocatalytic decomposition of gaseous water over TiO_2 and TiO_2—RuO_2 surfaces, *Chem. Phys. Lett.*, 72(1):87–89, 1980. doi:10.1016/0009-2614(80)80247-8.

139. Uekert, T. et al., Plastic waste as a feedstock for solar-driven H_2 generation, *Energy Environ. Sci.*, 2018. doi:10.1039/c8ee01408f.

140. Xie, Y. et al., Photocatalyzing waste plastics into C2 fuels under simulated natural environments, *Angew. Chem. Int. Ed. Engl.*, 2020. doi:10.1002/anie.201915766.

141. Chen, J. et al., How to build a microplastics-free environment: Strategies for microplastics degradation and plastics recycling, *Adv Sci (Weinh).*, 9(6):e2103764, 2022. doi:10.1002/advs.202103764.

142. Xie, A. et al., Photocatalytic technologies for transformation and degradation of microplastics in the environment: Current achievements and future prospects, *Catalysts*, 13:846, 2023. doi:10.3390/catal13050846.

143. Poerio, T., Piacentini, E., and Mazzei, R., Membrane processes for microplastic removal, *Molecules*, 24(22):4148, 2019. doi:10.3390/molecules24224148.

144. Joo, S.H. et al., Microplastics with adsorbed contaminants: Mechanisms and treatment, *Environ. Chall.*, 3:100042, 2021. doi:10.1016/j.envc.2021.100042.

145. Brienza, M., and Katsoyiannis, I., Sulfate radical technologies as tertiary treatment for the removal of emerging contaminants from wastewater, *Sustainability*, 9(9):1604, 2017. doi:10.3390/su9091604.

146. Lee, Q.Y., and Li, H., Photocatalytic degradation of plastic waste: A mini review, *Micromachines*, 12(8):907, 2021. doi:10.3390/mi12080907.

147. Mandade, P., Introduction, basic principles, mechanism, and challenges of photocatalysis. *Handbook Nanomaterials Wastewater Treat.*, 137–154, 2021. doi:10.1016/b978-0-12-821496-1.00016-7.

148. Nabi, I., et al., Complete photocatalytic mineralization of microplastic on TiO_2 nanoparticle film, *ISCIENCE*, 2020. doi:10.1016/j.isci.2020.101326.

149. Zhao, X.U., et al., Solid-phase photocatalytic degradation of polyethylene plastic under UV and solar light irradiation, *J. Mol. Catal. A Chem.*, 268(1–2):101–106, 2007. doi:10.1016/j.molcata.2006.12.012.

150. Thomas, R.T., Nair, V., and Sandhyarani, N., TiO_2 nanoparticle assisted solid phase photocatalytic degradation of polythene film: A mechanistic investigation, *Colloids Surf., A Physicochem. Eng. Asp.*, 422:1–9, 2013. doi:10.1016/j.colsurfa.2013.01.017.

151. Thomas, R.T., and Sandhyarani, N., Enhancement in the photocatalytic degradation of low density polyethylene–TiO_2 nanocomposite films under solar irradiation, *RSC Adv.*, 3(33):14080, 2013. doi:10.1039/c3ra42226g.

152. Llorente-García, B.E., et al., First insights into photocatalytic degradation of HDPE and LDPE microplastics by a mesoporous N–TiO_2 coating: Effect of size and shape of microplastics, *Coatings*, 10(7):658, 2020. doi:10.3390/coatings10070658.

153. Liang, W., et al., High photocatalytic degradation activity of polyethylene containing polyacrylamide grafted TiO_2, *Polym. Degrad. Stab.*, 98(9):1754–1761, 2013. doi:10.1016/j.polymdegradstab.2013.05.027.

154. Tofa, T.S., Ye, F., Kunjali, K.L., and Dutta, J., Enhanced visible light photodegradation of microplastic fragments with plasmonic platinum/zinc oxide nanorod photocatalysts, *Catalysts.*, 9(10):819, 2019. doi:10.3390/catal9100819.

155. Jiang, R., et al., Microplastic degradation by hydroxy-rich bismuth oxychloride, *J. Hazard. Mater.*, 124247, 2020. doi:10.1016/j.jhazmat.2020.124247.

156. Berr, M.J., et al., Hole scavenger redox potentials determine quantum efficiency and stability of Pt-decorated CdS nanorods for photocatalytic hydrogen generation, *Appl. Phys. Lett.*, 100(22):223903, 2012. doi:10.1063/1.4723575.

157. Nandy, S., Savant, S.A., and Haussener, S., Prospects and challenges in designing photocatalytic particle suspension reactors for solar fuel processing, *Chem. Sci.*, 12(29):9866–9884, 2021. doi:10.1039/d1sc01504d.

158. Klaewkla, R., Arend, M., and Hoelderich, F.W., A review of mass transfer controlling the reaction rate in heterogeneous catalytic systems, *Mass Transfer—Advanced Aspects*, 2011. doi:10.5772/22962.

2 Microplastic Abundance in the Indian Environs
A Review

Pankaj R. Sutkar and Vinayak P. Dhulap

2.1 INTRODUCTION

Through rapid industrialization, population, development and some global monetary turns of events, plastics are being brought into the sea, soil and air through a few pathways, resulting in related natural, ecological and medicinal issues (Abbasi et al., 2019). Worldwide plastic creation has been expanded from 1.5 million tons in the year 1950 to 359 million tons produced in 2018 (Ajith et al., 2020), and plastics arrive at the seashore and deep sea through many pathways (Anbumani & Kakkar, 2018). There are numerous examinations of plastic wastes in the sea that have been directed since the 1970s (Ashwini & Vargese, 2019). The microplastics which are produced from plastics are also causing adverse impressions on ecosystems and parts within it. In the last few years, attention to microplastic (< 5 mm polymer molecule, currently as microplastic) concentration and contamination has been expanded dramatically (Ballent et al., 2016; Bellas et al., 2016; Bhattacharya & Khare, 2020).

Microplastics (MPs), a new class of pollutants, have been produced as a result of increasing plastics manufacture and usage. By examining the available scientific literature, this chapter seeks to convey current understanding of MP contamination in India's aquatic systems, terrestrial systems, atmosphere and human consumables. Additionally, it points up knowledge gaps and the direction of future study (Vaid et al., 2021).

These microplastics keep the coherent separation along with standard worldwide (SI) unit terminology; that is, the size of microplastics is less than 5 mm to 1 μm (Bridson et al., 2020). Microplastics are an unavoidable and determined ecological foreign substance which adversely impacts fresh water, earthly and marine biological systems across the globe (Bridson et al., 2020; Choudhary et al., 2020; Chubarenko et al., 2018; Dahms et al., 2020). There are two principal characterizations of microplastics: essential microplastics and optional microplastics. Essential microplastics straightforwardly enter the environment in a tiny size (< 5 mm in width) and are created through expulsion or pounding, either as a feedstock for assembling items (Daniel et al., 2020) or, sometimes, for its direct use (Dantas et al., 2012). For instance, in cleaning items (Dowarah et al., 2020), a micro part of plastics in beautifying agents enter into the air medium (Dris et al., 2017). The optional microplastics are obtained from bigger plastic flotsam and jetsam (Dowarah et al., 2020). The problem of microplastics in the briny milieu is one among the significant foreign substances,

DOI: 10.1201/9781003449133-2

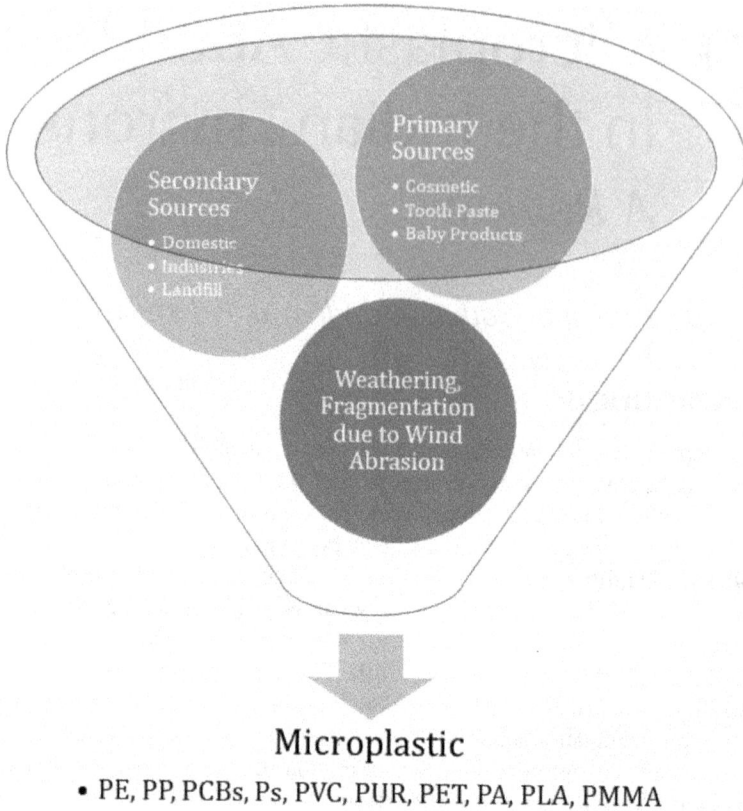

Secondary Sources
• Domestic
• Industries
• Landfill

Primary Sources
• Cosmetic
• Tooth Paste
• Baby Products

Weathering, Fragmentation due to Wind Abrasion

Microplastic

• PE, PP, PCBs, Ps, PVC, PUR, PET, PA, PLA, PMMA

FIGURE 2.1 Sources and types of microplastic.

since marine living beings misidentify microplastics as prey, and microplastics can be harmful or even deadly to them when ingested by their bodies (Eriksen et al., 2018). Additionally, now, microplastics are becoming vectors for moving poisonous synthetics (particularly persevering natural toxins and metals) from climate to biota (Ferreira et al., 2020). Its concentrations in all types of living and non-living matters are increasing and impacting bio-chemical processes within it.

2.2 HISTORICAL BACKGROUND

With the Arabian Sea in the west and the Bay of Bengal in the east, India has a long coastline that measures around 7,500 kilometers (Karuppasamy et al., 2020). There are 13 seashore states and associate domains on the coastline side, and there are many different types of ecosystems there, counting mangrove woodlands, coral ridges, wetlands, silt hills, mud flats and rugged and grimy coasts (James et al., 2020). Due to overfishing, growing urbanization, industrial activity and the overture of non-local genera, the Indian coastline is in a precarious state (Dowarah et al., 2020). Concurring to the CPCB, 62 metric tons of strong trash were created in India

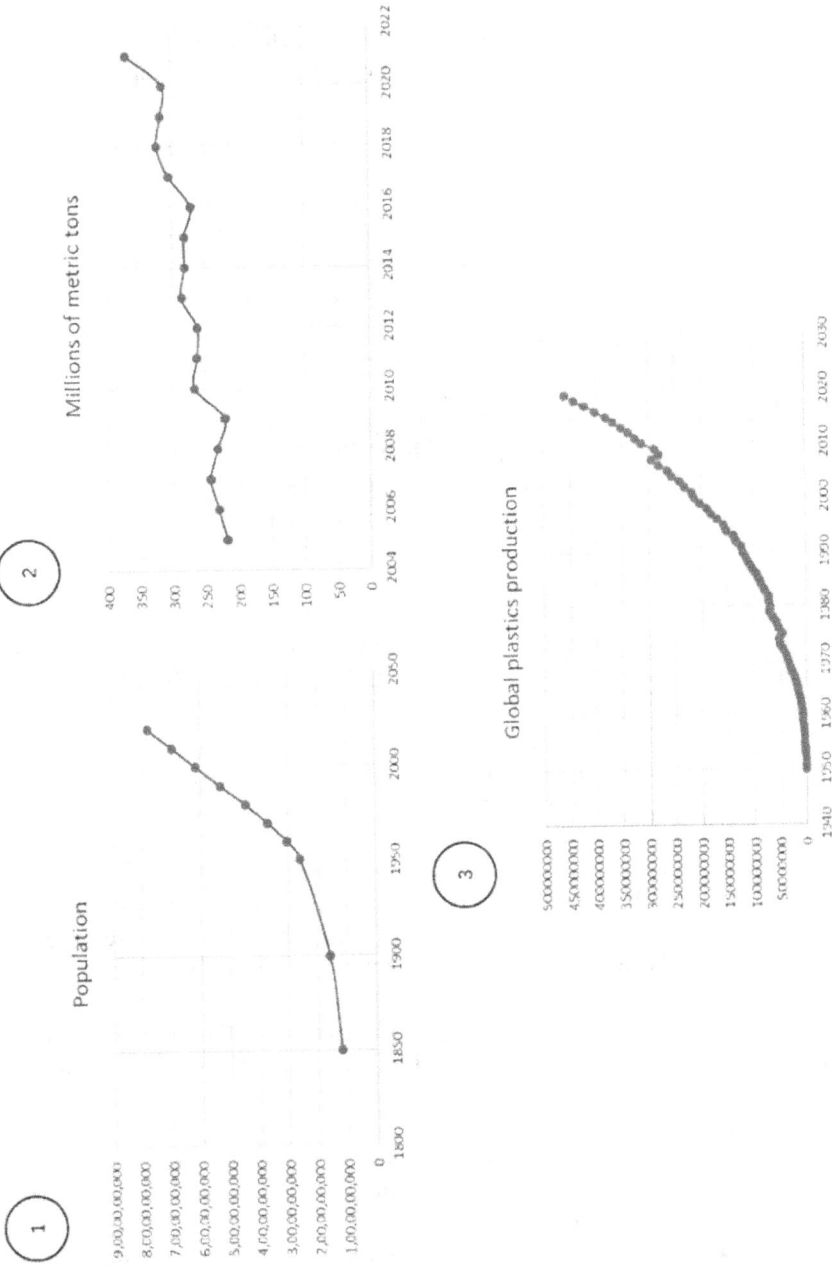

FIGURE 2.2 The historical trends of (1) population in a million (2) exports of plastics in a million tons and (3) plastic concentration in billion tons.

in 2015, with 82% of it being collected garbage and 18% being litter. Due to leachate discharge and unregulated aquatic streams, this open unloading contaminated the surface waters of the sea and fresh water (Sharma & Ghosh, 2019). The nation's expanding population has triggered an augmentation in aquatic rubble, which is primarily made up of plastic garbage. Over the past 30 years, a significant amount of the plastic debris that floats in the area along the coast may have come from land-based sources (Jayasiri et al., 2015).

2.3　REVIEW OF LITERATURE

According to the author, there are 4.8 to 12.7 million tons of plastic wastes which are generated from land and have gone into the sea and other water flows. There are ten countries out of the main 20 nations constantly delivering plastic, discarded solids to the seas which have seaboards on the "Indian Ocean" (Karthik et al., 2018). In addition, it is assessed that a plastic weight between 1.15 and 2.41 million tons presently moves from worldwide rivers in a framework obsessed by the sea consistently (Ganesan et al., 2019). Further, the top 20 contaminating waterways are situated in Asia and represent multiple thirds of the worldwide yearly information (Ríos et al., 2020). The River Ganges, considered worldwide, displays large concentrations of plastics, and is the second biggest contributor of plastic wastes into the sea (Chubarenko et al., 2018; Sarkar et al., 2019). Along with plastic pollution, there is also a developing interest in the pollution of microplastics in various ecological lattices in India (Sathish et al., 2020). The current microplastic survey articles and book sections distributed in India chiefly state the ecotoxicological and ecological effects of microplastics on the milieu (Bellas et al., 2016; Goswami et al., 2020; Gündoğdu, 2018).

Due to its cheap manufacturing cost, versatility, water resistance, high strength-to-weight ratio and thermal and electrical insulating competencies, plastic construction has increased significantly over the past 70 years and has become an essential component of contemporary life (GESAMP, 2015). The threats to the environment and to human health connected to the manufacture and disposal of plastics are causing concern (Sedlak, 2017; Thompson et al., 2009). Some authors (Machado et al., 2018) have shown how single-use plastic objects, such as throwaway face masks during the COVID-19 epidemic, might present a variety of issues resulting from the misapplication of plastic commodities.

Despite the fact that a couple of survey articles tended to the insightful strategies for microplastic contamination in the "Pacific, Arctic, Atlantic and Antarctic Oceans", no audit has been directed to the "Indian Ocean" (Goswami et al., 2020). So, this is the primary survey assessing the condition of the applied strategies for recognizable proof and measurement of microplastics in different ecological lattices (silt, water, biota, climatic residue and salt) in India, giving also a new view to future exploration needs (Karuppasamy et al., 2020).

Different examining and logical strategies have been embraced to survey the degree of microplastic contamination in India through different ecological networks (Imran et al., 2019). In any case, because of irregularity in these strategies, the spatial and worldly correlations are fairly troublesome. In this foundation, this chapter focuses on (i) summing up the present status of information concerning microplastic concentrations in various natural sources in India, (ii) evaluating and understanding the preferences and limits in different microplastic testing and logical techniques, (iii) examining the conveyance, sources and collaboration between microplastics and (iv) giving suggestions for the normalization and variation of acknowledged scientific strategies from public to the worldwide levels for successful microplastic contamination checking (Robin et al., 2020).

2.4 DATA COLLECTION

The microplastic study area in India includes its occurrences in "silt", "water", "biota", "salt", "air" and "residue" (distributed till May 30, 2020) (Ashwini & Vargese, 2019). Then all past and present researches were screened by its study zone and given examination globally and in India, including its seashores, estuaries, seaward and air. An aggregate of 41 examination articles has been considered for this investigation. The current review article includes (i) microplastic testing strategies, (ii) microplastic extraction measures, (iii) distinguishing proof procedures, (iv) presence and identification of microplastics and (v) physical, chemical and analytical techniques used for microplastic identification.

2.5 MICROPLASTIC POLLUTION IN INDIA

India is a country in southern Asia that links to the Indian Ocean through the Arabian Sea and the Bay of Bengal. It has a 7,517-km-long shoreline and is positioned amid latitudes 8°4′ and 37°6′ N and longitudes 68°7′ and 97°25′ E (Kumar et al., 2006).

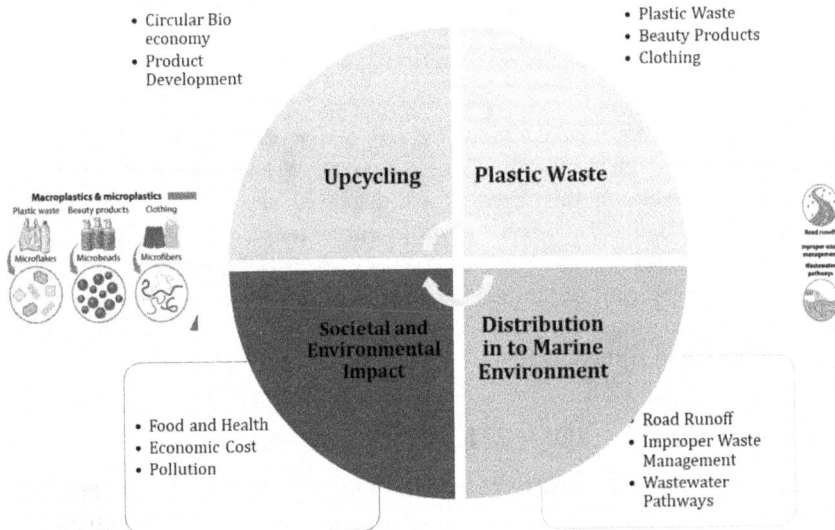

FIGURE 2.3 Sources of microplastic.

FIGURE 2.4 Locations taken into considerations for this study.

The present review was created after a thorough search of 64 studies in various areas of the Indian environment utilizing Google Scholar, Web of Science and SciFinder databases. The results of this research are displayed graphically. Southern India accounted for the bulk of the research in maritime habitats, followed by Western India, Eastern India, Bihar and West Bengal, and Northern India. As these resources are correspondingly vital to humankind, research interventions in freshwater systems, terrestrial systems, the atmosphere, and human consumables are required (Veerasingam et al., 2016; Sarkar et al., 2019; Pandey et al., 2021).

2.6 MICROPLASTIC IDENTIFICATION METHODS

The following table describes the location of microplastics found in India; plastic type, extraction and detection methods that are used for microplastic identification in "sediment, water, biota, salt, and atmospheric dust" along India's east and west coasts are shortened henceforth (Jayasiri et al., 2013).

2.6.1 SEDIMENT

By placing a wooden or metal edge on the silt surface, removing the stuff and seizing the analyzed microplastics with a brace dollop, microplastics in dregs from the seashores and seaside regions were collected (Sarkar et al., 2019). In India, various casing sizes were employed, and submerged residue tests were accumulated, depleting a van Veen or a Peterson size microplastic taster. Pellets of microplastic were occasionally collected by handpicking or filtering. No investigation was conducted to collect center silt for vertical microplastic distribution.

2.6.2 WATER

Seven studies looked on the manifestation of microplastics near the Indian shore. In collecting experiments, manta fishnets with lattice sizes ranging from 112, 200, 300, 333 and 335 m were utilized (Kosuth et al., 2018). The depths ranged from 20 cm to 3–5 m. To gauge the amount of water that would be put to the test, a stream meter was attached to the manta fishnets' entry point (Gautam & Anbumani, 2020). However, certain examinations have not yet established if a stream meter was installed on the net, and if the net is not hampered by an excessive amount of suspended and drifting objects, considerable errors might be provided by calculating merely the distance during examination (Kazour et al., 2019).

TABLE 2.1
Microplastic Identified in Sample Type of Sediment

Sr. No.	Location	Abundance	Extraction Method	Reference
1	Ganga River	107.57 to 409.86 objects/kg	Riddling	(Sarkar et al., 2019)
2	Chennai	304 (afore swamp), 896 (later swamp)	Cherry-pick	(Veerasingam et al., 2016)
3	Chennai and South Coast	not applicable	Stainless steel tweezers	(Ogata et al., 2009)
4	Southest Coast of India	1323 mg/m^2 (abnormal wave); 178 to 261 mg/m^2 (gentle wave)	Riddling	(Karthik et al., 2018)
5	Mumbai	194.33 ± 46.32 items/m^2	Riddling	(Jayasiri et al., 2013)
6	Rameshwaram	403 stuffs	Riddling	(Vidyasakar et al., 2018)
7	Goa	1655 (SW rainy season), 1345 (NE rainy season)	Cherry-pick	(Veerasingam et al., 2016)
8	Kerala	252.80 ± 25.76 items/m^2	Riddling	(Sruthy & Ramasamy, 2017)
9	Lakshdweep	603	Cherry-pick	(Mugilarasan et al., 2015)
10	Gujrat	81 mg items per kg	Riddling	(Reddy et al., 2006)

TABLE 2.2

Microplastic Identified for Sample Type—Water

Sr. No.	Location	Abundance	Extraction Method	Reference
1	Chennai	2 to 11 items/L	Filtration	(Ganesan et al., 2019)
2	Port Blair	0.93 ± 0.59 items/m^3	Riddling	(Goswami et al., 2020)
3	Bay of Bengal	16107 ± 47077.63 items/km^2	Riddling	(Eriksen et al., 2018)
4	Tutucorin	12.14 ± 3.11 to 31.05 ± 2.12 items/L	Riddling	(Patterson et al., 2019)
5	Tutucorin	3.1 ± 2.3 to 23.7 ± 4.2 items/L	Riddling	(Sathish et al., 2020)
6	Kerala Coast	1.25 ± 0.88 items/m^3	Riddling	(Robin et al., 2020)
7	Kochin	10 to 80%	Uphold and Riddling	(James et al., 2020)

2.6.3 BIOTA

There have been 11 investigations into the issue of microplastics in diverse oceanic biota (fish, mussels, shellfish, bivalves and spineless organisms) as well as 11 investigations into research facilities in India. Fish is the biota that is most frequently used in these studies (a total of seven studies) to examine microplastic ingestion (Anbumani & Kakkar, 2018). All biota samples were collected by hand, fishing nets or other means of confinement. Moreover, biota tests that were bought on the market were used. The entire collection of biota tests was then frozen at -20 °C till the subsequent step (Anbumani & Kakkar, 2018).

2.6.4 SALT

Four different tests were run to determine whether microplastics were present in different salt brands available in local Indian general stores. In "Tuticorin, Tamil Nadu", 25 briny mock-ups from various briny dishes (Patterson et al., 2019; Sathish et al., 2020) and 14 distinct briny brands generated from saltwater and water from deep well (Kumar et al., 2018) were examined. Seth and Shriwastav (2018) examined eight business brand salts produced on the west seaboard of India after Kim et al. (2018) acquired three business brands of ocean salts that were sold in Indian general stores.

2.6.5 RESIDUE

In India, there have been two investigations into the movement of microplastics in airborne dust. The KB-120F sort smart center stream suspended barometrical particulate sampler was used in the east Indian Ocean to collect environmental residue samples along the boat track at a rate of 100 0.1 L/min. In 2014, 33 indoor residue samples were collected from Patna City areas by Zhang et al. (2020). Tests with residue were homogenized, collected and stored at 4°C for study after being sieved

TABLE 2.3

Microplastic Identified in Any Sample Type Other Than Sediment and Water

Sr. No.	Location	Sample Type	Abundance	Reference
1	Kochi, Kerala	Benthic invertebrates	NA	(Naidu, 2019)
2	East Indian Ocean	Atmospheric Dust	0.4 to 0.6 objects/100 m^3	(Wang et al., 2020)
3	Patna	Indoor Dust	55e6800 mg (PET); <0.11e530 mg (PC)	(Zhang et al., 2020)
4	Tutucorin	Sea Salt	35 ± 15 to 72 ± 40 items/kg	(Sathish et al., 2020)
5	Tuticorin	Bore-well Salt	2 ± 1 to 29 ± 11 items/kg	(Sathish et al., 2020)
6	Gulf of Mannar	Fish	Out of 40 fish samples, 12 fish showed signs of consumption	(Kumar et al., 2018)
7	Chennai	Green Mussel	0.9 to 0.3 objects/10 g to 3.2 objects/10 g	(Naidu, 2019)
8	Andaman Island	Zooplankton	90%	(Goswami et al., 2020)
9	Pondicherry	Bivalves	0.18 to 1.84 objects/g; 0.50 to 4.8	(Dowarah et al., 2020)
10	Tuticorin	Oyster	5.21 ± 4.85 to 9.74 ± 8.92 items/ individual	(Patterson et al., 2019)

through a 150 m strainer. During an examination, field gaps were created by releasing aluminium foil into the air.

2.7 MICROPLASTIC EXTRACTION STRATEGIES

The isolation of microplastics from silt, water, biota, salt and residue tests was led in physical, chemical and analytical ways (Kanhai et al., 2017). The bigger-size microplastics were analyzed outwardly and gotten utilizing tweezers, while the little-size microplastics were removed using thickness-division and filtration techniques (Ballent et al., 2016).

2.7.1 Residue

Microplastics were extracted from the residue using NaCl, $ZnCl_2$, HNO_3, $CaCl_2$ and 30% H_2O_2 after the dried dregs tests were sieved using various size strainers. The supernatant was separated through a factor network size channel paper after thickness partition and processing, and it was then dried normally or in a broiler for further evaluation under microscopic and spectroscopic examination (Mugilarasan et al., 2015).

2.7.2 WATER

Seven samples were among the gathered water tests that were divided or sieved for size determination. Whereas NaCl (McEachern et al., 2019) and HNO$_3$ were utilized to extract the microplastics from water samples using thickness division, 30% H$_2$O$_2$ assimilation was employed to remove the natural matter (Nigam, 1982). During 12 to 72 hours, the absorption was allowed to proceed at room temperature or 75 °C. The supernatant was passed across channel panes with different lattice extents, such as 0.7M and 0.8M, subsequently thickness division and absorption. The channel sheets were then stored in petri plates after being desiccated at atmospheric temperature to 55°C (Robin et al., 2020; Sathish et al., 2020).

2.7.3 BIOTA

Before evaluation, the icy biota experiments were defrosted at atmospheric temperature. The biota's dimensions and mass were noted. The fragile networks from the armors of mussels and bivalves as well as the whole gastrointestinal tract of fish were examined. Several lattice-size channel sheets (0.7 m, 0.8 m, 5 m, and 11 m) were used for filtering after the tests were treated with 10% KOH and 30% H$_2$O (Naido et al., 2015). The channel sheets were then transferred to petri plates and dried for additional investigation.

2.7.4 SALT

To process the natural matter, 200–250 g of salt and H$_2$O$_2$ were combined, and the concoction was then stored at atmospheric temperature for 48 hours before being placed in the hatchery for 24 hours at 65°C (Kim et al., 2018; Selvam et al., 2020). 1 L of clean water was added to the sample after natural matter absorption to liquefy the Brakish mixture. The mixture was centrifuged, and the salt supernatant was divided using channel papers with various thicknesses (0.2 mm, 0.45 mm, 0.8 mm and 2.7 mm). The channel sheets were then transferred to petri plates and dried at that point (Selvam et al., 2020).

2.8 INFORMATION CURRENTLY AVAILABLE ON MICROPLASTICS IN VARIOUS ENVIRONMENTAL MATRICES

2.8.1 ABUNDANCE AND DISTRIBUTION

India's silt, water, biota, salt and residue studies yielded substantial results for the presence of microplastics. The east coastline side of India (ECI) was found to have a greater concentration of microplastic than the west coast. The analysis of microplastic dispersion along water in India's east (four investigations) and west (two investigations) beaches revealed that quantitative impacts were introduced in units such as things/l, things/km^2, things/m^3, and%. "Things/individual" and "Things/g" are the illuminating units for microplastics in biota. Also, it was discovered that the amount of surface silt on the remote island (Andaman Nicobar) was greater than that of the major stream in Chennai (Ganga) (Kazour et al., 2019). Mumbai had the highest microplastics deliberation (220 50 things/kg), but Chennai (3.1 2.63 things/km²) and Tuticorin (12.14 3.12 things/l) had lower echelons of microplastics in the surface water. The microplastic study identified 16,107 47,077.63 items/km² seaward of the "Bay of Bengal" (James et al., 2020).

The consumption of microplastics by fish was extensively studied on India's east and west shorelines (Bellas et al., 2016). According to the microplastic research that was focused on India, microplastic accumulated along seashores, close to the coast and seaward locations, notably close to stream mouths (Bhattacharya & Khare, 2020). This could raise concerns because of the fishes' capacity to ingest the marine food chain and its incorporation into animal life cycles (Maharana et al., 2020).

2.9 DIFFERENCES BETWEEN TERRESTRIAL AND AQUATIC SYSTEMS AND EMERGING CONCERNS ABOUT MICROPLASTIC POLLUTION

Less scientific research has been done on terrestrial systems than on aquatic ones. However, microplastic pollution on land may be 4–23 times more than in the water (Horton et al., 2017). In fact, compared to marine basins, agricultural soils may contain more microplastics (Nizzetto et al., 2016). Given that particulate material's quantity, composition and physico-chemical surface characteristics are typical in terrestrial and continental settings, more thorough assessments of microplastic pollution

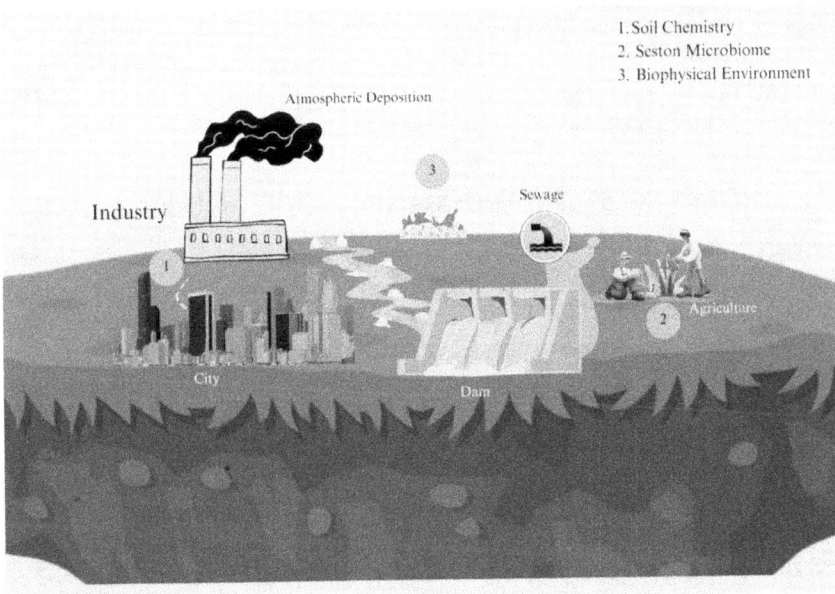

FIGURE 2.5 Microplastic fate in the terrestrial environment.

to terrestrial biodiversity are required (Dubovik et al., 2002). Microplastics are man-made and have xenobiotic structural characteristics that set them apart from natural material. More scientific research is required to comprehend their consequences in continental ecosystems in order to conserve terrestrial biodiversity (Machado et al., 2018). Thus, similar to how pollen tubes go unidirectionally to the ovules, plastic beads of appropriate pollen size may also do so (and occasionally intercellularly). However, it is still too early to estimate the extent to which microplastics may have a negative influence on crucial plant and pollinator ecological processes (Sanders & Lord, 1989).

2.10 MICROPLASTIC EFFECTS IN CONTINENTAL ECOSYSTEMS

As they can be loaded with harmful and opportunistic organisms and enter freshwater systems, plastics can have a detrimental effect on ecosystem functioning (Kirstein et al., 2016). When microplastics are discharged from sewage treatment facilities into continental waterways, they blend with a microbiome that is different from that of natural particles and may serve as a vehicle for the spread of illness. Future studies are required to fully comprehend how microplastics affect terrestrial microbiomes (Kirstein et al., 2016). Researchers in Sydney, Australia discovered that topsoils near roadways and industrial areas can have up to 7% by weight of microplastics, which can leach non-volatile organochlorines from PVC and other chlorinated microplastics and alter the geochemistry of soils (Fuller & Gautam, 2016).

When there are microplastics present in honey, for example, the biophysical habitat of terrestrial creatures might alter. A search for the causes of this contamination led researchers to hypothesize that microplastics would be found in the inflorescences of several species (Liebezeit & Liebezeit, 2015). Because of this, even in the absence of obvious signs of ingestion, microplastic-exposed springtails exhibited altered gut microbiota, a changed isotopic signature (15N and 13C) and showed negative impacts on development and reproduction (Zhu et al., 2018).

2.11 INDIA'S PLASTIC WASTE MANAGEMENT AND REGULATION

MoEF-CC, "National Environmental Engineering Research Institute" (NEERI), CPCB, SPCBs and the Ministry of Urban Development are responsible for managing plastic trash in India (MoUD) (Prakash et al., 2020). The management of plastic trash is challenging because of the age of the material, illegal offloading, a lack of structure and hesitation in forming organizations. 60% of municipal solid garbage is unloaded in an open area, which is a significant quantity of rubbish (MSW).

The techniques used to remove trash in India are bulk landfilling, thoughtless unloading of debris and disorganized landfilling. Plastic waste administration laws, which were put in place in 2016 to consolidate the current conventions for plastics holding, forbid single use of plastics as of January 1, 2019. In India, about 60% of recyclable plastic gets repurposed, but there is a lack of commitment to the elaboration and execution of plans for how single-use plastics are handled. The provincial activity strategy for India on marine rubble included strategies to address oceangoing plastics, including categorizing the stoolies of acreage and ocean-centered plastic waste and controlling the actual birthplace, raising public awareness of microplastics contamination and supporting innovative and scientific work (Liu et al., 2019).

2.12 CONCLUSION

The current chapter summarizes microplastic abundance and its identification as well as extraction methods which were used globally and with special preference to India. The current information on microplastics in different natural frameworks laterally from the east and west shorelines of India has been studied by Indian researchers. The article has also focused on microplastics occurrences in "silt", "water", "biota", "salt", "air" and "residue" in India. This chapter also summarizes physical, chemical and analytical techniques that were used for microplastic identification process (globally and in India). It covers all scientific strategies for microplastic detection that makes direct examination of microplastics in different natural sources. In this reread, the authors normalized characterization of the extent of microplastics and normalized conventions for checking microplastics in silt, water and biota to defeat a portion of the difficulties recognized in microplastic research around the world and in India, specifically. The increasing microplastic mixtures and their related entry in Indian water bodies, particularly in freshwater, cannot be ignored. It is precarious to estimate the degree of microplastics in the climatic residue for hazard assessment, though there is a need to study the toxicological impressions of the altered outlines of microplastics on lifeforms and their enrichment in the living processes and food webs, which gives us the freedom for future research and exploration.

REFERENCES

Abbasi, S. et al., 2019. Distribution and potential health impacts of microplastics and microrubbers in air and street dusts from Asaluyeh County, Iran. *Environmental Pollution*, Volume 244, pp. 153–164.

Ajith, N. et al., 2020. Global distribution of microplastics and its impact on marine environment—a review. *Environmental Science and Pollution Research*, Volume 27, pp. 25970–25986.

Anbumani, S.,& Kakkar, P., 2018. Ecotoxicological effects of microplastics on biota: A review. *Environmental Science and Pollution Research*, Volume 25, pp. 14373–14396.

Ashwini, S. K.,& Vargese, G., 2019. Environmental forensic analysis of the microplastic pollution at "Nattika" Beach, Kerala Coast, India. *Environmental Forensics*, 03 Dec, Volume 21(1), pp. 21–36.

Ballent, A. et al., 2016. Sources and sinks of microplastics in Canadian Lake Ontario nearshore, tributary and beach sediments. *Marine Pollution Bulletin*, Sept, Volume 110(1), pp. 383–395.

Bellas, J. et al., 2016. Ingestion of microplastics by demersal fish from the Spanish Atlantic and Mediterranean coasts. *Marine Pollution Bulletin*, Volume 109(1), pp. 55–60.

Bhattacharya, A.,& Khare, S. K., 2020. Ecological and toxicological manifestations of microplastics: Current scenario, research gaps, and possible alleviation measures. *Journal of Environmental Science and Health, Part C*, Volume 38(1), pp. 1–20.

Bridson, J. H. et al., 2020. Microplastic contamination in Auckland (New Zealand) beach sediments. *Marine Pollution Bulletin*, Volume 151, p. 110867.

Choudhary, D., Kurien, C.,& Kumar, A., 2020. Microplastic contamination and life cycle assessment of bottled drinking water. *Advances in Water Pollution Monitoring and Control*, pp. 41–48.

Chubarenko, I. et al., 2018. Three-dimensional distribution of anthropogenic microparticles in the body of sandy beaches. *Science of the Total Environment*, Volume 628–629, pp. 1340–1351.

Dahms, H. T., Rensburg, G. J. V.,& Greenfield, R., 2020. The microplastic profile of an urban African stream. *Science of the Total Environment*, Volume 731, p. 138893.

Daniel, D. B., Ashraf, P. M.,& Thomas, S. N., 2020. Abundance, characteristics and seasonal variation of microplastics in Indian white shrimps (Fenneropenaeus indicus) from coastal waters off Cochin, Kerala, India. *Science of the Total Environment*, Volume 737, p. 139839.

Dantas, D. V., Barletta, M.,& Costa, M. F. D., 2012. The seasonal and spatial patterns of ingestion of polyfilament nylon fragments by estuarine drums (Sciaenidae). *Environmental Science and Pollution Research*, Volume 19, pp. 600–606.

Dowarah, K. et al., 2020. Quantification of microplastics using Nile Red in two bivalve species Perna viridis and Meretrix meretrix from three estuaries in Pondicherry, India and microplastic uptake by local communities through bivalve diet. *Marine Pollution Bulletin*, Volume 153, p. 110982.

Dris, R. et al., 2017. A first overview of textile fibers, including microplastics, in indoor and outdoor environments. *Environmental Pollution*, Volume 221, pp. 453–458.

Dubovik, O. et al., 2002. Variability of absorption and optical properties of key aerosol types observed in worldwide locations. *Journal of Atmosperic Sciences*, Volume 59(3), pp. 590–608.

Eriksen, M. et al., 2018. Microplastic sampling with the AVANI trawl compared to two neuston trawls in the Bay of Bengal and South Pacific. *Environmental Pollution*, Volume 232, pp. 430–439.

Ferreira, M. et al., 2020. Presence of microplastics in water, sediments and fish species in an urban coastal environment of Fiji, a Pacific small island developing state. *Marine Pollution Bulletin*, Volume 153, p. 110991.

Fuller, S.,& Gautam, A., 2016. A procedure for measuring microplastics using pressurized fluid extraction. *Environmental Science & Technology*, Volume 50, pp. 5774–5780.

Ganesan, M., Nallathambi, G.,& Srinivasalu, S., 2019. Fate and transport of microplastics from water sources. *Current Science*, Volume 117(11), pp. 1879–1885.

Gautam, K.,& Anbumani, S., 2020. Ecotoxicological effects of organic micro-pollutants on the environment. *Current Developments in Biotechnology and Bioengineering*, pp. 481–501.

GESAMP, 2015. *GESAMP*. [Online] Available at: gesamp.org/publications/reports-and-studies-no-90.

Goswami, P., Vinithkumar, N. V.,& Dharani, G., 2020. First evidence of microplastics bioaccumulation by marine organisms in the Port Blair Bay, Andaman Islands. *Marine Pollution Bulletin*, Volume 155, p. 111163.

Gündoğdu, S., 2018. Contamination of table salts from Turkey with microplastics. *Food Additives & Contaminants: Part A*, Volume 35(8), pp. 1006–1014.

Horton, A. A. et al., 2017. Microplastics in freshwater and terrestrial environments: Evaluating the current understanding to identify the knowledge gaps and future research priorities. *Science of the Total Environment*, Volume 586, pp. 127–141.

Imran, M., Das, K. R.,& Naik, M. M., 2019. Co-selection of multi-antibiotic resistance in bacterial pathogens in metal and microplastic contaminated environments: An emerging health threat. *Chemosphere*, Volume 215, pp. 846–857.

James, K. et al., 2020. An assessment of microplastics in the ecosystem and selected commercially important fishes off Kochi, south eastern Arabian Sea, India. *Marine Pollution Bulletin*, Volume 154, p. 111027.

Jayasiri, H., Purushothaman, C. S., & Vennila, A. (2013). Quantitative analysis of plastic debris on recreational beaches in Mumbai, India. Marine Pollution Bulletin, Volume 77(1–2), pp. 107–112. https://doi.org/10.1016/j.marpolbul.2013.10.024

Jayasiri, H. B., Purushothaman, C. S.,& Vennila, A., 2015. Bimonthly variability of persistent organochlorines in plastic pellets from four beaches in Mumbai coast, India. *Environmental Monitoring and Assessment*, Volume 187.

Kanhai, L. D. K. et al., 2017. Microplastic abundance, distribution and composition along a latitudinal gradient in the Atlantic Ocean. *Marine Pollution Bulletin*, Volume 115(1–2), pp. 307–314.

Karthik, R. et al., 2018. Microplastics along the beaches of southeast coast of India. *Science of the Total Environment*, Volume 645, pp. 1388–1399.

Karuppasamy, P. et al., 2020. Baseline survey of micro and mesoplastics in the gastro-intestinal tract of commercial fish from Southeast coast of the Bay of Bengal. *Marine Pollution Bulletin*, Volume 153, p. 110974.

Kazour, M. et al., 2019. Microplastics pollution along the Lebanese coast (Eastern Mediterranean Basin): Occurrence in surface water, sediments and biota samples. *Science of the Total Environment*, Volume 696, p. 133933.

Kim, J.-S., Lee, H.-J., Kim, S.-K.,& Kim, H.-J., 2018. Global pattern of microplastics (MPs) in commercial food-grade salts: Sea salt as an indicator of seawater MP pollution. *Environmental Science Technology*, Volume 52(21), pp. 12819–12828.

Kirstein, I. V. et al., 2016. Dangerous hitchhikers? Evidence for potentially pathogenic Vibrio spp. on microplastic particles. *Marine Environmental Research*, Volume 120, pp. 1–8.

Kosuth, M., Mason, S. A.,& Wattenberg, E. V., 2018. Anthropogenic contamination of tap water, beer, and sea salt. *PLoS One*, Volume 13(4).

Kumar, V. E., Ravikumar, G., & Jeyasanta, K. I., 2018. Occurrence of microplastics in fishes from two landing sites in Tuticorin, South east coast of India. *Marine Pollution Bulletin*, Volume 135, pp. 889–894.

Kumar, V. S. et al., 2006. Coastal processes along the Indian coastline. *Current Sciecne*, Volume 91(4), pp. 530–536.

Liebezeit, G., & Liebezeit, E., 2015. Origin of synthetic particles in honeys. *Polish Journal of Food and Nutrition Sciences*, Volume 65, pp. 143–147.

Liu, K., Wang, X., Fang, T., Xu, P., Zhu, L., & Li, D. (2019). Source and potential risk assessment of suspended atmospheric microplastics in Shanghai. Science of the Total Environment, 675, 462–471. https://doi.org/10.1016/j.scitotenv.2019.04.110

Machado, A. A. D. S. et al., 2018. Microplastics as an emerging threat to terrestrial ecosystems. *Global Change Biology*, Volume 24(4), pp. 1405–1416.

Maharana, D. et al., 2020. Assessment of micro and macroplastics along the west coast of India: Abundance, distribution, polymer type and toxicity. *Chemosphere*, Volume 246, p. 125708.

McEachern, K. et al., 2019. Microplastics in Tampa Bay, Florida: Abundance and variability in estuarine waters and sediments. *Marine Pollution Bulletin*, Volume 148, pp. 97–106.

Mugilarasan, M., Venkatachalapathy, R., Sharmila, N., & Kaliyamoorthi, G., 2015. Occurrence of microplastic resin pellets from Chennai and Tinnakkara Island: Towards the establishment of background level for plastic pollution. *Indian Journal of Geo-Marine Sciences*, Volume 46(6).

Naido, T., Smit, A. J., & Glassom, D., 2015. Plastic ingestion by estuarine mullet Mugil cephalus (Mugilidae) in an urban harbour, KwaZulu-Natal, South Africa. *African Journal of Marine Science*, Volume 38(1), pp. 145–149.

Naidu, S. A., 2019. Preliminary study and first evidence of presence of microplastics and colorants in green mussel, Perna viridis (Linnaeus, 1758), from southeast coast of India. *Marine Pollution Bulletin*, Volume 140, pp. 416–422.

Nigam, R., 1982. Plastic pellets on the Caranzalem beach sands, Goa, India. *Mahasagar*, Volume 15(2), pp. 125–127.

Nizzetto, L., Futter, M., & Langaas, S., 2016. Are agricultural soils dumps for microplastics of urban origin? *Environmental Science Technology*, Volume 50(20), pp. 10777–10779.

Ogata, Y. et al., 2009. International pellet watch: Global monitoring of persistent organic pollutants (POPs) in coastal waters. 1. Initial phase data on PCBs, DDTs, and HCHs. *Marine Pollution Bulletin*, Volume 58(10), pp. 1437–1446.

Pandey, D., Singh, A., Ramanathan, A., & Kumar, M., 2021. The combined exposure of microplastics and toxic contaminants in the floodplains of north India: A review. *Journal of Environmental Management*, Volume 279, p. 111557.

Patterson, J. et al., 2019. Profiling microplastics in the Indian edible oyster, Magallana bilineata collected from the Tuticorin coast, Gulf of Mannar, Southeastern India. *Science of the Total Environment*, Volume 691, pp. 727–735.

Prakash, V. et al., 2020. Occurrence and ecotoxicological effects of microplastics on aquatic and terrestrial ecosystems. *Microplastics in Terrestrial Environments*, Volume 95, pp. 223–243.

Reddy, M. S., Basha, S., Adimurthy, S., & Ramachandraiah, G., 2006. Description of the small plastics fragments in marine sediments along the Alang-Sosiya ship-breaking yard, India. *Estuarine, Coastal and Shelf Science*, Volume 68(3–4), pp. 656–660.

Ríos, M. F., Hernández-Moresino, R. D., & Galván, D. E., 2020. Assessing urban microplastic pollution in a benthic habitat of Patagonia Argentina. *Marine Pollution Bulletin*, Volume 159, p. 111491.

Robin, R. et al., 2020. Holistic assessment of microplastics in various coastal environmental matrices, southwest coast of India. *Science of the Total Environment*, Volume 703, p. 134947.

Sanders, L. C., & Lord, E. M., 1989. Directed movement of latex-particles in the gynoecia of 3 species of flowering plants. *Science*, Volume 243, pp. 1606–1608.

Sarkar, D. J. et al., 2019. Spatial distribution of meso and microplastics in the sediments of river Ganga at eastern India. *Science of the Total Environment*, Volume 694, p. 133712.

Sathish, N., Jeyasanta, I., & Patterson, J. (2020). Microplastics in salt of Tuticorin, southeast coast of India. *Archives of Environmental Contamination and Toxicology*, Volume 79(1), pp. 111–121. https://doi.org/10.1007/s00244-020-00731-0

Sedlak, D., 2017. Three lessons for the microplastics voyage. *Environmental Science and Technology*, Volume 51(14), pp. 7747–7748.

Selvam, S. et al., 2020. Microplastic presence in commercial marine sea salts: A baseline study along Tuticorin Coastal salt pan stations, Gulf of Mannar, South India. *Marine Pollution Bulletin*, Volume 150, p. 110675.

Seth, C. K., & Shriwastav, A., 2018. Contamination of Indian sea salts with microplastics and a potential prevention strategy. *Environmental Science and Pollution Research*, Volume 35, pp. 30122–30131.

Sharma, G., & Ghosh, C., 2019. Microplastics: An unsafe pathway from aquatic environment to health—a review: Case studies from India. *Emerging Issues in Ecology and Environmental Science*, pp. 67–72.

Sruthy, S., & Ramasamy, E., 2017. Microplastic pollution in Vembanad Lake, Kerala, India: The first report of microplastics in lake and estuarine sediments in India. *Environmental Pollution*, Volume 222, pp. 315–322.

Thompson, R. C., Swan, S. H., Moore, C. J., & Saal, F. S. V., 2009. Our plastic age. *Biological Sciences*, Volume 364(1526).

Vaid, M., Mehra, K., & Gupta, A., 2021. Microplastics as contaminants in Indian environment: A review. *Environmental Science and Pollution Research*, Volume 28, pp. 68025–68052.

Veerasingam, S., Saha, M., Suneel, V., Vethamony, P., Rodrigues, A. C., Bhattacharyya, S., & Naik, B. G. (2016). Characteristics, seasonal distribution and surface degradation features of microplastic pellets along the Goa coast, India. *Chemosphere*, Volume 159, pp. 496–505. https://doi.org/10.1016/j.chemosphere.2016.06.056

Vidyasakar, A. et al., 2018. Macrodebris and microplastic distribution in the beaches of Rameswaram Coral Island, Gulf of Mannar, Southeast coast of India: A first report. *Marine Pollution Bulletin*, Volume 137, pp. 610–616.

Wang, X. et al., 2020. Atmospheric microplastic over the South China Sea and East Indian Ocean: Abundance, distribution and source. *Journal of Hazardous Materials*, Volume 389, p. 121846.

Zhang, J., Wang, L., & Kannan, K., 2020. Microplastics in house dust from 12 countries and associated human exposure. *Environment International*, Volume 134, p. 105314.

Zhu, D. et al., 2018. Exposure of soil collembolans to microplastics perturbs their gut microbiota and alters their isotopic composition. *Soil Biology and Biochemistry*, Volume 116, pp. 302–310.

3 Toxic Effect of Food-Borne Microplastics on Human Health

Riashree Mondal, Subarna Bhattacharyya and Punarbasu Chaudhuri

3.1 INTRODUCTION

MPs have become an omnipresent pollutant in all spheres of the planet. MPs are defined as small, suspended particles with a diameter under 5 mm. These are divided mainly into two categories: primary and secondary. Primary MPs are used intentionally in textiles, sandblasting media, and facial or body scrubs, and secondary MPs are generated from the degradation of larger plastics (Udovicki et al., 2022). These are now existing in the marine and freshwater systems, soil, air, as well as in the food chain. Studies have shown that microplastic exposure might turn into toxic effects on humans through inhalation and ingestion. Therefore, the objective of this chapter is to find out the transmission process of MPs involved in interacting with the food chain, foods affected by MPs, which cause food-borne toxicity in the digestive system, and finally, stipulate some preventive measures and sustainable alternatives to cope with the plastics issue.

3.2 TRANSFER AND INTERACTION OF MICROPLASTICS (MPS) INTO THE FOOD CHAIN

Plastics can be sustained for years in the environment because they are very resistant to degradation. The presence of MPs has been observed in both terrestrial and aquatic systems. Aquatic animals can mistakenly consume MPs, as they resemble their food items. Paropa, polychaetes, and pinworms can ingest MPs such as polystyrene, which may be consumed by higher-food-chain animals such as shrimp (Saeedi, 2023). Marine algae can be contaminated by MPs that produce mycotoxins. These can be transferred to the seafood consumed by humans (Mamun et al., 2023). The first study was documented by Guo et al., 2020 to prove MP accumulation and transfer in the food chain by demonstrating transfer from moss to zooplankton and then to a fish species (Figure 3.1). Adverse metabolic changes such as weight loss, altered triceraldehyde-to-choleserol ratio, and change in muscle and liver cholesterol distribution levels have been monitored in the fish. Further, these will possibly induce harmful human health effects directly or indirectly as fish is consumed as food.

DOI: 10.1201/9781003449133-3

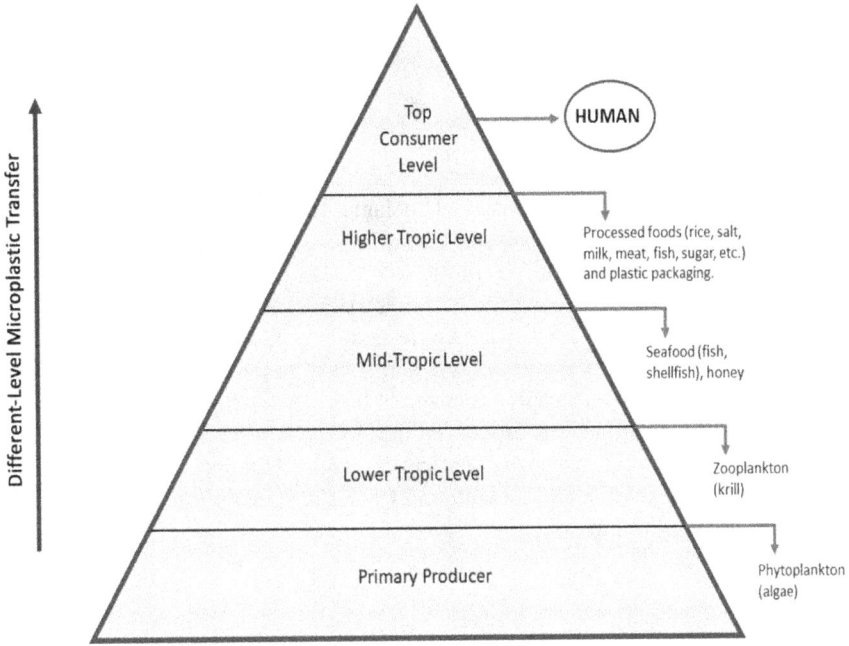

FIGURE 3.1 Microplastics food chain.

The soil ecosystem is a necessary part of the terrestrial system, as it plays an important role in food production and supply. Reports have shown that around 79% of plastic litter is deposited in landfills, accumulating in the agricultural soil and plant system (Guo et al., 2020). Studies have shown that the presence of MPs in soil makes plants undergo a loss of growth, inhibits water and nutrient absorption, lowers height and fruit number in plants, and may also affect the photosynthesis process (Li et al., 2023). Hence, there is a high chance of migration of MPs into the human body via the affected stems, fruits, and other edible sections of the plants when consumed. Further, MPs consumed by predator animals and animals like cows and goats remain a longer time in their organs from ingestion to excretion, enhancing the transfer of MPs to a higher level of the food chain and causing bio-magnification in the ecosystem (Saeedi, 2023).

3.3 PRESENCE AND TRANSMISSION OF MPS IN THE FOODS CONSUMED BY HUMANS

There are mainly three different routes through which MPs contaminate the human system; namely, inhalation (both indoor and outdoor air), ingestion (different foods and drinking water), and dermal contact (dust, clothing, and personal care products) (Prata et al., 2020). Here, ingestion is a significant pathway of contamination which includes foods such as fish, fruits, vegetables, meat, cereal, legumes, and drinking water.

3.3.1 FISH

Several studies have demonstrated the presence of MPs in freshwater and marine fish species, including edible, demersal, pelagic, and reef fishes worldwide. The diverse group of shellfish and crustaceans such as crabs, lobsters, shrimp, crayfish, and prawns play an important role in the food chain and food security. MPs have been detected in cupped oysters and Japanese carpet shells ranging from 0.18 to 3.84 particles/g and 0.9 to 2.5 particles/g, respectively. Canned fish can possibly be contaminated during processing, storage, and transportation. A study reported that four of 20 branded canned sardines and sprats from 13 countries have spotted MPs. Reefs built by corals are also an important part of the marine ecosystem, and medicine prepared from corals has shown anticancer, anti-inflammatory properties, bone repair, and neurological benefits (Mamun et al., 2023). MPs infecting these coral reefs have impaired the mesenteric tissue and gut cavity, which may further impact the human body.

3.3.2 SUGAR

The contamination of MPs in sugar might have different reasons, including processing, purification, refinement, drying, packaging, and other sources such as dust from aprons, gloves, and equipment used. There are very few studies indicating microplastic particles in sugar, and commercial, unpacked, unbranded, and unlabelled sugar samples in Bangladesh have been spotted with 343.7 MP particles/kg (Afrin et al., 2022). MPs have also been detected and quantified in honey samples in Germany with 40–660 fibers/kg and 0–38 fragments/kg (Liebezeit and Liebezeit, 2015). The possible reason for contamination may be through flowers or transport by bees or during processing and storing with plastic equipment (Toussaint et al., 2019).

3.3.3 SALT

The main source of contamination in salt is the fresh water and ocean water accumulated with plastic debris. MPs have been detected in sea salts with 350 particles/gm (Sivagami et al., 2021), lake salts at 75.2 particles/gm, rock salts at 39.03 particles/gm, and also in crystal/powder salts, table salts, and well salts (Danopoulos et al., 2020). Here, contamination during processing and storage is also an unavoidable factor.

3.3.4 DRINKING WATER

The human body requires adequate water to maintain normal body functions. Drinking water is also used for domestic use, food processing, and other recreational activities. Contaminated drinking water can transmit various diseases. Several studies have reported plastic contamination in drinking water. Many countries such as Germany, India, and China have demonstrated very high levels of MPs such as 118 particles/liter in mineral water, 39 items/liter in bottled water, and 440 MP/liter in tap water (Schymanski et al., 2018; Mason et al., 2018; Tong et al., 2020). These data indicate toxicity, as drinking water can get easily contaminated through various sources and pathways.

3.3.5 FRUITS AND VEGETABLES

Fruits and vegetables are an essential part of a healthy diet, as they provide vitamins and minerals. 400–500 gm of fruits and vegetables should be consumed by an individual daily. Although high concentrations of MPs have been investigated in daily consumed fruits like apples, pears, and vegetables such as broccoli, lettuce, carrot, and potato, the average number of MPs found in them is 132,740 particles/gm (Gea et al., 2020).

3.3.6 MILK

Milk is a vital nutrient for humans, especially for infants. It is very important for their growth and development, as it is rich in components such as proteins, vitamins, and minerals. Different processing, packaging systems and insufficient hygiene may cause MP contamination in milk and dairy products. A study done in Mexico has shown 150 particles of microplastics in 23 different dairy milk samples (Pironti et al., 2021). Furthermore, human breast milk is also getting contaminated with MPs, assuming that MPs may pass through the cell membranes and translocate to different glands and organs. A study done in Italy has revealed MPs in 26 of 34 individual milk samples (Ragusa et al., 2022).

3.3.7 RICE

Rice is a widely consumed staple food. Around 70% of total calories comes from rice. MP contamination has been known to occur in the transfer of MPs from the soil to the paddy plant (Wu et al., 2022) and also in store-brought branded rice, in which washed rice contained 3.7 mg/100 gm and unwashed rice consisted of 2.8 mg/100 gm (Dessì et al., 2021).

3.3.8 MEAT

There are very few studies that have declared the presence of MPs in meat – e.g., chicken meat has been contaminated with MPs from soil (Huerta Lwanga et al., 2017), and packaged meat has been shown to be contaminated with extruded polystyrene (XPS) food tray and sealing film (Kedzierski et al., 2020).

Other sources include plastic packaging, which apparently releases plastics into foods, such as rice cooking bags, ice-cube bags, nylon tea bags, disposable plastic cups, disposable paper cups, and polypropylene feeding bottles (Mamun et al., 2023).

3.4 MECHANISM OF FOOD-BORNE TOXICITY

Ingestion of food is currently considered a significant pathway of microplastic contamination, and toxicity induced by food mostly affects the digestive system. MP exposure can be explored in various levels of the human body, from cellular and tissue levels to organ levels. Organs in the body work together to maintain the body's normal functioning. So, the threat to one organ will impact others, consequently, causing toxicity in the digestive, respiratory, endocrine, reproductive, and immune systems.

The MPs may interact with bio-molecules such as lipids, carbohydrates, nucleic acids, ions, and water after ingestion. The intestine is a very crucial part of the digestive system, as it plays important roles such as digestion, absorption, and metabolism. MPs of 5 to 10 μm are capable of reaching the intestinal barrier (Figure 3.2). After entering the intestine, the particles can disrupt the cellular layer, irritate the intestinal tissue, release toxic substances, and cross the epithelium through the paracellular route; or some particles may not cross the bloodstream; they stay inside the intestinal cells and release into the gut lumen after the end of the lifecycle of the intestinal cells (Paul et al., 2020). The intestinal flora regulate different biochemical reactions, including protein synthesis and mineral absorption, but the existence of any exogenous toxins like MPs or NPs can disrupt the microbial ecology, causing a disturbance in metabolic activity, inflammation, and immune regulation (Mamun et al., 2023). Table 3.1 refers to the data regarding the studies on the gastrointestinal system affected by MPs.

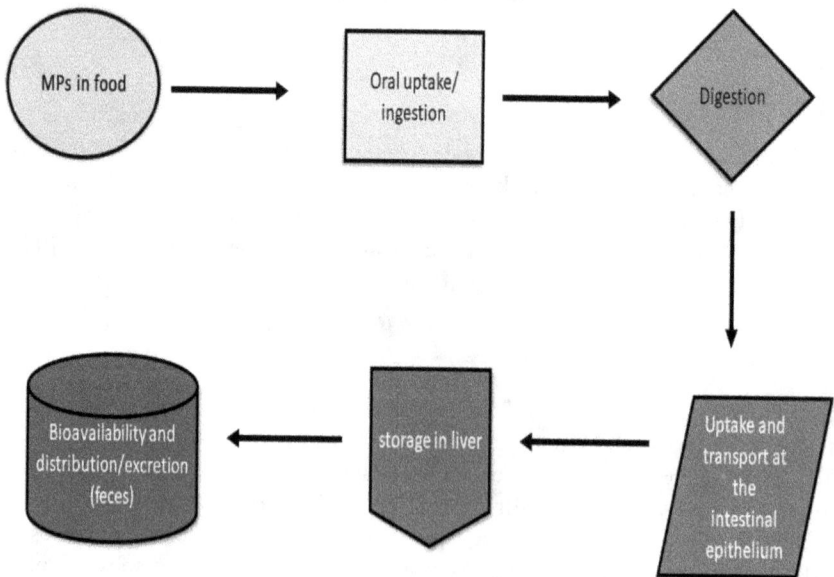

FIGURE 3.2 Pathway of MPs in the digestive system of the human body.

TABLE 3.1

Studies on the Gastrointestinal System Affected by MPs

Type of Samples	Size/types of Plastics	Organs/Tissues Examined	Biological Findings	Reference
Zebrafish (*Danio rerio*)	Polyamides, Polyethylene, Polypropylene, Polyvinyl chloride, Polystyrene particles. Size: ~70 μm- Polyamide, Polyethylene, Polypropylene, and Polyvinyl Chloride. 0.1, 1.0, and 5 μm- Polystyrene.	Intestine, gill, liver, kidney	- Cracking of the villi and splitting of enterocytes were observed. - Tissue damage from oxidative stress.	(Lei et al., 2018)
Mice	Polystyrene Size: 0.5 and 50 μm.	Colon	- Decrease in the secretion of mucin. - Gut microbiota dysbiosis.	(Lu et al., 2018)
Mice	Green fluorescent micro-polystyrene Size: 5 μm	Liver	- Vascular degeneration, chronic inflammatory infiltration, and hepatocellular oedema. - Oxidative stress, apoptosis, and inflammation in the hepatocytes.	(Li et al., 2021)
Human	Polyethylene, Polypropylene	The gastrointestinal tract, human stools	- Toxicity, inflammation through endocytosis by M cells of Peyer's patches and then into the circulatory system.	(Liao and Yang, 2020)

(Continued)

TABLE 3.1 (Continued)
Studies on the Gastrointestinal System Affected by MPs

Type of Samples	Size/types of Plastics	Organs/Tissues Examined	Biological Findings	Reference
Mice	Polystyrene microplastics	Intestine	- Intestinal immune imbalance and increase the expression of inflammation factors (TNF-α, IL-1β, and IFN-γ). - Histopathological damage in colonic mucosa.	(Liu et al., 2022)
Cell cultures: intestinal epithelial cell lines: LS174T, HT-29, and Caco-2	Nanoparticles Size: ~60 nm	Gastrointestinal system	- Positive NPs affected cell viability and cell line apoptosis.	(Inkielewicz-Stepniak et al., 2018)
Human intestinal epithelium in vitro	Polystyrene nanoparticles Size: 20–40 nm	Intestine	- The nanoparticle-induced apoptosis in individual cells propagated across the cell monolayer through "bystander killing effects." - Decrease in cell viability.	(Thubagere and Reinhard, 2010)
human stool	Polypropylene and polyethylene terephthalate Size: 50–500 μm	-	- Suggests unintentional ingestion of plastics from different sources.	(Schwabl et al., 2019)

3.5 PREVENTIVE MEASURES AGAINST PLASTIC POLLUTION

Strategies and policies should be created on plastic waste management and governance to eliminate the risk of harmful impacts on humans and the overall ecosystem. Both behavioral and natural science should be applied to reduce the amplification of plastic issues. Countries like Hong Kong have regulated a scheme for disposable plastic tableware, and China has prohibited the sale of chemical products containing microbeads by the end of 2022 (Yang et al., 2022). In India, the Ministry of Environment, Forest, and Climate Change has adopted policies to phase out single-use plastic carry bags which are 50 to 75 microns and have increased to 120 microns from 31 December 2022 (Government Notifies the Plastic Waste Management Amendment Rules, 2021, Prohibiting Identified Single Use Plastic Items by 2022, 2021). Moreover, strategies are adopted for alternatives with biodegradable constituents, such as bioplastics made of polymers from biological sources. Scientists across India are working on this to reduce the impact of plastic pollution.

3.6 CONCLUSION

Due to their lightweight, inexpensive, degradation-resistant nature and vigorous usage, plastics have contaminated the food web, causing immense harm to the human body through ingestion. The toxicity can mostly occur in the digestive system through ingestion. Studies can conclude that MPs in foods with harmful additives reach the gut epithelial layer, causing an obstruction for nutrient uptake. Some may enter the bloodstream, releasing toxic substances throughout the body. This eventually results in oxidative stress, immune disruption, metabolic disturbance, and overall breakdown of the whole system. In this situation, degradable bioplastics and keratin-based polymers not only fulfill plastic demands but also ensure the reduction of MP-based toxicity also.

REFERENCES

Afrin, S., Rahman, Md. M., Hossain, Md. N., Uddin, Md. K., Malafaia, G., 2022. Are there plastic particles in my sugar? A pioneering study on the characterization of microplastics in commercial sugars and risk assessment. Sci Total Environ 837, 155849. https://doi.org/10.1016/j.scitotenv.2022.155849.

Danopoulos, E., Jenner, L., Twiddy, M., Rotchell, J.M., 2020. Microplastic contamination of salt intended for human consumption: A systematic review and meta-analysis. SN Appl Sci 2. https://doi.org/10.1007/s42452-020-03749-0.

Dessì, C., Okoffo, E.D., O'Brien, J.W., Gallen, M., Samanipour, S., Kaserzon, S., Rauert, C., Wang, X., Thomas, K.V., 2021. Plastics contamination of store-bought rice. J Hazard Mater 416. https://doi.org/10.1016/j.jhazmat.2021.125778.

Gea, O. C., Ferrante, M., Banni, M., Favara, C., Nicolosi, I., Cristaldi, A., Fiore, M., Zuccarello, P., 2020. Micro- and nano-plastics in edible fruit and vegetables. The first diet risks assessment for the general population. Environ Res 187, 109677. https://doi.org/10.1016/j.envres.2020.109677.

Government Notifies the Plastic Waste Management Amendment Rules, 2021. Prohibiting identified single use plastic items by 2022, Controller of Publications, Delhi. pib.gov.in.

Guo, J.-J., Huang, X.-P., Xiang, L., Wang, Y.-Z., Li, Y.-W., Li, H., Cai, Q.-Y., Mo, C.-H., Wong, M.-H., 2020. Source, migration and toxicology of microplastics in soil. Environ Int 137, 105263. https://doi.org/https://doi.org/10.1016/j.envint.2019.105263.

Huerta Lwanga, E., Mendoza Vega, J., Ku Quej, V., Chi, J. de los A., Sanchez del Cid, L., Chi, C., Escalona Segura, G., Gertsen, H., Salánki, T., van der Ploeg, M., Koelmans, A.A., Geissen, V., 2017. Field evidence for transfer of plastic debris along a terrestrial food chain. Sci Rep 7, 14071. https://doi.org/10.1038/s41598-017-14588-2.

Inkielewicz-Stepniak, I., Tajber, L., Behan, G., Zhang, H., Radomski, M., Medina, C., Santos-Martinez, M., 2018. The role of mucin in the toxicological impact of polystyrene nanoparticles. Materials 11, 724. https://doi.org/10.3390/ma11050724.

Kedzierski, M., Lechat, B., Sire, O., Le Maguer, G., Le Tilly, V., Bruzaud, S., 2020. Microplastic contamination of packaged meat: Occurrence and associated risks. Food Package Shelf Life 24, 100489. https://doi.org/https://doi.org/10.1016/j.fpsl.2020.100489.

Lei, L., Wu, S., Lu, S., Liu, M., Song, Y., Fu, Z., Shi, H., Raley-Susman, K.M., He, D., 2018. Microplastic particles cause intestinal damage and other adverse effects in zebrafish danio rerio and nematode Caenorhabditis elegans. Sci Total Environ 619–620, 1–8. https://doi.org/https://doi.org/10.1016/j.scitotenv.2017.11.103.

Li, S., Shi, M., Wang, Y., Xiao, Y., Cai, D., Xiao, F., 2021. Keap1-nrf2 pathway up-regulation via hydrogen sulfide mitigates polystyrene microplastics induced-hepatotoxic effects. J Hazard Mater 402, 123933. https://doi.org/https://doi.org/10.1016/j.jhazmat.2020.123933.

Li, Z., Yang, Y., Chen, X., He, Y., Bolan, N., Rinklebe, J., Lam, S.S., Peng, W., Sonne, C., 2023. A discussion of microplastics in soil and risks for ecosystems and food chains. Chemosphere 313, 137637. https://doi.org/https://doi.org/10.1016/j.chemosphere.2022.137637.

Liao, Y., Yang, J., 2020. Microplastic serves as a potential vector for Cr in an in-vitro human digestive model. Sci Total Environ 703, 134805. https://doi.org/https://doi.org/10.1016/j.scitotenv.2019.134805.

Liebezeit, G., Liebezeit, E., 2015. Origin of synthetic particles in honeys. Pol J Food Nutr Sci 65, 143–147. https://doi.org/10.1515/pjfns-2015-0025.

Liu, S., Li, H., Wang, J., Wu, B., Guo, X., 2022. Polystyrene microplastics aggravate inflammatory damage in mice with intestinal immune imbalance. Sci Total Environ 833, 155198. https://doi.org/https://doi.org/10.1016/j.scitotenv.2022.155198.

Lu, L., Wan, Z., Luo, T., Fu, Z., Jin, Y., 2018. Polystyrene microplastics induce gut microbiota dysbiosis and hepatic lipid metabolism disorder in mice. Sci Total Environ 631–632, 449–458. https://doi.org/https://doi.org/10.1016/j.scitotenv.2018.03.051.

Mamun, A.A., Prasetya, T.A.E., Dewi, I.R., Ahmad, M., 2023. Microplastics in human food chains: Food becoming a threat to health safety. Sci Total Environ 858, 159834. https://doi.org/https://doi.org/10.1016/j.scitotenv.2022.159834.

Mason, S.A., Welch, V.G., Neratko, J., 2018. Synthetic polymer contamination in bottled water. Front Chem 6. https://doi.org/10.3389/fchem.2018.00407.

Ministry of Environment, Forest and Climate Change, 2021. Government notifies the plastic waste management rules, 2021, prohibiting identified single use items by 2022 [www document].

Paul, M.B., Stock, V., Cara-Carmona, J., Lisicki, E., Shopova, S., Fessard, V., Braeuning, A., Sieg, H., Böhmert, L., 2020. Micro- and nanoplastics—current state of knowledge with the focus on oral uptake and toxicity. Nanoscale Adv 2, 4350–4367. https://doi.org/10.1039/d0na00539h.

Pironti, C., Ricciardi, M., Motta, O., Miele, Y., Proto, A., Montano, L., 2021. Microplastics in the environment: Intake through the food web, human exposure and toxicological effects. Toxics 9. https://doi.org/10.3390/toxics9090224.

Prata, J.C., Da Costa, J.P., Lopes, I., Duarte, A.C., Rocha-Santos, T., 2020. environmental exposure to microplastics: An overview on possible human health effects. Sci Total Environ 702, 134455. https://doi.org/https://doi.org/10.1016/j.scitotenv.2019.134455.

Ragusa, A., Notarstefano, V., Svelato, A., Belloni, A., Gioacchini, G., Blondeel, C., Zucchelli, E., De Luca, C., D'avino, S., Gulotta, A., Carnevali, O., Giorgini, E., 2022. Raman microspectroscopy detection and characterisation of microplastics in human breastmilk. Polymers (Basel) 14. https://doi.org/10.3390/polym14132700.

Saeedi, M., 2023. How microplastics interact with food chain: A short overview of fate and impacts. J Food Sci Technol. https://doi.org/10.1007/s13197-023-05720-4.

Schwabl, P., Köppel, S., Königshofer, P., Bucsics, T., Trauner, M., Reiberger, T., Liebmann, B., 2019. Detection of various microplastics in human stool. Ann Intern Med 171, 453–457. https://doi.org/10.7326/m19-0618.

Schymanski, D., Goldbeck, C., Humpf, H.U., Fürst, P., 2018. Analysis of microplastics in water by micro-Raman spectroscopy: Release of plastic particles from different packaging into mineral water. Water Res 129, 154–162. https://doi.org/10.1016/j.watres.2017.11.011.

Sivagami, M., Selvambigai, M., Devan, U., Velangani, A.A.J., Karmegam, N., Biruntha, M., Arun, A., Kim, W., Govarthanan, M., Kumar, P., 2021. Extraction of microplastics from commonly used sea salts in India and their toxicological evaluation. Chemosphere 263. https://doi.org/10.1016/j.chemosphere.2020.128181.

Thubagere, A., Reinhard, B.M., 2010. Nanoparticle-induced apoptosis propagates through hydrogen-peroxide-mediated bystander killing: Insights from a human intestinal epithelium in vitro model. ACS Nano 4, 3611–3622. https://doi.org/10.1021/nn100389a.

Tong, H., Jiang, Q., Hu, X., Zhong, X., 2020. Occurrence and identification of microplastics in tap water from China. Chemosphere 252. https://doi.org/10.1016/j.chemosphere.2020.126493.

Toussaint, B., Raffael, B., Angers-Loustau, A., Gilliland, D., Kestens, V., Petrillo, M., Rio-Echevarria, I.M., Van Den Eede, G., 2019. Review of micro- and nanoplastic contamination in the food chain. Food Addit Contam Part A 36, 639–673. https://doi.org/10.1080/19440049.2019.1583381.

Udovicki, B., Andjelkovic, M., Cirkovic-Velickovic, T., Rajkovic, A., 2022. Microplastics in food: Scoping review on health effects, occurrence, and human exposure. Int J Food Contam 9, 7. https://doi.org/10.1186/s40550-022-00093-6.

Wu, X., Hou, H., Liu, Y., Yin, S., Bian, S., Liang, S., Wan, C., Yuan, S., Xiao, K., Liu, B., Hu, J., Yang, J., 2022. Microplastics affect rice (Oryza sativa l.) quality by interfering metabolite accumulation and energy expenditure pathways: A field study. J Hazard Mater 422, 126834. https://doi.org/https://doi.org/10.1016/j.jhazmat.2021.126834.

Yang, X., Man, Y., Wong, M., Owen, R., Chow, K.L., 2022. Environmental health impacts of microplastics exposure on structural organization levels in the human body. Sci Total Environ 825, 154025. https://doi.org/10.1016/j.scitotenv.2022.154025.

4 Microplastics in the Atmosphere

Moumita Sharma

4.1 INTRODUCTION

Globally, there is a growing fret over microplastics (Rochman et al., 2019). Microplastics (MPs) are frequently described as plastic fragments with a size range of 5 mm to 100 nm (Thompson et al., 2004). Nanoplastics, which are predicted to be as common as their bulk counterparts, are of special concern because of the evolving research into plastic particles. 75.9% of this garbage was dumped as trash between 1950 and 2015, which is a worry for the ecosystem (Geyer et al., 2017). About 12,000 million metric tonnes of plastic wastes are anticipated to be residing in landfills by 2050 (Jambeck et al., 2015). Our oceans accumulate 8.75 million metric tonnes of plastic from land-based sources on an annual basis (Barboza et al., 2019; Song et al., 2017). Nanoplastics are typically interpreted as plastic fragments having size lower than 1 μm and are of paramount importance for the safety of seafood as well as for improving the spread of contaminants in the environment and posing significant health concerns to humans (Bank and Hansson, 2019; Alimi et al., 2018). Environmental microplastics have not yet been extensively quantified, however. The precise plastic particles that this chapter will focus on are called microplastics.

Microplastics have been detected in an extensive variety of media—including soils, aquatic systems (such as oceans, shorelines, rivers, swamps), and inside the alimentary canal of both vertebrate and invertebrate animals (Prata et al., 2019; Auta et al., 2017). The maritime environment has received the majority of emphasis in study up to this point, but other environmental compartments are receiving a growing amount of attention (Horton and Dixon, 2018). Many suspended elements are transported locally or globally via the atmosphere, which is a significant pathway. Recent investigations have shown that atmospheric microplastic particles can travel into distant locations, and even into ocean surface air (Zhang et al., 2019; Klein and Fischer, 2019). Numerous phenomena, including wind direction and speed, updrafts and downdrafts, turbulence, and convection lift, are present in the atmosphere. As a result, they are regarded as significant conduits for influencing microplastic transport as well as the source-sink dynamics and flow mechanism of the plastic pollution in marine as well as terrestrial environments (Zhang et al., 2019; K. Liu, Wu et al., 2019). Microplastics are currently considered to be a significant source of air pollution due to their ability to absorb and combine with other contaminants (such mercury or PAHs) (Barboza et al., 2019; C. Liu et al., 2019).

Microplastics fall within the primary or secondary plastics categories. When macroplastics, such as fibres from synthetic fabrics, break down and fragment,

DOI: 10.1201/9781003449133-4

secondary microplastics are created (GESAMP, 2016). Primary microplastics are made specifically to fit specific applications (like micro beads). Due to the shape difference that may have an impact on its aerodynamics and, ultimately, atmospheric transport, such a differentiation may be crucial for the research of atmospheric transport. There is compelling evidence that microplastics are entering the environment at all stages of the lifecycle of a plastic product, from producers to waste management, with the potential for trophic transmission and human health exposure (Bank and Hansson, 2019).

Considering how many studies there are on microplastics in marine and terrestrial settings, research on air microplastics has only lately gathered interest (Alimba and Faggio, 2019; Prata et al., 2019). The majority of research that has been published thus far concentrates on atmospheric deposition, which is the passive accumulation of debris at a chosen area. Just over four dozen researches have examined the presence of microplastics in atmospheric aerosols. Microplastics can be inhaled by humans through aerosols, which has been identified as a significant pathway (Enyoh et al., 2019; Vianello et al., 2019). They can travel up to 95 km, per a recent study (Allen et al., 2019). Microplastics are released into the air through a variety of processes, including clothing washing and drying, emissions generated from the synthetic textile industry, tyre erosiveness from synthetic rubber, vinyl chloride, as well as polyvinyl chloride (PVC) industry emissions, worsening of household furniture, and defilation from city dust (Browne et al., 2011; Henry et al., 2019; Dris et al., 2016). Additionally, significant amounts of plastics are burned every day in open landfills, which causes the volatilization of numerous toxic substances that unavoidably mix with the atmospheric aerosol (Velis and Cook, 2021).

Size, abundance, shapes, and components have all been studied for suburban, urban, and also distant locations (Dris et al., 2015; C. Liu et al., 2019; Klein and Fischer, 2019). It is vital to gather and evaluate recent research findings, assess the state of knowledge, and contrast the characteristics of atmospheric microplastics with those of microplastics from other settings in order to better comprehend the current situation of atmospheric microplastics. A growing global issue is the possible effects of airborne microplastics on the transportation, deposition, and exposure of humans and remote locations via food webs. This chapter summarises the current level of knowledge in the study of air microplastic pollution with an emphasis on its advancement, information gaps, and suggestions to encourage future research that is standardised and comparable.

4.2 METHODS OF MICROPLASTIC ANALYSIS

4.2.1 SAMPLE COLLECTION

Early investigations collected a variety of wet and dry deposits for various times and amounts of precipitation using non-standard collection equipment. However, due to recent advancements in the procedure of passive sampling in atmospheric deposition, Norwegian Institute for Air Research has created a harmonised metallic/glass system. This technology offers a passive atmospheric deposition method that is plastic-free and standardised, making it perfect for microplastic research. These bulk-deposition, also known as total-deposition, samplers offer the benefits of being easy to use, standardised processes and not requiring electricity at the study site. This standardised sample method allows for the very low cost, very efficient adherence to a standardised protocol for collection of data gathering in distant locations with limited infrastructure. The volume of blow by (wind driving particles out of the collection funnel before entrapment), which allows comparison to other deposited material in addition to previous plastic studies, is another justification for employing a standardised sampler. Different sampling techniques, such as sweeping, actively pumping the dust, and vacuuming was used to sample indoor as well as outdoor dust, which makes it challenging to compare the data. To collect indoor dust deposition, C. Liu et al., 2019, employed hog bristle brushes; the quantity of material retained inside the brush is unknown. The collected dust deposition was then transferred as completely as possible to sample bags. It is straightforward to replicate this process, but it is arduous to estimate the relative volume of the air that is sampled or whether the MPs that were collected were simply the consequence of atmospheric deposition.

4.2.1.1 Active Sampling

The aerosol sample is drawn into the active collectors using a tool, usually a vacuum pump. Some of these pumps, referred to as high-volume air samplers, can draw hundreds of litres of air per minute, while others, referred to as low-volume air samplers, can only pull a few litres (Uddin, 2016). The active sampling technique has the advantages of drawing an investigated volume of aerosol and typically collecting a significant volume of sample reasonably rapidly. Similar data across the spatiotemporal domain will be produced. The method also makes it possible to calculate concentrations per unit volume. For years, active sampling has been used to monitor air contaminants, including microorganisms, carbon, lead, and mercury (Dommergue et al., 2019). This method is well-established and well-accepted. In comparison to passive samplers, active samplers are more expensive to purchase, operate, and require a power supply and some basic skills to use. A media/collection substrate is used by the majority of active samplers in use; just a few authors have acknowledged using glass-fibre, quartz, or cellulose filters.

4.2.1.2 Passive Sampling

A time-integrated sample from passive sampling is often obtained through air deposition over a defined area, such as a surface which can be both dry and wet. Simple equipment like funnels, bottles, planar surfaces, and rain gauges that allow for atmospheric deposition are just a few examples of the different samplers that are available. These samplers can estimate the amount of microplastics per unit area but not the concentrations of aerosol per unit volume. The local weather conditions, height, and angle of deployment have a significant impact on passive samplers, despite being cheaper and simpler to deploy. Additionally, they cannot be used for deployments lasting only a few minutes or hours. The selection of a sampler must take into account the scientific objective(s), which will help establish for the purpose of calculating the long-term mass accumulation rates of microplastics or their concentration in aerosol.

4.2.2   Criteria for Identifying Microplastics Visually

Globally, a huge proportion of plastics are produced from nonrenewable fossil fuels. When viewed using a stereomicroscope, plastic particles bigger than 500 µm are normally identified visually based on their shape and colour, with chemical analysis being utilised for confirmation (Hidalgo-Ruz et al., 2012). Because of the weathering and particle size, there are some differences in the technique used to identify air microplastics. But when identifying air microplastics, the following recommendations are frequently followed:

- Plastics cannot include any biogenic (organic or biological) formations (Dris et al., 2015).
- The removal of biofilms and other organic or inorganic adhesion from the microplastic particles is necessary to avoid artefacts that obstruct precise and accurate identification (Löder and Gerdts, 2015).
- Fibres are anticipated to exhibit three-dimensional bending and to have a reasonably even or constant thickness along their whole length (Dris et al., 2015).
- Films and fragments are required to display a degree of transparency or clarity and reasonably homogeneous coloration. Extremely worn particles, however, may exhibit prominent interior coloration "spots" along with colour loss or bleaching at the particle corners and exterior (Löder and Gerdts, 2015).
- Aged plastic has embrittled and weathered surfaces, uneven forms, and broken and sharp edges, as would be expected in the environmental samples (Hidalgo-Ruz et al., 2012). According to Zhou et al. (2018), pitting, gouging, and surface scratches or tears can also be seen in worn plastics.
- Colours that range from clear and different shades of white through vibrant orange, greens, blues, and purples through to black can be used to identify plastics (Löder and Gerdts, 2015; Hidalgo-Ruz et al., 2012). It's important to properly inspect transparent, red, and green threads to determine their composition (Dris et al., 2015). From the study, it is noticed that, throughout the procedure of sample preparation (Hydrogen peroxide [H_2O_2] digestion), biogenic and plastic material become bleached, which makes coloured plastic particles less noticeable and harder to distinguish from leftover (after digestion) biogenic material.

4.2.3   Organic Matrix Elimination

For the purpose of analysing microplastics, sample preparation is constantly evolving and improving. Since the focus of microplastic research has shifted away from straightforward sample matrices and also the analytical techniques have advanced to permit for smaller particle analysis, it is now crucial to isolate small microplastics from the residual sample material. It is particularly critical when sample materials contain a lot of organic material, since the organic material impedes spectrographic analysis by increasing noise in the spectra, bio-coating, and screening of plastic particulates (Löder and Gerdts, 2015). Numerous techniques, such as KOH, HNO$_3$, NaOH, HCl, H$_2$O$_2$ + Fe, H$_2$O$_2$, H$_2$O$_2$ + H$_2$SO$_4$, and enzymatic procedures, were used to remove organic material (Renner et al., 2018).

4.2.4 ANALYTICAL METHODS

Several researchers have tried to characterise the microplastics as polymeric materials using methods including μ-FTIR, HPLC-MSMS, ATR-FTIR, μ -RS, and thermal desorption, combined with thermo-gravimetric investigations. It is crucial to point out the fundamental contrasts between these strategies. Fibres with a particle size of up to 10 and 50 μm can be studied using the ATR-FTIR and μ-FTIR spectroscopies, respectively (Vianello et al., 2019). With its ability to recognise MPs up to 1 μm in size, μ-RS has a technological advantage. The FTIR method has, however, been preferred in research because of its quick-throughput and large-polymer-spectrum libraries. The use of PTFE, polycarbonate, and nylon filters by some researchers has hindered the utilisation of μ-FTIR or μ-RS because they interfere with MP spectra. A silver filter with a 0.2 μm pore size or an aluminium oxide is the preferable filter for both μ-FTIR as well as μ-RS analysis. Other techniques for identifying MPs include Py-GCMS, TD, and HPLC-MS-MS; however, they cannot reveal the quantity, shape, or size of MPs; instead, they can only provide the concentration of a certain polymeric particle in the sample.

4.3 SHAPE, SIZE, COMPOSITION, AND COLOUR OF ATMOSPHERIC MICROPLASTICS

4.3.1 COMPOSITION

The most important factor in determining plastic contamination is its chemical composition (Hartmann et al., 2019). Additionally comprised of many distinct kinds of polymer are microplastics (Rochman et al., 2019). Numerous polymers have been generated and used in commercial as well as residential applications. The constitution of the plastic polymers, which are divided into thermoplastics and thermosets, governs the features of plastics, both chemical and physical. Low-density polyethylene (17.5%), polypropylene (19.3%), high-density polyethylene (12.3%), polyurethane (7.7%), polyvinyl chloride (10.2%), polystyrene (6.6%), and polyethylene terephthalate (7.4%) are the polymer types with the highest demand for plastics and highest production rates. According to published research on the polymer composition of seawater, polyethylene is the most predominant polymer, followed by polypropylene and polystyrene (PlasticsEurope, 2018).

4.3.2 COLOUR

A variety of colours have been observed in microplastics, including orange, green, red, yellow, tan, brown, grey, off-white, blue, white, and others. (Rochman et al., 2019; Bergmann et al., 2019). Blue and red fibres have been reported the most frequently. Visual inspection could under represent particles that are dark, white, clear, or translucent (Hartmann et al., 2019). When preparing samples, colour can be used to identify the potential contaminations as well as sources of plastic waste. Polypropylene has been linked to clear and transparent products; polyethylene, to white; and LDPE, to opaque colours (Hartmann et al., 2019). But based just on colour, a plastic particle cannot be classified as to what kind or where it originates from. Importantly, since brighter hues are easier to see when inspected visually, colour information might be misinterpreted (Rochman et al., 2019).

4.3.3 SIZE AND SHAPE OF THE MICROPLASTICS

The interaction of a plastic object with biota and its fate in the environment are significantly influenced by plastic particle size (Besseling et al., 2017). Typically, the collection and analysis procedure operationally defines the microplastic size limitations (Hartmann et al., 2019). There are many different sizes of microplastics; however, they are typically believed to be between 1 mm and 5 mm long. Microplastics in the air are far smaller than those in sediment and water (Prata et al., 2019; Auta et al., 2017). For instance, the samples from the Pyrenees Mountains exhibited a larger proportion of fragments (size ranging between 50 m composed of 70% of microplastic particulates), with plastic fibres typically being smaller than 300 metres in length (50%) in those samples (Allen et al., 2019).

The shapes and sizes of microplastics in the environment are quite diverse. Microplastic forms are frequently described as spheres, pellets, beads, flakes, foam, pieces, fibres, and films (Hidalgo-Ruz et al., 2012). These morphologies are influenced by the main microplastics' starting state, the erosive and degradative processes that take place on their surfaces, and the time that the plastic particles have spent in the environment. According to research, degraded microplastics with sharp edges signify a recent entry into the environment, whereas those with smooth edges signify a long-term presence. The air microplastics have been identified in a range of morphologies, including fibre, fragment, foam, and film.

4.4 TRANSPORT OF MICROPLASTICS IN THE ATMOSPHERE

In freshwater, marine, terrestrial, and now air environmental compartments, microplastic pollution is pervasive (Horton and Dixon, 2018; Bergmann et al., 2019). There is a connection between these ecosystems, and a complex web of source-pathway-sink relationships can affect how microplastics move through and are retained in these environmental matrices. The deposition of microplastics in terrestrial or aquatic habitats has been linked to atmospheric transmission of microplastics, which has recently been recognised as a significant vector (Zhang et al., 2019). Such mobility, particularly the movement between the terrestrial and marine environments, has a significant impact on the source-sink behaviour of plastic pollution in many ecosystems (Bank and Hansson, 2019). The microplastics found in the oceans could eventually originate from terrestrial habitats, according to a study based on air microplastics in the Shanghai region and the west Pacific Ocean (K. Liu, Wang et al., 2019). According to K. Liu, Wu et al. (2019), suspended air microplastics, including pollution from textile microfibres, may be a substantial origin of the ocean's microplastic pollution. Atmospheric mobility has a considerable impact on the migration of microplastics and their potential environmental sinks (Horton and Dixon, 2018). The transport of microplastic particles will be significantly impacted by their density and form. The mechanisms governing the transfer of microplastics inside air are, however, poorly understood (Allen et al., 2019). More specifically, it is unknown how much air fallout contaminates the ground and water. More studies are required in this field, particularly those that consider source-pathway-sink processes, transport characteristics, and weather-related factors spatially.

4.5 HEALTH IMPLICATIONS CAUSED BY ATMOSPHERIC MICROPLASTICS

A schematic representation of the health hazard caused by different microplastics is shown in Figure 4.1.

Different kinds of organisms, including those commonly found in the human diet, can consume microplastics that are present in the environment (Rochman et al., 2019; Prata et al., 2019). The newest scientific research studies on air microplastics underline the extensive spatiotemporal dimensions of the techniques that influence the origins, fate, movement, and consequences of microplastics on living beings and

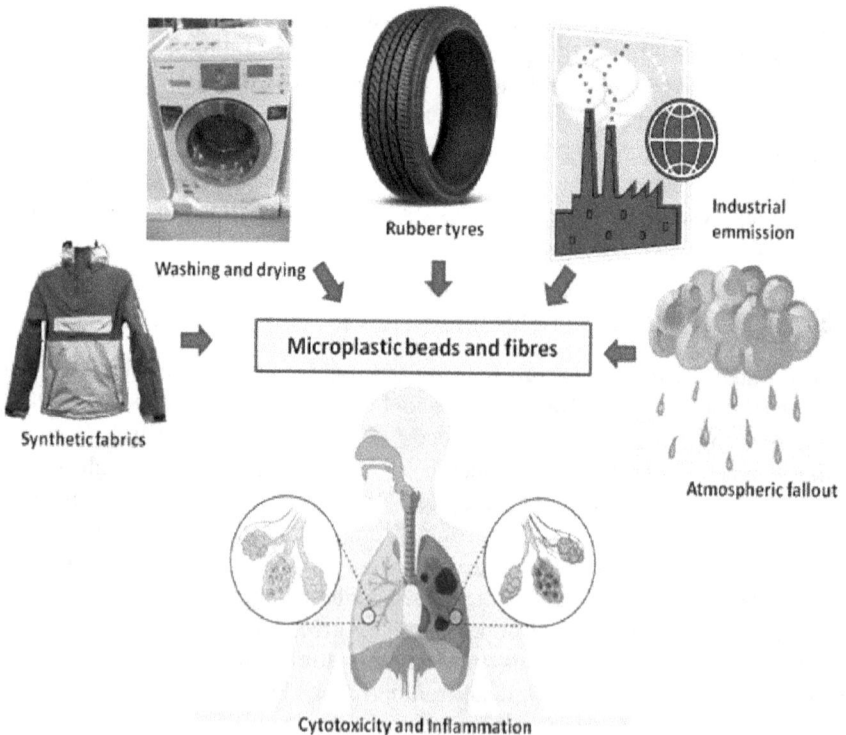

FIGURE 4.1 Health hazard caused by various atmospheric microplastics.

the environment, including people (Bank and Hansson, 2019). Airborne MPs have the potential to carry particle-reactive contaminants to human lungs, including PAHs (Polycyclic Aromatic Hydrocarbons), metals produced by vehicle emissions, as well as microorganisms (Gasperi et al., 2018). As a result, breathing in these MPs could be hazardous to one's health and result in long-lasting illnesses. Several contaminants— viz. plasticisers, antimicrobials, flame retardants, and Bisphenol A—may be produced during the breakdown of MPs (Berkner et al., 2004). According to epidemiological research, air pollution from ambient atmospheric particles has a negative impact on the heart and lungs. There is a chance of exposure through the absorption of dust, especially for little newborns, despite the argument that the plainly apparent microplastic threads are too big to be ingested (Wright and Kelly, 2017).

Previous research found cellulosic as well as plastic fibres inside human lungs (lung tumours that had been excised and lung biopsies) as well as in the respiratory and health issues that workers in factories that processed plastics displayed (coughing, wheezing, dyspnea, occupational asthma) (Pauly et al., 1998). Furthermore, it has been demonstrated that microplastic particulates (>100 µm) are bio-persistent and can pass past the digestive tract's epithelium. Based on the air microplastic content, it may be possible to determine the human exposure to microplastics, particularly through dust consumption. Aerosol also contains other pollutants, like phthalates (Wang et al., 2018). However, it is still unclear how much harm these MPs may actually do to people's health. There is a connection with previously published studies showing that fibre accumulation in alveoli, granulomas, and terminal bronchioles caused persistent inflammation (Amato-Lourenço et al., 2020). In a different investigation, silica, asbestos, and diesel exhaust fibres were linked to secondary genotoxicity as a result of ongoing reactive oxygen species generation (Schins, 2002). Inhaling MPs was linked to an increased risk of interstitial lung disease, but, as these associations were predicated on tests using nonenvironmental amounts, there is some doubt as to whether these correlations are accurate.

4.6 CONCLUSION

Today, microplastics are recognised as air pollutants and particulates. The atmosphere and atmospheric deposition of urban, rural, and remote locations include microplastics, according to recent study. As an air pollutant, microplastic pollution has the potential to travel great distances and affect areas that are distant from the sources of the pollution.

Studies that have been published on the relative abundance of air microplastics show that these particles can have a wide variety of characteristics and concentrations, depending on the region. Fibres and fragments are the microplastic structures that are most frequently seen in the atmosphere. It is difficult to draw conclusions on size distribution from these tests because of the variability in the targeted particle sizes. Atmospheric microplastics can be carried to isolated locations and deposited through dry or wet deposition due to their light weight, resilience, and other inherent qualities. Atmospheric microplastics from different sources to the ocean as well as land surface are significantly impacted by wind, snowfall, and weathering. More needs to be learned about the geographical and temporal variability of atmospheric microplastic depositions. Understanding the interactions between microplastics in the atmosphere, ecosystems, various chemicals, and human exposure will require more investigation.

REFERENCES

Alimba, C.G., Faggio, C., 2019. Microplastics in the marine environment: Current trends in environmental pollution and mechanisms of toxicological profile. Environ. Toxicol. Pharmacol. 68, 61–74. https://doi.org/10.1016/j.etap.2019.03.001.

Alimi, O.S., Farner Budarz, J., Hernandez, L.M., Tufenkji, N., 2018. Microplastics and nanoplastics in aquatic environments: Aggregation, deposition, and enhanced contaminant transport. Environ. Sci. Technol. 52, 1704–1724. https://doi.org/10.1021/acs.est.7b05559.

Allen, S., Allen, D., Phoenix, V.R., Le Roux, G., Durántez Jiménez, P., Simonneau, A., Binet, S., Galop, D., 2019. Atmospheric transport and deposition of microplastics in a remote mountain catchment. Nat. Geosci. 12, 339–344. https://doi.org/10.1038/s41561-019-0335-5.

Amato-Lourenço, L.F., dos Santos Galvão, L., de Weger, L.A., Hiemstra, P.S., Vijver, M.G., Mauad, T., 2020. An emerging class of air pollutants: Potential effects of microplastics to respiratory human health? Sci. Total Environ. 749, 141676. https://doi.org/10.1016/j.scitotenv.2020.141676.

Auta, H.S., Emenike, C.U., Fauziah, S.H., 2017. Distribution and importance of microplastics in the marine environment: A review of the sources, fate, effects, and potential solutions. Environ. Int. 102, 165–176. https://doi.org/10.1016/j.envint.2017.02.013.

Bank, M.S., Hansson, S.V., 2019. The plastic cycle: A novel and holistic paradigm for the Anthropocene. Environ. Sci. Technol. 53, 7177–7179. https://doi.org/10.1021/acs.est.9b02942.

Barboza, L.G.A., Cózar, A., Gimenez, B.C.G., Barros, T.L., Kershaw, P.J., Guilhermino, L., 2019. Macroplastics pollution in the marine environment, 305–328. https://doi.org/10.1016/B978-0-12-805052-1.00019-X.

Bergmann, M., Mützel, S., Primpke, S., Tekman, M.B., Trachsel, J., Gerdts, G., 2019. White and wonderful? Microplastics prevail in snow from the Alps to the Arctic. Sci. Adv. 5, eaax1157. https://doi.org/10.1126/sciadv.aax1157.

Berkner, S., Streck, G., Herrmann, R., 2004. Development and validation of a method for determination of trace levels of alkylphenols and bisphenol A in atmospheric samples. Chemosphere 54, 575–584. https://doi.org/10.1016/S0045-6535(03)00759-8.

Besseling, E., Quik, J.T.K., Sun, M., Koelmans, A.A., 2017. Fate of nano- and microplastic in freshwater systems: A modeling study. Environ. Pollut. Barking Essex 1987 220, 540–548. https://doi.org/10.1016/j.envpol.2016.10.001.

Browne, M.A., Crump, P., Niven, S.J., Teuten, E., Tonkin, A., Galloway, T., Thompson, R., 2011. Accumulation of microplastic on shorelines worldwide: Sources and sinks. Environ. Sci. Technol. 45, 9175–9179. https://doi.org/10.1021/es201811s.

Dommergue, A., Amato, P., Tignat-Perrier, R., Magand, O., Thollot, A., Joly, M., Bouvier, L., Sellegri, K., Vogel, T., Sonke, J.E., Jaffrezo, J.-L., Andrade, M., Moreno, I., Labuschagne, C., Martin, L., Zhang, Q., Larose, C., 2019. Methods to investigate the global atmospheric microbiome. Front. Microbiol. 10.

Dris, R., Gasperi, J., Rocher, V., Mohamed, S., Tassin, B., 2015. Microplastic contamination in an urban area: A case study in greater Paris. Environ. Chem. 12. https://doi.org/10.1071/EN14167.

Dris, R., Gasperi, J., Saad, M., Mirande, C., Tassin, B., 2016. Synthetic fibers in atmospheric fallout: A source of microplastics in the environment? Mar. Pollut. Bull. 104, 290–293. https://doi.org/10.1016/j.marpolbul.2016.01.006.

Enyoh, C.E., Verla, A.W., Verla, E.N., Ibe, F.C., Amaobi, C.E., 2019. Airborne microplastics: A review study on method for analysis, occurrence, movement and risks. Environ. Monit. Assess. 191, 668. https://doi.org/10.1007/s10661-019-7842-0.

Gasperi, J., Wright, S.L., Dris, R., Collard, F., Mandin, C., Guerrouache, M., Langlois, V., Kelly, F.J., Tassin, B., 2018. Microplastics in air: Are we breathing it in? Micro Nanoplastics Ed. Dr Teresa AP Rocha-St. 1, 1–5. https://doi.org/10.1016/j.coesh.2017. 10.002.

GESAMP, 2016. Sources, Fate and Effects of Microplastics in the Marine Environment (Part 2 of a Global Assessment). GESAMP Reports and Studies Series No. 93. International Maritime Organization, London, p. 220.

Geyer, R., Jambeck, J.R., Law, K.L., 2017. Production, use, and fate of all plastics ever made. Sci. Adv. 3, e1700782. https://doi.org/10.1126/sciadv.1700782.

Hartmann, N.B., Hüffer, T., Thompson, R.C., Hassellöv, M., Verschoor, A., Daugaard, A.E., Rist, S., Karlsson, T., Brennholt, N., Cole, M., Herrling, M.P., Hess, M.C., Ivleva, N.P., Lusher, A.L., Wagner, M., 2019. Are we speaking the same language? Recommendations for a definition and categorization framework for plastic debris. Environ. Sci. Technol. 53, 1039–1047. https://doi.org/10.1021/acs.est.8b05297.

Henry, B., Laitala, K., Klepp, I.G., 2019. Microfibres from apparel and home textiles: Prospects for including microplastics in environmental sustainability assessment. Sci. Total Environ. 652, 483–494. https://doi.org/10.1016/j.scitotenv.2018.10.166.

Hidalgo-Ruz, V., Gutow, L., Thompson, R.C., Thiel, M., 2012. Microplastics in the marine environment: A review of the methods used for identification and quantification. Environ. Sci. Technol. 46, 3060–3075. https://doi.org/10.1021/es2031505.

Horton, A.A., Dixon, S.J., 2018. Microplastics: An introduction to environmental transport processes. Wiley Interdiscip. Rev. Water 5. https://doi.org/10.1002/wat2.1268.

Jambeck, J.R., Geyer, R., Wilcox, C., Siegler, T.R., Perryman, M., Andrady, A., Narayan, R., Law, K.L., 2015. Marine pollution. Plastic waste inputs from land into the ocean. Science 347, 768–771. https://doi.org/10.1126/science.1260352.

Klein, M., Fischer, E.K., 2019. Microplastic abundance in atmospheric deposition within the Metropolitan area of Hamburg, Germany. Sci. Total Environ. 685, 96–103. https://doi.org/10.1016/j.scitotenv.2019.05.405.

Liu, C., Li, J., Zhang, Y., Wang, L., Deng, J., Gao, Y., Yu, L., Zhang, J., Sun, H., 2019. Widespread distribution of PET and PC microplastics in dust in urban China and their estimated human exposure. Environ. Int. 128, 116–124. https://doi.org/10.1016/j.envint.2019.04.024.

Liu, K., Wang, X., Fang, T., Xu, P., Zhu, L., Li, D., 2019. Source and potential risk assessment of suspended atmospheric microplastics in Shanghai. Sci. Total Environ. 675, 462–471. https://doi.org/10.1016/j.scitotenv.2019.04.110.

Liu, K., Wu, T., Wang, X., Song, Z., Zong, C., Wei, N., Li, D., 2019. Consistent transport of terrestrial microplastics to the ocean through atmosphere. Environ. Sci. Technol. 53, 10612–10619. https://doi.org/10.1021/acs.est.9b03427.

Löder, M.G.J., Gerdts, G., 2015. Methodology used for the detection and identification of microplastics—a critical appraisal. In: Bergmann, M., Gutow, L., Klages, M. (Eds.), Marine Anthropogenic Litter. Springer International Publishing, Cham, pp. 201–227. https://doi.org/10.1007/978-3-319-16510-3_8.

Pauly, J.L., Stegmeier, S.J., Allaart, H.A., Cheney, R.T., Zhang, P.J., Mayer, A.G., Streck, R.J., 1998. Inhaled cellulosic and plastic fibers found in human lung tissue. Cancer Epidemiol. Biomarkers Prev. 7(5), 419–428.

PlasticsEurope, 2018. European plastic waste: Recycling overtakes landfill for the first time. URL www.plasticseurope.org/en/newsroom/press-releases/archive-press-releases-2018/european-plastic-waste-recycling-overtakes-landfill-first-time (accessed 7.13.20).

Prata, J.C., da Costa, J.P., Girão, A.V., Lopes, I., Duarte, A.C., Rocha-Santos, T., 2019. Identifying a quick and efficient method of removing organic matter without damaging microplastic samples. Sci. Total Environ. 686, 131–139. https://doi.org/10.1016/j.scitotenv.2019.05.456.

Renner, G., Schmidt, T.C., Schram, J., 2018. Analytical methodologies for monitoring micro(nano)plastics: Which are fit for purpose? Micro Nanoplastics Ed. Dr Teresa AP Rocha-St. 1, 55–61. https://doi.org/10.1016/j.coesh.2017.11.001.

Rochman, C.M., Brookson, C., Bikker, J., Djuric, N., Earn, A., Bucci, K., Athey, S., Huntington, A., McIlwraith, H., Munno, K., De Frond, H., Kolomijeca, A., Erdle, L., Grbic, J., Bayoumi, M., Borrelle, S.B., Wu, T., Santoro, S., Werbowski, L.M., Zhu, X., Giles, R.K., Hamilton, B.M., Thaysen, C., Kaura, A., Klasios, N., Ead, L., Kim, J., Sherlock, C., Ho, A., Hung, C., 2019. Rethinking microplastics as a diverse contaminant suite. Environ. Toxicol. Chem. 38, 703–711. https://doi.org/10.1002/etc.4371.

Schins, R.P.F., 2002. Mechanisms of genotoxicity of particles and fibers. Inhal. Toxicol. 14, 57–78. https://doi.org/10.1080/089583701753338631.

Song, Y.K., Hong, S.H., Jang, M., Han, G.M., Jung, S.W., Shim, W.J., 2017. Combined effects of UV exposure duration and mechanical abrasion on microplastic fragmentation by polymer type. Environ. Sci. Technol. 51, 4368–4376. https://doi.org/10.1021/acs.est.6b06155.

Thompson, R.C., Olsen, Y.S., Mitchell, R.P., Davis, A., Rowland, S.J., John, A.W.G., Mcgonigle, D.F., Russell, A.E., 2004. Lost at sea: Where is all the plastic? Science (New York, N.Y.). 304, 838.

Uddin, S., 2016. Use of satellite images to map spatio-temporal variability of PM2.5 in air. Athens J. Sci. 3, 183–198. https://doi.org/10.30958/ajs.3-3-1.

Velis, C.A., Cook, E., 2021. Mismanagement of plastic waste through open burning with emphasis on the global south: A systematic review of risks to occupational and public health. Environ. Sci. Technol. 55, 7186–7207. https://doi.org/10.1021/acs.est.0c08536.

Vianello, A., Jensen, R.L., Liu, L., Vollertsen, J., 2019. Simulating human exposure to indoor airborne microplastics using a breathing thermal manikin. Sci. Rep. 9, 8670. https://doi.org/10.1038/s41598-019-45054-w.

Wang, Y., Ding, D., Shu, M., Wei, Z., Wang, T., Zhang, Q., Ji, X., Zhou, P., Dan, M., 2018. Characteristics of indoor and outdoor fine phthalates during different seasons and haze periods in Beijing. Aerosol Air Qual. Res. 19. https://doi.org/10.4209/aaqr.2018.03.0114.

Wright, S.L., Kelly, F.J., 2017. Plastic and human health: A micro issue? Environ. Sci. Technol. 51(12), 6634–6647.

Zhang, Y., Gao, T., Kang, S., Sillanpää, M., 2019. Importance of atmospheric transport for microplastics deposited in remote areas. Environ. Pollut. Barking Essex 1987 254, 112953. https://doi.org/10.1016/j.envpol.2019.07.121.

Zhou, Q., Zhang, H., Fu, C., Zhou, Y., Dai, Z., Li, Y., Tu, C., Luo, Y., 2018. The distribution and morphology of microplastics in coastal soils adjacent to the Bohai Sea and the Yellow Sea. Geoderma. 322, 201–208.

5 Microplastics in the Atmosphere
Identification, Sources and Transport Pathways

Rahul Arya, Jaswant Rathore, Ajay Kumar Mishra and Arnab Mondal

5.1 INTRODUCTION

Plastic is a synthetic organic polymer having a wide range of applications, from domestic commodities to industrial usage, from its very inception. It is used extensively in packaging industry as bags, bottles and wraps. The popularity of plastics has sky-rocketed since its inception due to its inexpensive manufacturing techniques, lightweight nature, durability, inertness and resistivity to environmental factors like moisture, microbial degradation, etc. A study in 2019 revealed that the packaging sector had the largest share of plastic generation, followed by the construction industry, the textile industry and the consumer delivery sector (Geyer et al., 2017; Kumar et al., 2021). Their resistance to biodegradation ensures that plastic decomposition is almost impossible, indicating most of the produced plastic is still available in nature in one form or another since inception. This suggests that the presence of plastics indicates the Anthropocene, and thus, can be identified as an environmental pointer for the era (Zalasiewicz et al., 2016). Overexploitation of plastics, along with its longevity and resistivity to degradation post-usage, has led us to a serious environmental crisis.

Only 9% of the plastic materials are recycled, whereas the rest gets accumulated in the surroundings. About three-fourth of the 8.75 billion tonnes of plastic produced between 1950 and 2015 were dumped as waste (Habibi et al., 2022). Going by the current trend as presented in Figure 5.1, the global production of plastics is anticipated to reach 25 billion tonnes by the year 2050, of which 12,000 million tonnes of plastic debris will be in landfills (Wright et al., 2020).

Though plastic doesn't decompose naturally, it is subjected to breakdown due to erosion by solar radiation, biotic and abiotic degradation. The term 'microplastics' (MP) to denote fragmented particles of plastic was first used in the year 2004 (Thompson et al., 2004). Later in 2008, National Oceanic and Atmospheric Administration (NOAA) redefined it as fragmented plastics with size range of less than 5 mm (Arthur et al., 2009). In some contemporary redefinitions, attempts were made to put in place a lower limit of 1 μm (Frias and Nash, 2019; Hartmann et al.,

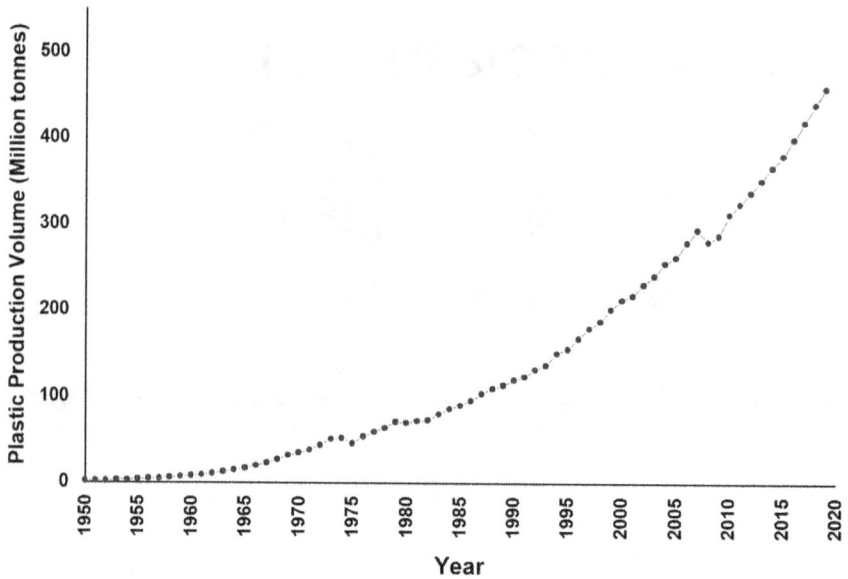

FIGURE 5.1 Global production of plastics annually (1951–2019).

2019; Zhang et al., 2021). However, till date, there is no standard definition of micro-plastics' size range, but the most commonly accepted size range is from 1 μm to 5 mm, which will used in this chapter hereafter. Microplastics less than the size of 1 μm are further classified as nanoplastics and pose a greater potential hazard to human wellbeing (Yuan et al., 2022). However, due to limited literature and lack of detailed quantification of environmental nanoplastics, this chapter focuses primarily on environmental microplastics. Given their size, microplastics can be ingested by marine or freshwater creatures. Microplastics have undesirable impacts on living beings, and their impact upon bioaccumulation in organisms is currently being stud-ied (Miller et al., 2020). Recent studies on MPs are primarily fixated on aquatic and terrestrial ecosystems (Qi et al., 2020; Xu et al., 2020). The study of air-borne micro-plastics became consistent only after the detection of microplastics in atmospheric fallout during a study conducted in 2015 (Abad López et al., 2023; Chen et al., 2020; Dris et al., 2015; Prata, 2018). While marine microplastic pollution are well docu-mented, the sources and pathways of atmospheric microplastic contaminants largely remains unknown (Dris et al., 2016; Su et al., 2022). Therefore, the present chapter attempts to deliver an insight of microplastics in the atmosphere, its sources, types, transport pathways and an overview of the present measurement techniques.

5.2 MICROPLASTIC SOURCES

The presence or occurrence of the airborne microplastics varies in the atmosphere contingent upon the geographic location, type of microplastics, transboundary move-ment and meteorological parameters such as mixing process, dilution and diffusion of the air, rainfall, temperature, etc. Furthermore, occurrence, or presence, of airborne MPs also depends upon their sources, which depend upon the biodegradation of plastics. Plastics are not produced naturally, so the sources of microplastics are anthropogenic only, and they are found in the atmosphere, both indoors and outdoors, through the decomposition of plastics. Some plastics easily deteriorate when exposed to solar radiation, whereas some plastics deteriorate less frequently, contingent upon the molecular bond of the plastic material. Thus, it becomes prominent to study the sources of microplastics to better estimate their impact on the atmosphere. Identifying the sources of airborne microplastics is vital for creating focused strategies that minimize emissions and reduce exposure levels for humans and the environment. Various causes, including industrial processes, tire degradation from automobiles, atmospheric deposition and residential activities, have been effectively determined by scientific investigations. Policymakers, stakeholders and the public may promote cleaner air and better ecosystems by implementing initiatives to prevent the release of airborne microplastics by recognizing these sources. Based on the literature, some of the key sources of airborne microplastics are discussed in the following sections.

5.2.1 Industrial Emission

Microplastics get released into the environment due to industrial activities, particularly those that involve the manufacturing, processing and recycling of plastics. During the analysis of airborne MPs around industrial facilities, Rillig et al. (2019) and Li et al. (2018) highlighted the production and dispersion of microplastic particles during manufacturing operations, waste management and combustion.

5.2.2 Vehicle Tire Wear and Road Dust

The atmosphere's microplastics attributed to road traffic are a critical source. Microplastic particles are released due to road dust and tire-wear emissions. Massive quantities of airborne microplastics caused by tire wear and road abrasion have been identified during tests conducted by Sommer et al. (2018) close to roads and highways.

5.2.3 Transboundary Movement of Microplastics

Through currents in the air, microplastics may traverse quite a distance before being dumped onto terrestrial and aquatic habitats. Urban locations, open dumps and maritime environments are sources of air microplastics. Chen et al. (2020) looked at deposition of MPs in pristine areas and pinpointed the movement of airborne microplastics from far-off sources.

5.2.4 Domestic Activities and Indoor Environments

Domestic activities, including cleaning, cooking and using plastic products indoors, can potentially emit microplastics into the air. Cooking oil has been identified as a potential source of airborne microplastics by Rist et al. (2019), who studied microplastic emissions from cooking activities. The prevalence of microplastics in indoor air can also be attributed to indoor environments with plastic-containing items. Additionally, the sources of microplastics vary in both indoor and outdoor environments, and Figure 5.2 shows some key sources of both indoor and outdoor microplastics sources.

5.2.5 OUTDOOR SOURCES

The sources of air-borne microplastics in the outdoor environment are abundant. Synthetic clothing is one of the predominant sources of microplastic fibers (Dris et al., 2016). Apart from this, the incomplete combustion of plastic wastes in landfills, scaffolding mesh used in construction sites, abrasion from rubber tires and synthetic textiles, degradation of municipal solid wastes and dust storms are potential sources of outdoor microplastic contamination. Meteorological occurrences such as dust storms and monsoons promote the transboundary movement of microplastics (Abad López et al., 2023). Climatic factors like winds and rains positively correlate with microplastic concentrations (Ding et al., 2022). Microplastic deposition and size (due to hygroscopic growth) is higher during the wet periods (rainfall and snowfall), as compared to the dry periods. Increase in microplastic deposition during the dry periods highly correlates with the increase in wind velocity and relative humidity. The deposited microplastics onto the land or water during the dry periods may get resuspended in air, leading to a cyclical exchange of microplastics via air, water and land (Abad López et al., 2023). Vegetation is also known to temporarily store microplastics in its leaves. Earlier studies have reported that suspended atmospheric microplastics is one of the major contributors of microplastics in ocean.

5.2.6 INDOOR SOURCES

The primary sources of indoor microplastic contaminants are tiny fibers from furniture, curtains, cushions, personal care products and daily wear (Qiu et al., 2020). Like the outdoor environment, microplastic concentrations in the indoor environment, too, are influenced by temperature and humidity. The use of wall paints, plastic containers and wrappers, too, influence the concentration of microplastics in the indoor environment. The air conditioners resuspend sediment particles/fibers and redistribute them in the indoor air (Abad López et al., 2023). Previous research has established that the concentration of microplastics in the indoor environment is higher than outdoor air within same area. Microplastics in the indoor environment also get released from wear and tear of daily wear during washing and drying as well (Habibi et al., 2022).

In addition to these local sources, microplastics also travel from distant sources under the influence of global air-circulation patterns and regional winds. Microplastics, due to their smaller size and light weight, can remain suspended in the atmosphere for long time-frames and come to the surface as atmospheric fallout. Atmospheric fallout is the sedimentation of the dust or fine particles from the atmosphere to the surface of the earth. The atmospheric fallout of microplastics has been quantified in megacities like Paris (France) and Dongguan (China). The average

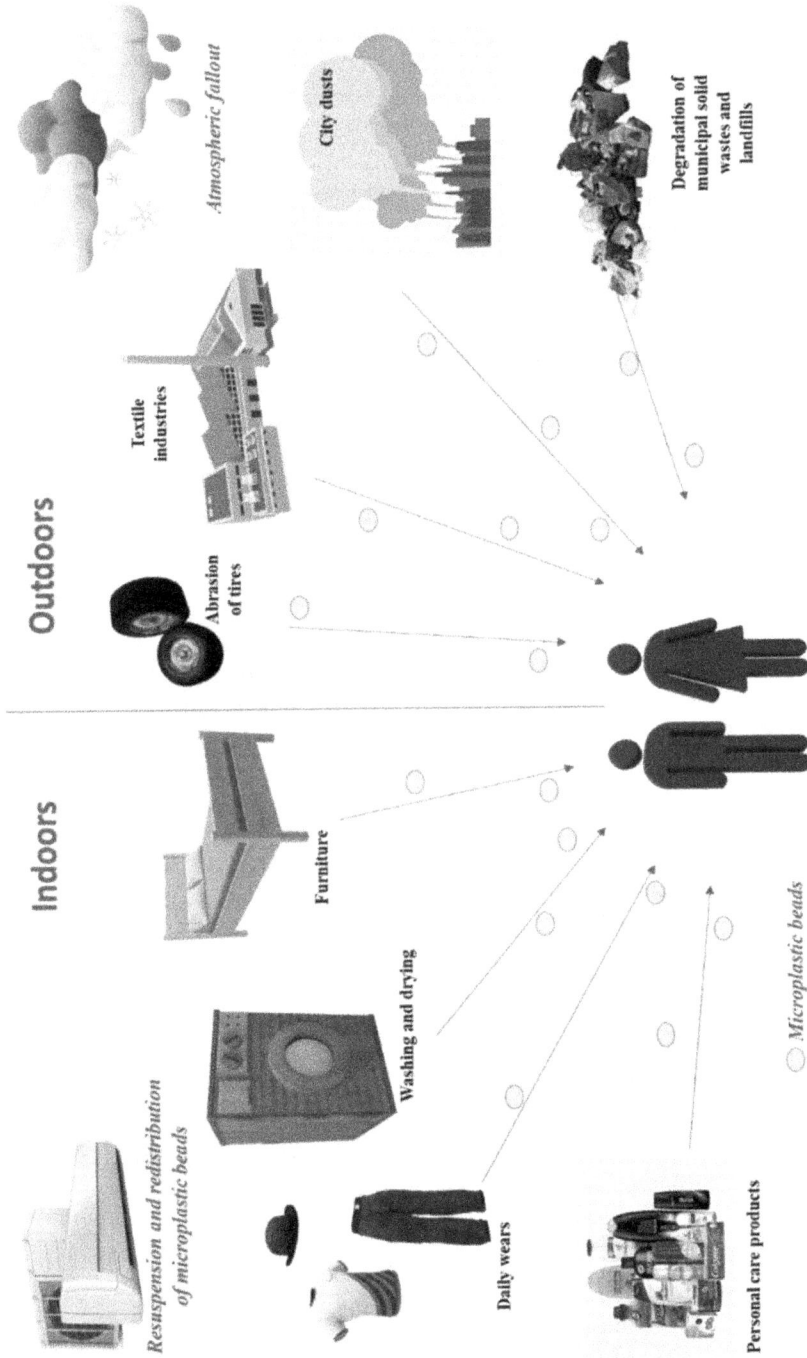

Outdoors

Atmospheric fallout

City dusts

Degradation of municipal solid wastes and landfills

Textile industries

Abrasion of tires

Indoors

Furniture

Washing and drying

Resuspension and redistribution of microplastic beads

Daily wears

Personal care products

○ Microplastic beads

FIGURE 5.2 Pictorial representation of indoor and outdoor sources of microplastics in the atmosphere.

deposition rate of microplastics was found to be 110 ± 96 m^{-2} day^{-1} for the urban site and 53 ± 38 m^{-2} day^{-1} for the suburban site of Paris, France (Dris et al., 2015; Wright et al., 2020). Similarly, the average deposition rate of microplastics reported was 36 ± 7 m^{-2} day^{-1} for the city of Dongguan, China (Cai et al., 2017). Interestingly, the average deposition rate of microplastic fallout at a remote, pristine mountain area of Pyrenees (France) was reported to be 365 ± 69 m^{-2} day^{-1} (Dris et al., 2015). The presence of microplastics in far-flung areas indicates that microplastics can rapidly get transported through the wind away from their place of origin. The air-mass back trajectory models can help to identify the source of origin of the microplastics in pristine regions. A recent study has revealed that the microplastics can travel up to a distance of ~95 kms before fallout (Habibi et al., 2022).

5.3 TYPES OF MICROPLASTICS

Frias and Nash (2019) have defined MPs as 'any synthetic solid (regular/irregular shape) particle matrix with size ranging between 1 µm – 5 mm which is insoluble in water'. Hartmann et al. (2019) have projected an outline for the identification of types of plastics with descriptions like (a) chemical composition; (b) color; (c) shape (*spheres, pellets, foams, fibers, fragments, films*). Based on the chemical composition of microplastics, they are classified as *polyethylene, polypropylene, polyethylene terephthalate, polystyrene, etc.* Based on the nature of origin, MPs are classified as primary or secondary. The primary MPs are manufactured as is to be used in cosmetic products, whereas the secondary MPs are formed due to the fragmentation and degradation of plastic materials as well as fibers and synthetic textiles (Abad López et al., 2023). Primary microplastics such as microbeads are used in cosmetic products like face wash and face scrubs (Napper et al., 2015). The contribution of secondary microplastics to global microplastic concentrations is maximal, as compared to primary microplastics (Duis and Coors, 2016).

Kooi and Koelmans (2019) proposed the definition of microplastics in terms of a three-dimensional (3D) probability distribution, considering size, shape and density as dimensions. The characteristics of microplastics such as their size, abundance, shapes and components have been greatly studied and reported for urban, suburban and remote areas.

5.3.1 ANALYTICAL MEASUREMENTS OF MICROPLASTICS

In this section, we discuss the identification and analysis methodologies of microplastics in the atmospheric air. This includes aerosol sample collection, separation of microplastics from remaining aerosols, followed by the analysis of microplastics using different analytical techniques.

5.3.2 SAMPLE COLLECTION

There are two widely used sampling techniques for the measurement of airborne microplastics—i.e., active and passive sampling. In both these techniques, a sample of atmospheric aerosols is collected. In active sampling, a pump with a known flow rate is used to suck the ambient air, and the aerosols drawn with the air are deposited on a substrate, or filter. The deposited aerosol volume is then used for further analysis. The advantages of active sampling are known aerosol volume and large sample collection in quick time. In passive sampling, the sample is collected via wet or dry deposition at any location—e.g., settling over a surface after wet or dry deposition. For passive sampling, funnels, rain gauges, bottles and planar surfaces which permit the deposition of aerosols can be used. The passive samplers can't be used to assess per-unit-volume concentration but can provide an idea about microplastics per unit area. These are less expensive, easy to use but suffer due to local weather and deployment specifications. Additionally, they can't be used for short-term monitoring. Hence, the sampling choice depends upon the research objectives, resources and duration.

5.3.3 SAMPLE PREPARATION

After each collection of atmospheric fallout, the funnel is then rinsed with distilled water three times, consecutively, for the recovery of the particles adhered on the surface of the funnel, as the fourth rinse shows the number of microplastics to be equivalent to the blank sample. All the glassware used must be washed once with 70% ethanol solution and thrice with ultrapure water. The samples collected after the rinsing are then covered to avoid further contamination from indoor air. Using the values of the collecting surface area and the sampling period, the atmospheric fallout is expressed as the number of particles per square meter per day.

5.4 ANALYSIS

5.4.1 VISUAL OBSERVATIONS

The samples are then filtered on the Whatman quartz fiber GF/A filters with pore size of 1.6 μm. Filters are then observed under the fluorescence stereomicroscope. All the fiber samples are counted in bright light, while the non-fibrous materials (considered as potential microplastics) are observed under the UV lights ($\lambda = 475$ nm).

5.4.2 FTIR ANALYSIS

Chemical characterization of the total fibers collected from the sampling can be analyzed using Fourier-Transform-Infrared (FTIR) micro-spectroscopy, coupled with an ATR (Attenuated Total Reflectance). The characterization of the total fibers is useful in determining the proportion of synthetic and natural fibers and also helps in the identification of predominant plastic polymers in the sample. The smallest atmospheric microplastic particle identified using μFTIR is 11 μm (Zhang et al., 2020)

5.4.3 RAMAN SPECTROSCOPY

Raman spectroscopy uses a higher-frequency laser ($\lambda = 532$ nm) for excitation for the release of a photon from the surface of the material. The release of photons is generally along the line of the laser (Rayleigh scattering), but a few photons are emitted at right angles (Raman scattering). The theoretical limitation of Raman spectroscopy are submicron particles less than 10 μm (Zhang et al., 2020).

5.4.4 MICROPLASTIC QUANTIFICATION

To quantify the presence of microplastics in the deposited atmospheric fallout samples, a correction factor is applied based on the FTIR results.

$$Microplastics\ m^{-2}day^{-1} = \frac{Potential\ microplastics \times \%identified \times 31.85}{nd}$$

where 'nd' is the sampling duration (number of sampling days) and '31.85' is a standardized factor based on the sampling area. The microplastic deposition during a fallout is expressed in terms of n m^{-2} day^{-1} (n = *number of microplastics*), which is a standard metric for geographical comparisons.

5.5 PHYSIOGNOMY OF MICROPLASTICS

Identifying microplastics' sources, behavior and possible ecological impact requires analyzing them based on their size, shape and color. Shape analysis allows for the classification of different types of plastic waste; size analysis evaluates their transit and bioavailability; and color analysis helps identify their sources. Combining other analysis techniques helps us comprehend microplastics more thoroughly. Improvements in monitoring, mitigation tactics and regulatory measures to address the global problem of microplastic pollution will result from continued research in this area.

5.5.1 SHAPE

Shape analysis of microplastics is crucial for identification, as various types of plastic debris exhibit distinct morphologies. Recent studies have employed microscopy and image-analysis techniques to characterize microplastic shapes. For instance, Shi et al. (2022) utilized scanning electron microscopy (SEM) to categorize microplastics into fragments, fibers, films and beads. Such classification facilitates the identification of potential sources, as certain shapes are associated with specific plastic products or degradation processes.

5.5.2 SIZE

The sizes of microplastics fluctuate enormously, from submicron particles to higher fragments. Understanding their motion, accumulation and potential interactions with biota depends on knowing their size distribution. Techniques including sieving, flow cytometry and laser diffraction are frequently employed to measure the size of microplastics. Microscopy and sieving were used to analyze size distribution by Conkle et al. (2018), who found that marine sediment samples contained a lot of microplastics with a size range of 50–500 micrometers. Assessing the potential for organism ingestion and entry into various environmental compartments is made easier with the help of size analysis.

5.5.3 COLOR

Microplastics can exhibit a range of colors, depending on their composition and weathering processes. Color analysis provides valuable information about the type and source of microplastics. Spectroscopic techniques, such as Fourier-transform infrared spectroscopy (FTIR) and Raman spectroscopy, enable the identification and classification of microplastics based on their color spectra. A study by Dümichen et al. (2017) employed FTIR spectroscopy to differentiate microplastics by color, enabling the identification of different polymer types. Color analysis assists in tracing the origin of microplastics, such as those from specific consumer products or industrial processes.

5.5.4 COMBINED ANALYSIS

Analyses of shape, size and color improve understanding of microplastic characteristics. For instance, Dümichen et al. (2017) classified microplastics according to their shape, size and color using a combination of microscopy, FTIR spectroscopy and image analysis. Table 5.1 presents the different characteristics of microplastics in the atmosphere. This integrated technique enables a more thorough identification of microplastics, enabling researchers to identify particular types, gauge their sources and assess their environmental impacts more correctly.

5.6 TRANSPORT PATHWAYS OF MICROPLASTICS IN AIR

Airborne particles are transported globally or locally in the atmosphere, owing to upward and downward drafts, wind speed and wind direction, atmospheric disturbance/turbulence and convection. Microplastics have smaller density due to which they remain in the air for a longer period and get transported to distant places. The investigation of sources and transport of microplastics is indispensable and, till date, the knowledge about their transport in the atmosphere is lacking. The movement of atmospheric aerosols or microplastics is driven by three mechanisms, which are transportation, dispersion and deposition. Transportation occurs due to ambient wind speed and wind direction, dispersion occurs due to atmospheric turbulence and deposition results from the downdraft of airborne particles due to scavenging, precipitation and sedimentation.

   All these mechanisms are aided by the size, shape and length of the particle (Zhou et al., 2017). As particle size increases, their atmospheric residence time decreases, or the smaller particles remain for a longer period in the atmosphere and get transported to great distances (Dris et al., 2016). The shape and molecular density of the particles also play a key role in their residence time and transboundary movement (Horton and Dixon, 2018); however, the substantiation of this requires further studies (Rochman et al., 2019). The morphology of the particle can also govern

TABLE 5.1
Summary of Characteristics of Microplastics in the Atmosphere from Various Studies

Study Area	Sources	Size	Shape	Colors	Method	References
Paris, France	Total fallout (dry & wet)	50–1000 μm	≥90% Fibers ⁓10% Fragments	Blue Red	SM FTIRs	Dris et al., 2015
Paris, France	Fallout	50–1400 μm	Fibers	N/A	SM FTIRs	Dris et al., 2016
Paris, France	Indoor and Outdoor air	50–450 μm	Fibers	N/A	SM FTIRs	Dris et al., 2017
Hamburg, Germany	Atmospheric deposition	63–5000 μm	Fragment: ≥90% Fibers: ≤10	N/A	μ-Raman	Klein and Fischer, 2019
Tehran, Iran	Urban dust	100–1000 μm	Granule Dominant: 60% Fibers: 35% Sphere: 5%	Yellow, White, Orange, Black, Grey, Green, Blue, Pink, Red	FM SEM/EDS	Dehghani et al., 2017
Asaluyeh, Iran	Suspended dust, urban dust	100–1000 μm	Fibers Granules	White–Transparent:70% Yellow, Blue, Orange, Green	Fluorescence microscopy SEM/EDS	Abbasi et al., 2019
Dongguan, China	Fallout (dry & wet)	200–4200 μm	Fibers Foam Films	Blue Red Transparent Grey	SM FTIRs	Cai et al., 2017

Location	Source	Size	Shape	Color	Method	Reference
Shanghai, China	Microplastics (suspended)	23–5000 µm	Fibers Fragment Granules	Blue, Black, Red Transparent, Brown, Green, Yellow and Grey	SM FTIRs	Liu, Wang K et al., 2019
Yntai, China	Atmospheric deposition	100–500 µm	Fiber, Foam, Film	N/A	SM FTIRs	Zhou et al., 2017
39 major cities in China	Indoor and outdoor dust	N/A	Fibers, Granules, Cellulose and Rayon	N/A	FTIRs	Liu C et al., 2019
Pyrenees mountains Europe	Fallout (dry & wet)	10–5000 µm	Fragments, Fibers, Films	N/A	SM µ-Raman	Allen et al., 2019
Europe and Arctic	European & Arctic snow (wet)	11–475 µm	Fibers	N/A	µ-Raman FTIR-imaging	Bergmann et al., 2019
West Pacific Ocean	Suspended atmospheric microplastics	20–200 µm	Fibers, Fragment, Granule, Microbead	Yellow, Purple, Brown, Transparent White, Orange, Black, Grey, Green, Blue, Pink, Red	SM FTIRs	Liu, Wu K et al., 2019

SM: Stereomicroscope; FM: Fluorescence Microscope; FTIRs: Fourier transform Infrared Spectroscopy.

their long-range transport as settling speed of the particle change with the particle morphology (Allen et al., 2019). In addition to these, topography of the location, its climate and weather conditions such as rainfall, pressure, temperature, wind speed and direction, snowfall also have substantial influence on the movement of the atmospheric particulates (Allen et al., 2019).

Previous studies have reported the far-off movement of MPs from land to marine and far remote places (Klein and Fischer, 2019; Zhang et al., 2019), polar regions (Liu et al., 2019) and glaciers (Ambrosini et al., 2019; Jiménez-Vélez et al., 2009). Bergmann et al. (2019) identified microplastics from snow particles at the poles. They revealed that microplastics move through the atmosphere, passing the clouds, coagulate with the snow and get deposited, with a scavenging effect. The clouds can scavenge the microplastics along with other particles and gases and then precipitate them to different terrestrial and aquatic biomes. Klein and Fischer (2019) reported that the microplastics found in the countryside and municipal areas of Hamburg, Germany was due to long-range transport. Selvam et al. (2020) couldn't identify the source of microplastics in the sea salt; however, they didn't disregard the transboundary movement of microplastics from remote places. To investigate the transport of microplastics, backward trajectory analysis can be an appropriate technique. These trajectories can identify the movement of particles by taking into account the wind speed and direction, settling speed, height of the mixing layer, etc. Lagrangian atmospheric models such as HYSPLIT, LAGRANTO, etc., may deliver evidence about the source of the fragmented plastics along with the elevation, mixing and distance information (Allen et al., 2019).

In a previous study (Allen et al., 2019), movement of microplastics over the French Pyrenees was investigated using calculations of settling speeds. Based on the settling speeds of the microplastics, wind speed during the event along with the direction and mixing layer depth, backward trajectory analysis was performed. The results revealed that the source of microplastics was 28 km away from the study area in the northwest to southwest direction. Similarly, Zhang et al. (2019) reported the long-range transport of microplastics for more than 100 km at high-altitude glaciers in the Tibetan Plateau.

5.7   HEALTH RISKS OF MICROPLASTICS IN AIR

Aerosols are the significant pathways for inhalation of microplastics by living beings. MPs suspended in the air are hazardous to ecosystems, animals and human beings. Upon inhalation or ingestion, airborne microplastics can cause respiratory problems, ecological disruption and detrimental effects on animal health. The potential dangers are also increased by the movement of substances linked to microplastics. Ongoing research is essential for creating mitigation methods to reduce their release and exposure and fully understanding the health effects of airborne microplastics.

5.7.1   HUMAN HEALTH RISKS

Humans that inhale airborne microplastics may experience adverse health impacts. According to studies, microplastics can build up in the respiratory system and may result in inflammation and respiratory issues. Microplastics can cause lung toxicity in human lung cells, according to a study by Prata (2018), underscoring the possible dangers of breathing them in.

5.7.2 ECOSYSTEM IMPACTS

Microplastics in the air can be deposited on land and water, harming the environment. According to research, animals such as insects, birds and marine species can consume airborne microplastics. This can damage these creatures' health, reproductive efficiency and survival. Numerous creatures have been shown to ingest airborne microplastics, according to Wright et al. (2020), underscoring the potential for ecological damage.

5.7.3 ANIMAL HEALTH

Microplastics in the air can harm the health of animals. When inhaled or ingested, microplastics can cause organ damage, vitamin shortages and gastrointestinal blockages. According to a study by Ma et al. (2020), fish can experience physical and behavioral alterations from microplastics that can influence their behavior toward food and general health.

5.7.4 TRANSFER OF CHEMICALS

Harmful substances like heavy metals and POPs (persistent organic pollutants) are transported via microplastics. These substances can increase the health risks to organisms when inhaled or consumed. When evaluating the health effects of microplastics, a study by Lusher et al. (2022) underlined the potential for microplastics to absorb and transport pollutants. It emphasized the necessity to take into account chemical interactions.

5.8 CONCLUSION

Airborne microplastics are now acknowledged and regarded as particulate matter and potential pollutants. Though in an early stage, their prevalent presence in the atmospheric air has initiated concern among researchers, policymakers and NGOs across the globe. The latest studies have shown the occurrence of airborne microplastics in urban as well as rural areas. They are present in the atmosphere, both indoors and outdoors; however, indoor airborne microplastics are rather neglected but poses a greater risk due to increased human exposure. This chapter summarizes different characteristics of airborne microplastics, including their types, occurrence, sources, transport pathways, sampling procedures, impacts, etc. Based on the published literature, it is found that the airborne microplastics show a broad array of characteristics and concentrations across the globe. They are found in different size ranges and have a varied range of sources, both indoor and outdoor. The sampling tools for atmospheric microplastics include the use of a sampling pump or a rain gauge or fallout collector without a vacuum pump. Their occurrence in the atmosphere is governed by transportation, dispersion and deposition. Additionally, meteorological conditions play a major role in their presence. Being less-dense aerosols, they demonstrate the potential of transboundary movement and are even found in remote areas like polar snow, glaciers and farmlands. Their tinier size scatters them in inhalable air, which allows them to enter human respiratory system and/or the pores in human skin, thus augmenting their dangers to human health.

5.8.1 FUTURE WORKS

Based on the aforementioned literature, we suggest the following investigations to better comprehend the distribution of airborne microplastics and their implications to human health.

- To carry out global-scale measurements to comprehend their spatio-temporal distribution worldwide, especially in areas with very high population density.
- To standardize the sampling methods for detection of airborne microplastics, which will allow intercomparison of different circumstances and provide a worldwide information about the status of airborne microplastics. Additionally, it will also provide improved-quality data.
- Use of remote sensing techniques such as light detection and ranging (lidar), which are used to monitor microplastics in terrestrial and even marine ecosystems. The use of different types of lidar systems can provide information about vertical distributions of airborne microplastics. With help of other remote-sensing techniques and in-situ measurements, we can estimate their impact on air quality, their radiative effects, etc.
- To carry out studies on chemical characterization, size distribution, number concentration and source apportionment of airborne microplastics to do a better assessment of the risks for human health.
- To study the complex interplay between MPs suspended in air and human health along with its role in other ecological systems.
- Identification and quantification of microplastics in human lung biopsies to better comprehend their impact on human respiratory system.
- Development of models to approximate the distribution of airborne MPs and to assess the ecological implications of airborne microplastics.

REFERENCES

Abad López, A.P., Trilleras, J., Arana, V.A., Garcia-Alzate, L.S., Grande-Tovar, C.D., 2023. Atmospheric microplastics: Exposure, toxicity, and detrimental health effects. RSC Adv. 13, 7468–7489. https://doi.org/10.1039/D2RA07098G.

Abbasi, S., Keshavarzi, B., Moore, F., Turner, A., Kelly, F.J., Dominguez, A.O., Jaafarzadeh, N., 2019. Distribution and potential health impacts of microplastics and microrubbers in air and street dusts from Asaluyeh County. Iran. Environ. Pollut. 244, 153–164. https://doi.org/10.1016/j.envpol.2018.10.039.

Allen, S., Allen, D., Phoenix, V.R., Le Roux, G., Durántez Jiménez, P., Simonneau, A., Binet, S., Galop, D., 2019. Atmospheric transport and deposition of microplastics in a remote mountain catchment. Nat. Geosci. 12, 339–344. https://doi.org/10.1038/s41561-019-0335-5.

Ambrosini, R., Azzoni, R.S., Pittino, F., Diolaiuti, G., Franzetti, A., Parolini, M., 2019. First evidence of microplastic contamination in the supraglacial debris of an alpine glacier. Environ. Pollut. 253, 297–301. https://doi.org/10.1016/j.envpol.2019.07.005.

Arthur, C., Baker, J.E., Bamford, H.A., 2009. Proceedings of the international research workshop on the occurrence, effects, and fate of microplastic marine debris, September 9–11, 2008, University of Washington Tacoma, Tacoma, WA, USA. NOAA technical memorandum NOS-OR&R 30.

Bergmann, M., Mützel, S., Primpke, S., Tekman, M.B., Trachsel, J., Gerdts, G., 2019. White and wonderful? Microplastics prevail in snow from the Alps to the Arctic. Sci. Adv. 5, eaax1157. https://doi.org/10.1126/sciadv.aax1157.

Cai, L., Wang, J., Peng, J., Tan, Z., Zhan, Z., Tan, X., Chen, Q., 2017. Characteristic of microplastics in the atmospheric fallout from Dongguan city, China: Preliminary research and first evidence. Environ. Sci. Pollut. Res. 24, 24928–24935. https://doi.org/10.1007/s11356-017-0116-x.

Chen, G., Feng, Q., Wang, J., 2020. Mini-review of microplastics in the atmosphere and their risks to humans. Sci. Total Environ. 703, 135504. https://doi.org/10.1016/j.scitotenv.2019.135504.

Conkle, J.L., Báez Del Valle, C.D., Turner, J.W., 2018. Are we underestimating microplastic contamination in aquatic environments? Environ. Manage. 61, 1–8. https://doi.org/10.1007/s00267-017-0947-8.

Dehghani, S., Moore, F., Akhbarizadeh, R., 2017. Microplastic pollution in deposited urban dust, Tehran metropolis. Iran. Environ. Sci. Pollut. Res. 24, 20360–20371. https://doi.org/10.1007/s11356-017-9674-1.

Ding, J., Sun, C., He, C., Zheng, L., Dai, D., Li, F., 2022. Atmospheric microplastics in the Northwestern pacific ocean: Distribution, source, and deposition. Sci. Total Environ. 829, 154337. https://doi.org/10.1016/j.scitotenv.2022.154337.

Dris, R., Gasperi, J., Mirande, C., Mandin, C., Guerrouache, M., Langlois, V., Tassin, B., 2017. A first overview of textile fibers, including microplastics, in indoor and outdoor environments. Environ. Pollut. 221, 453–458. https://doi.org/10.1016/j.envpol.2016.12.013.

Dris, R., Gasperi, J., Rocher, V., Saad, M., Renault, N., Tassin, B., 2015. Microplastic contamination in an urban area: A case study in greater Paris. Environ. Chem. 12, 592. https://doi.org/10.1071/EN14167.

Dris, R., Gasperi, J., Saad, M., Mirande, C., Tassin, B., 2016. Synthetic fibers in atmospheric fallout: A source of microplastics in the environment? Mar. Pollut. Bull. 104, 290–293. https://doi.org/10.1016/j.marpolbul.2016.01.006.

Duis, K., Coors, A., 2016. Microplastics in the aquatic and terrestrial environment: Sources (with a specific focus on personal care products), fate and effects. Environ. Sci. Eur. 28, 2. https://doi.org/10.1186/s12302-015-0069-y.

Dümichen, E., Eisentraut, P., Bannick, C.G., Barthel, A.K., Senz, R., Braun, U., 2017. Fast identification of microplastics in complex environmental samples by a thermal degradation method. Chemosphere. 174, 572–584. https:// doi.org/10.1016/j.chemosphere.2017.02.010.

Frias, J.P.G.L., Nash, R., 2019. Microplastics: Finding a consensus on the definition. Mar. Pollut. Bull. 138, 145–147. https://doi.org/10.1016/j.marpolbul.2018.11.022.

Geyer, R., Jambeck, J.R., Law, K.L., 2017. Production, use, and fate of all plastics ever made. Sci. Adv. 3, e1700782. https://doi.org/10.1126/sciadv.1700782.

Habibi, N., Uddin, S., Fowler, S.W., Behbehani, M., 2022. Microplastics in the atmosphere: A review. J. Environ. Expo. Assess. https://doi.org/10.20517/jeea.2021.07.

Hartmann, N.B., Hüffer, T., Thompson, R.C., Hassellöv, M., Verschoor, A., Daugaard, A.E., Rist, S., Karlsson, T., Brennholt, N., Cole, M., Herrling, M.P., Hess, M.C., Ivleva, N.P.,

Lusher, A.L., Wagner, M., 2019. Are we speaking the same language? Recommendations for a definition and categorization framework for plastic debris. Environ. Sci. Technol. 53, 1039–1047. https://doi.org/10.1021/acs.est.8b05297.

Horton, A.A., Dixon, S.J., 2018. Microplastics: An introduction to environmental transport processes. WIREs Water 5. https://doi.org/10.1002/wat2.1268.

Jiménez-Vélez, B., Detrés, Y., Armstrong, R.A., Gioda, A., 2009. Characterization of African Dust (PM2.5) across the Atlantic Ocean during AEROSE 2004. Atmos. Environ. 43, 2659–2664. https://doi.org/10.1016/j.atmosenv.2009.01.045.

Klein, M., Fischer, E.K., 2019. Microplastic abundance in atmospheric deposition within the Metropolitan area of Hamburg, Germany. Sci. Total Environ. 685, 96–103. https://doi.org/10.1016/j.scitotenv.2019.05.405.

Kooi, M., Koelmans, A.A., 2019. Simplifying microplastic via continuous probability distributions for size, shape, and density. Environ. Sci. Technol. Lett. 6, 551–557. https://doi.org/10.1021/acs.estlett.9b00379.

Kumar, Rakesh, Verma, A., Shome, A., Sinha, R., Sinha, S., Jha, P.K., Kumar, Ritesh, Kumar, P., Shubham, Das, S., Sharma, P., Vara Prasad, P.V., 2021. Impacts of plastic pollution on ecosystem services, sustainable development goals, and need to focus on circular economy and policy interventions. Sustainability. 13, 9963. https://doi.org/10.3390/su13179963.

Li, J., Yang, D., Li, L., Jabeen, K., Shi, H., 2018. Microplastics in commercial bivalves from China. Environ. Pollution. 234, 114–121.

Liu, C., Li, J., Zhang, Y., Wang, L., Deng, J., Gao, Y., Yu, L., Zhang, J., Sun, H., 2019. Widespread distribution of PET and PC microplastics in dust in urban China and their estimated human exposure. Environ. Int. 128, 116–124. https://doi.org/10.1016/j.envint.2019.04.024.

Liu, K., Wang, X., Fang, T., Xu, P., Zhu, L., Li, D., 2019. Source and potential risk assessment of suspended atmospheric microplastics in Shanghai. Sci. Total Environ. 675, 462–471. https://doi.org/10.1016/j.scitotenv.2019.04.110.

Liu, K., Wu, T., Wang, X., Song, Z., Zong, C., Wei, N., Li, D., 2019. Consistent transport of terrestrial microplastics to the ocean through atmosphere. Environ. Sci. Technol. 1–12. https://doi.org/10.1021/acs.est.9b03427.

Lusher, A.L., Provencher, J.F., Baak, J.E., Hamilton, B.M., Vorkamp, K., Hallanger, I.G., Pijogge, L., Liboiron, M., Bourdages, M.P.T., Hammer, S., Gavrilo, M., 2022. Monitoring litter and microplastics in Arctic mammals and birds. Arctic Science, 8(4), 1217–1235. https://doi.org/10.1139/as-2021-0058.

Ma, H., Pu, S., Liu, S., Bai, Y., Mandal, S., Xing, B., 2020. Microplastics in aquatic environments: Toxicity to trigger ecological consequences. Environ. Pollut. 261, 114089. https://doi.org/10.1016/j.envpol.2020.114089.

Miller, M.E., Hamann, M., Kroon, F.J., 2020. Bioaccumulation and biomagnification of microplastics in marine organisms: A review and meta-analysis of current data. PLoS One. 15, e0240792. https://doi.org/10.1371/journal.pone.0240792.

Napper, I.E., Bakir, A., Rowland, S.J., Thompson, R.C., 2015. Characterisation, quantity and sorptive properties of microplastics extracted from cosmetics. Mar. Pollut. Bull. 99, 178–185. https://doi.org/10.1016/j.marpolbul.2015.07.029.

Prata, J.C., 2018. Airborne microplastics: Consequences to human health? Environ. Pollut. 234, 115–126. https://doi.org/10.1016/j.envpol.2017.11.043.

Qi, R., Jones, D.L., Li, Z., Liu, Q., Yan, C., 2020. Behavior of microplastics and plastic film residues in the soil environment: A critical review. Sci. Total Environ. 703, 134722. https://doi.org/10.1016/j.scitotenv.2019.134722.

Qiu, R., Song, Y., Zhang, X., Xie, B., He, D., 2020. Microplastics in urban environments: Sources, pathways, and distribution. In: He, D., Luo, Y. (Eds.), Microplastics in Terrestrial Environments, The Handbook of Environmental Chemistry. Springer International Publishing, Cham, pp. 41–61. https://doi.org/10.1007/698_2020_447.

Rillig, M.C., Lehmann, A., Mäder, P., Yang, X., 2019. Environmental hazard of microplastics in the context of soil fertility. Curr. Opin. Environ. Sci. Health. 7, 65–69.

Rist, S., Carney Almroth, B., Hartmann, N.B., Karlsson, T.M., Pizzol, L., Sonnemann, G., Wagner, M., 2019. Are we speaking the same language? Recommendations for a definition and categorization framework for plastic debris. Environ. Sci. Technol. 53(3), 1039–1047.

Rochman, C.M., Brookson, C., Bikker, J., Djuric, N., Earn, A., Bucci, K., Athey, S., Huntington, A., McIlwraith, H., Munno, K., De Frond, H., Kolomijeca, A., Erdle, L., Grbic, J., Bayoumi, M., Borrelle, S.B., Wu, T., Santoro, S., Werbowski, L.M., Zhu, X., Giles, R.K., Hamilton, B.M., Thaysen, C., Kaura, A., Klasios, N., Ead, L., Kim, J., Sherlock, C., Ho, A., Hung, C., 2019. Rethinking microplastics as a diverse contaminant suite. Environ. Toxicol. Chem. 38, 703–711. https://doi.org/10.1002/etc.4371.

Selvam, S., Manisha, A., Venkatramanan, S., Chung, S.Y., Paramasivam, C.R., Singaraja, C., 2020. Microplastic presence in commercial marine sea salts: A baseline study along Tuticorin Coastal salt pan stations, Gulf of Mannar, South India. Mar. Pollut. Bull. 150, 110675. https://doi.org/10.1016/j.marpolbul.2019.110675.

Shi, B., Patel, M., Yu, D., Yan, J., Li, Z., Petriw, D., Pruyn, T., Smyth, K., Passeport, E., Miller, R., Howe, J.Y., 2022. Automatic quantification and classification of microplastics in scanning electron micrographs via deep learning. Sci. Total Environ. 825, 153903. https://doi.org/10.1016/j.scitotenv.2022.153903.

Sommer, F., Dietze, V., Baum, A., Sauer, J., Gilge, S., Maschowski, C., Gieré, R., 2018. Tire abrasion as a major source of microplastics in the environment. Aerosol Air Qual. Res. 18(8), 2014–2028. https://doi.org/10.4209/aaqr.2018.03.0099.

Su, L., Xiong, X., Zhang, Y., Wu, C., Xu, X., Sun, C., Shi, H., 2022. Global transportation of plastics and microplastics: A critical review of pathways and influences. Sci. Total Environ. 831, 154884. https://doi.org/10.1016/j.scitotenv.2022.154884.

Thompson, R.C., Olsen, Y., Mitchell, R.P., Davis, A., Rowland, S.J., John, A.W.G., McGonigle, D., Russell, A.E., 2004. Lost at sea: Where is all the plastic? Science 304, 838–838. https://doi.org/10.1126/science.1094559.

Wright, S.L., Ulke, J., Font, A., Chan, K.L.A., Kelly, F.J., 2020. Atmospheric microplastic deposition in an urban environment and an evaluation of transport. Environ. Int. 136, 105411. https://doi.org/10.1016/j.envint.2019.105411.

Xu, B., Liu, F., Cryder, Z., Huang, D., Lu, Zhijiang, He, Y., Wang, H., Lu, Zhenmei, Brookes, P.C., Tang, C., Gan, J., Xu, J., 2020. Microplastics in the soil environment: Occurrence, risks, interactions and fate—A review. Crit. Rev. Environ. Sci. Technol. 50, 2175–2222. https://doi.org/10.1080/10643389.2019.1694822.

Yuan, Z., Nag, R., Cummins, E., 2022. Human health concerns regarding microplastics in the aquatic environment—From marine to food systems. Sci. Total Environ. 823, 153730. https://doi.org/10.1016/j.scitotenv.2022.153730.

Zalasiewicz, J., Waters, C.N., Ivar Do Sul, J.A., Corcoran, P.L., Barnosky, A.D., Cearreta, A., Edgeworth, M., Gałuszka, A., Jeandel, C., Leinfelder, R., McNeill, J.R., Steffen, W., Summerhayes, C., Wagreich, M., Williams, M., Wolfe, A.P., Yonan, Y., 2016. The geological cycle of plastics and their use as a stratigraphic indicator of the Anthropocene. Anthropocene 13, 4–17. https://doi.org/10.1016/j.ancene.2016.01.002.

Zhang, F., Wang, X., Xu, J., Zhu, L., Peng, G., Xu, P., Li, D., 2019. Food-web transfer of microplastics between wild caught fish and crustaceans in East China Sea. Mar. Pollut. Bull. 146, 173–182. https://doi.org/10.1016/j.marpolbul.2019.05.061.

Zhang, S., Sun, Y., Liu, B., Li, R., 2021. Full size microplastics in crab and fish collected from the mangrove wetland of Beibu Gulf: Evidences from Raman Tweezers (1–20 µm) and spectroscopy (20–5000 µm). Sci. Total Environ. 759, 143504. https://doi.org/10.1016/j.scitotenv.2020.143504.

Zhang, Y., Kang, S., Allen, S., Allen, D., Gao, T., Sillanpää, M., 2020. Atmospheric microplastics: A review on current status and perspectives. Earth-Sci. Rev. 203, 103118. https://doi.org/10.1016/j.earscirev.2020.103118.

Zhou, Q., Tian, C., Luo, Y., 2017. Various forms and deposition fluxes of microplastics identified in the coastal urban atmosphere. Chin. Sci. Bull. 62, 3902–3909. https://doi.org/10.1360/N972017-00956.

6 Microplastic in the Marine Environment

Sayan Mukherjee, Subhasis Ghosh, Sanket Roy,
Swastika Saha, Surajit Mondal and Papita Das

6.1 INTRODUCTION

Microplastics (MPs) are extremely small particles (≤ 5 mm), insoluble in the aqueous environment (Bergmann *et al.*, 2015). Overall, MPs are classified into the following categories according to their size: (a) large microplastics (5 mm -1 mm), (b) small microplastics (1 mm -1 µm), and (c) nano plastics (< 1 µm) (Crawford and Quinn, 2016). In the current scenario, pollution caused by MPs and nano plastics has become a pressing issue and has been reported to be present in the ocean, directly affecting its flora and fauna (Nguyen *et al.*, 2019). Apart from the classification as per size variations, MPs of various shapes are also to be found (Hidalgo-Ruz *et al.*, 2012; Wright *et al.*, 2013). As per sources, MPs can be of two types—i.e., primary microplastics (intentionally manufactured especially by cosmetic industries) and secondary microplastics (generated through degradation or weathering of residual plastics through wave action, bio-film growth, exposure to sunlight, mechanical shear, and thermal oxidation) (Arthur *et al.*, 2009; Andrady, 2017).

Along with soil and air, MPs are significantly present in marine environments, and also, very recent studies suggest their presence in bottles of potable water also (Desforges *et al.*, 2015; Gasperi *et al.*, 2018; Kosuth *et al.*, 2018; Schymanski *et al.*, 2018). In lower organisms, the major pathway of entry of MPs into their bodies is through food, as often, given their minute size and different colors (Shaw and Day, 1994), MPs get confused with food (Moore, 2008). Through this, MPs become bioavailable to different algae, zooplankton, fish, and finally, are introduced into the human body system (Cedervall *et al.*, 2012). Once inside the human body, MPs are found to get deposited in human tissue and possess the ability to cause internal blockings and disruption in the endocrine system (Wright *et al.*, 2013).

Despite all this information, a more detailed study is required for the WHO to classify MPs and MPs sorbed POPs (persistent organic compounds) as a potential threat to humanity (Pete Marsden *et al.*, 2019). This chapter deals with the emergence, effects, and threats posed by microplastics and the various methods of reorganization and separation of the microplastics.

6.2 INCEPTION OF MICROPLASTICS

6.2.1 THE CATASTROPHE: BEGIN TO END

Microplastic is mixture of particles of different diameters, from micrometers to millimeters, shapes, and colors. The shape of microplastics can be spherical, fragmented, or fiber. Over time these particles can be converted to even smaller particles, known as nano plastics. Hence, they are considered as an intermediate of nano and macro plastics (Hale *et al.*, 2020). Though the attention of researchers previously considered microplastic as an "ocean-centric" problem, in 2004, it was reported that they also act as atmospheric debris. Alongside the scientific research, there is growing interest by the media and policy makers. A considerable concentration of microplastics is also reported in the Arctic Sea. But, as mentioned previously, the contamination in water is actually a small fraction of a larger threat. From different sources, microplastics are being released into the environment and not only contaminating the abiotic factors in the ecosystem but also generating problems in the biome.

Different sources are responsible for producing microplastics. The origin can be broadly categorized as primary and secondary sources. The primary source includes direct release. On the other hand, secondary sources include the cause of fragmentation of larger particles (An *et al.*, 2020), though it is hard to detect an origin from the sample collected from the environment. The primary source of microplastic includes personal care products, plastic pellets, paints, etc. Plastic pellets are plastic granules with a regular shape. It is used to develop various plastic products. The products are from the clothing industry, household appliances, building materials, etc. According to previous reports from 2009 to 2017, the demand for plastic pellets increased from 5.11 million tons to 13.79 million tons in China. Also, a study in United Kingdom shows that the nation produces 5.3 billion tons of plastic every year (Cole and Sherrington, 2016).

Microbeads, a derivative of microplastic, is used in personal care product to replace synthetic pigments. Mainly two types of microbeads are used in these products—viz., thermoplastics and thermoset plastics (Bhattacharya, 2016). The so-called microbeads contain the major classes of plastics and co-polymers. These substances are discharged directly into the environment with water. An assumption was made that about 11% of plastic waste discharged to North Sea is contributed by these microbeads (Waldschläger *et al.*, 2020).

A secondary source includes the washing of clothes, landfills, the construction industry, littering, etc. Nowadays, polyester, along with nylon and acrylic fiber, is used to wash clothes. These fibers release fabric that serves as a secondary source of microplastic (De Falco *et al.*, 2018).

Though landfill is banned in many countries, it plays a significant role as the secondary source of microplastic. As plastics are non-biodegradable, after landfilling, they don't decompose but get fragmented. These plastic fragments are later converted to microplastics (Barnes *et al.*, 2009). Like landfill, two other sources of microplastic are construction and littering, though they are different sources but not totally separated. Construction workers use different plastic equipment, which can generate microplastics after improper disposal or littering. The dominant contributors of microplastic found in littering are bottles, bottle caps, cigarettes, rubbers, etc. Every year, 6 trillion cigarettes are consumed throughout the globe that left 4.5 trillion cigarette filters; they weigh almost 750,000 tons and contribute to generating microplastic (Novotny and Slaughter, 2014). Though the amount is very small, ports, fishing, and shipping also act as sources of microplastic generation. They contribute to the microplastic contamination of surface water (Barnes *et al.*, 2009). Also, sports grounds and vehicle tires contribute to this process. Tires contains styrene-butadiene rubber, which is responsible for microplastic pollution.

6.2.1.1   Distribution

As the microplastic enters the marine ecosystem, it slowly breaks down, forming different variants with various morphologies. These particles, being light in weight, are driven by the ocean waves and carried to remote areas and subsequently accumulate in different ocean gyres. With the movement of the gyres, the plastic particles continue to accumulate to a large extent, forming a phenomenon known as garbage patches, as shown in Figure 6.1.

Microplastics are polymers, and their distribution is determined by their morphology and chemical content. As a result, it is vital to investigate, both because their morphological structure permits the understanding of horizontal distribution patterns and because their chemical structure enables the understanding of vertical distribution patterns. Several investigations were conducted about the identification of polymers using their morphology and categorized as fibers, films, fragments, and foams (Anderson *et al.*, 2017). An investigation about microplastics in the Canadian Lake discovered the presence of some types in high concentrations, while some were found in lower concentrations. Identification of their specific origin may be difficult, as a single polymer have a wide range of sources (Ballent *et al.*, 2016).

Microplastics in the North Atlantic Ocean ranged in length from 0.41 to 420 mm and weighed less than 0.05 g. Researchers conducted studies in different oceans such as the Southern Ocean, Gulf of Guinea, North Atlantic Ocean, Mediterranean Sea (Van Cauwenberghe *et al.*, 2013), Pacific (Hirai *et al.*, 2011), Northeast Pacific Ocean (Desforges *et al.*, 2014), Straits of Malacca and Bay of Bengal (Ryan, 2013), Arctic Sea (Obbard *et al.*, 2014), and North Atlantic Ocean (Lusher *et al.*, 2014).

Some investigations described the types of polymers along with their source, whereas other researchers concentrated solely on the level of contamination. More research has focused on the Pacific and Atlantic Oceans, but relatively few have focused on the Arctic and Antarctic Oceans, and there have been few investigations along the Indian Ocean's coast. As a result, it is suggested that these undiscovered features be considered for future research.

6.2.1.2   Factors for Microplastic Ingestion

There are many entry points of primary and secondary microplastics into the environment, but the main path is wastewater treatment plants and sewage sludge, along with the wind (Allen *et al.*, 2019). The input of microplastic in the wastewater treatment plant depends on the catchment area. Wastewater is mainly treated in three different stages. The particle quantity in the inflow is 320 particle/lit ((Ding *et al.*, 2021). According to few studies, it was found that the concentration of microplastics

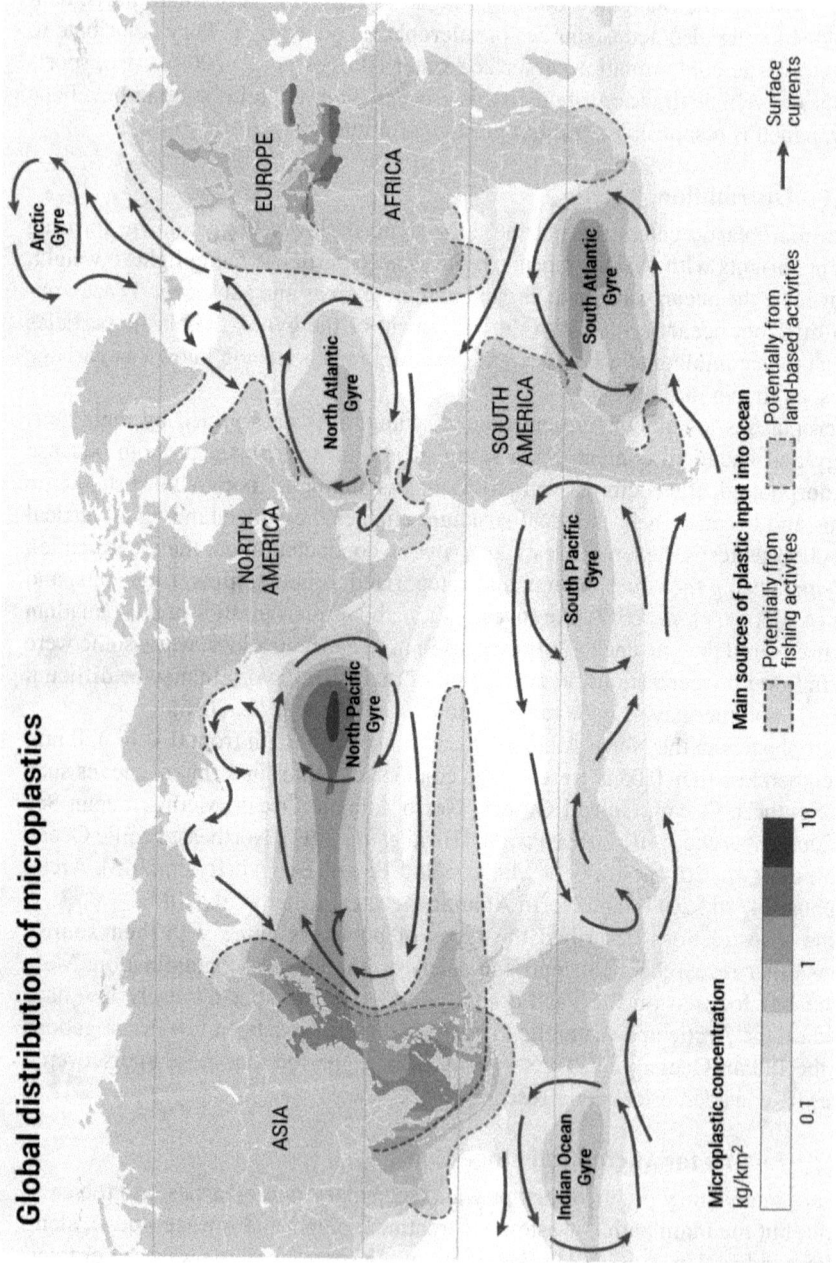

FIGURE 6.1 Global distribution of microplastics dispersed by the surface currents and gyres (www.grida.no; creator credit: Riccardo Pravettoni and Philippe Rekacewicz, global distribution of microplastics, 2019).

in the effluent was between < 1 particle/lit and 100 particle/lit. From this data, it is clear that the treatment plant contains a particle load between 95 and 99%.

This discharge, due to heavy rainfall, releases a mixture of wastewater and rainwater. This source of microplastics is not studied well, and quantification is not possible at present. Connecting with treatment plants, another entry point for microplastics is sewage sludge. It is the solid material that gets separated with the activated sludge from wastewater during treatment. The sludge can be used as fertilizer. Currently, the use of sewage sludge is maintained by the EU because of heavy metal content (Hudcová *et al.*, 2019), though no country has estimated thresholds of microplastic contamination so far. Recent studies have found that the microplastics' dry weight in sludge is between 1,000 and 20,000 particles/kg (Mintenig *et al.*, 2014). It was also observed that, in Europe, between 125 and 850 tons of microplastics are discharged in environment by sewage sludge.

There are different distribution routes of microplastic distribution into the environment. The sewage sludge can be washed out by landfill leachate that carries microplastics into the environment. When sewage sludge is applied to a field, wind and rain leads to further spreading of the contaminant. The particle can not only mix with aquatic ecosystem but also contaminate the soil and be stored there (Zubris and Richards, 2005). Figure 6.2. shows the journey of these microplastics from their sources to the tissues of various organisms.

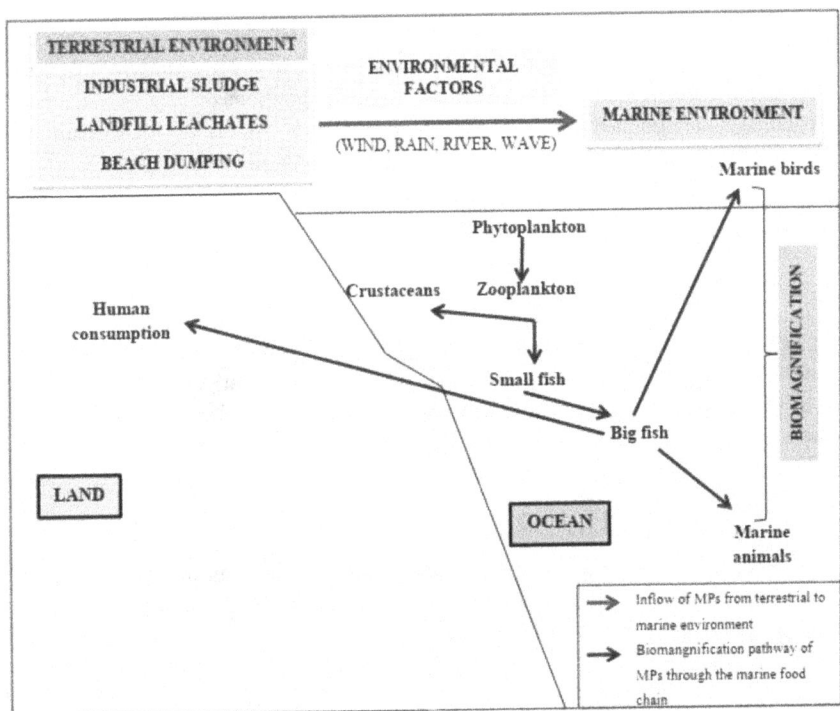

FIGURE 6.2 Transportation of microplastics into the environment.

6.2.1.3 Trophic Transfer

Recently, trophic transfer research concentrated on microplastics impact. Batel *et al.* (2016) created a freshwater food chain with *Danio rerio* and *Artemia sp. nauplii* to demonstrate the trophic transmission of microplastics. Virgin microplastics did not have any discernible effects.

Tosetto *et al.* (2017) also investigated the impact of microplastics (38–45 m) on a marine food chain and focused on amphipods in an Australian study. Plastics were placed in an urban bay for two months and aided in the sorption of PAHs from seawater at a relevant concentration and were given to the amphipods, which were then fed to gobies. Scientists concluded that there was no significant effect on fish behavior after monitoring behavioral changes in response to pollutants before and after exposure. These investigations give a fundamental understanding about the complex interactions that exist between biological entities and developing pollutants. Adsorption/desorption kinetics between plastics, contaminants, and substances and their persistent effects on organisms, especially top predators, is little known. Lacking reliable data to describe predators and prey, we do not know enough about the movement of contaminants to make an environmental statement or marine microplastic risk assessment about dietary effects on seafood.

6.2.1.4 Impacts on Biological Organisms

The bioaccumulation of microplastic in various seafoods and additives has emerged as a concern associated with human health. The risk associated with microplastics largely depends on the physical characteristics of contaminants such as surface area, shape, and size; also, the durability of the components and pathogenicity after accumulation. The source-to-sink transfer of microplastic happens largely via fish, as fishery products have been found to be the accumulator of microplastics. Consumption of these products can lead to a threat to humans (Alberghini *et al.*, 2023).

6.3 SAMPLING, IDENTIFICATION, AND QUANTIFICATION

6.3.1 SAMPLE PREPARATION METHODS

To understand the extent of effect of microplastics, it is imperative to understand the morphological and chemical characteristics of the pollutant. This calls for extensive screening and analysis of the samples procured from different zones of marine ecosystem—viz. sedimentary, aquatic, and biological, the methods of which have been shown in Figure 6.3.

6.3.1.1 Physical Method

It is the means of the initial microplastic separation technique. Techniques like sieving, filtration, and visual sorting followed by density separation are utilized, for which NaCl is the most recommended salt.

6.3.1.1.1 Sieving

It is perhaps the simplest process of the physical separation method. Here, the materials are left on the top of the sieving mesh, and the passed-over materials are

TABLE 6.1
The Effects of Microplastics on Different Biological Organisms

Organism	Organ of Deposition	Consumption Cause	Final Fate/Affect	Reference
Fish	Gut. Gastrointestinal tract.	Transfer from lower trophic level.	Eliminated by feces. Transferred to animal and human.	Koongolla et al., 2020; Possatto et al., 2011; Foekema et al., 2013; Farrell and Nelson, 2013.
Molluscs	Circulatory system. Gut. Muscle tissue. Intestine.	Accidental consumption with food.	Cause lysosomal commotion and oxidative stress damage.	Bouwmeester et al., 2015; Canesi et al., 2012; Farrell and Nelson, 2013; Von Moos et al., 2012.
Crustaceans	Stomach.	Accidental consumption with food.	Affect the swimming capabilities. Change enzyme activity. Responsible for oxidative stress damage.	Gambardella et al., 2017; Andrade and Ovando, 2017.
Marine mammals	Gut. Intestine.	Bioaccumulation by the consumption of planktons.	Serious injury. Malnutrition.	Fossi et al., 2014; Besseling et al., 2015; Baulch and Perry 2014.
Benthic invertebrates	Gut. Biopores.	Accidental consumption with food.	Inflammation in the tissue. Reduced feeding habit. Death.	Besseling et al., 2013; Kim and An 2019.
Microbiota	Tissue. Mid-gut region.	Ingestion with food such as microalgae.	Reduction in feeding. Inhibition of reproduction. Reduction in the size of eggs. Reduction of growth. Species extinction.	Cole et al., 2015; Davarpanah and Guilhermino, 2019; Hamer et al., 2014, Martins and Guilhermino, 2018.
Humans	Gastrointestinal tract. Cerebral cell. Epithelial cells.	Consumption with food. Migration with human placenta. Sea derived salts. Honey, beer. Drinking water. Facial scrubs.	Circular damage. Contraction in muscle cells. Production of reactive oxygen species. Endocrine disruptions. Arteriosclerosis.	Devriese et al., 2015; Romeo et al., 2015; Wick et al., 2010; Lind and Lind, 2011; Berntsen et al., 2010; Selvam et al., 2020; Mintenig et al., 2019; Lithner et al., 2011; Napper et al., 2015.

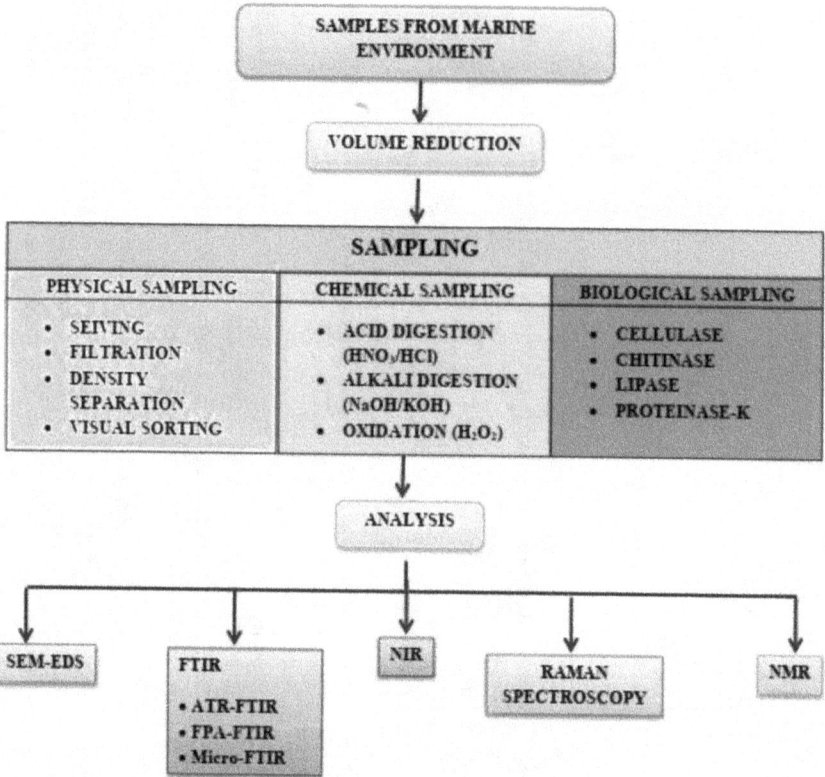

FIGURE 6.3 Steps for sampling, identification, and quantification of microplastics in marine samples.

discarded. It has been found out that, apart from usage of a single mesh, 3–4 meshes are also used for a size differentiation study. The most commonly used pore sizes, in this case, are 5 mm, 1 mm, 335 μm, 330 μm, 80 μm, 0.2 μm (Tiwari *et al.*, 2019).

6.3.1.1.2 Filtration

Filtration is the process of solid separation from liquid particles by a size-selective barrier that allows liquid particles to pass through. The most commonly used filtration agents are a funnel, vacuum pump, etc. (Crawford and Quinn, 2016). The most widely used pore sizes in the filter papers are 0.7 μm, 0.22 μm, 5 μm, 10 μm, 11 μm. But cloth filter paper can contaminate the process, as it releases cellulose. This is also the case for nylon filter paper. All of this contamination can lead to over- and/or underestimation of MPs (Lares *et al.*, 2019).

6.3.1.1.3 Visual Sorting

To separate non-MP particles like biological remains, wood, paint, shell fragments, seaweed, etc., with ease, visual identification and separation is crucial. For this, the structure and form of the particles as well as their hardness or elasticity are to be tested (Fries *et al.*, 2013). Studies also suggest the hot needle test—i.e. the process of prodding needles in particles to avoid misidentification (Shim *et al.*, 2017; De Witte *et al.*, 2014). Visual sorting can be a complex and time-consuming process (Wang *et al.*, 2017). Three major methods used in this procedure are (i) through naked eyes and (ii) optical microscopy, where tweezers are used as a pre-screening tool. But both of these methods involve the chance of misidentification. (iii) Fluorescence staining, where particles are stained with a salvatochromic dye named "nile red" and identified under orange, red, green filters through fluorescent spectroscopy. This process has a short incubation period and has a high recovery rate. It is to be mentioned that, in fluorescence staining, the MP particles are categorized on the basis of their hydrophobicity. But the difficulty in fiber staining and low fluorescent ability of some MP particles may create some hindrances.

6.3.1.1.4 Density Separation

The density of plastics is known to be lower than that of sediments, which makes it plausible for segregation from sediments and other inorganic substances, which are not damaged by chemical or enzymatic digestion (Stock *et al.*, 2019). This process involves the mixing of sediments in a saturated salt solution, where the supernatant containing the microplastic is separated and filtered (Hidalgo-Ruz *et al.*, 2012). Various parameters are responsible for determination of the microplastic density, such as additive concentration, polymer type, as well as adsorbed substances and organisms. NaCl is the most commonly used salt solution for microplastic separation due to its affordability and easy method of use. Polymers of low density such as polypropylene and polyamide are known to be easily separated using NaCl. However, polyvinyl chloride (PVC) and polyethylene terephthalate cannot be separated due to their higher density. Higher density salts like sodium iodide (NaI) and lithium metatungstate are favored alternatives for separation of high density microplastics (Stock *et al.*, 2019; Masura *et al.*, 2015). Zinc chloride is also considered to be favorable for extraction of polyamide, PVC, polyethylene terephthalate, etc., with maximum recovery and low interference. Other than salts, oils can also be used for microplastic separation. Plastics being lipophilic in nature, they have shown themselves to be easily segregated by canola oil (92–97% recovery rate). The excess oil can further be removed using ethanol in this process. However, the presence of canola oil in trace amounts have been reported to hamper the identification process. Olive oil, as an alternative, can be used to mitigate this issue (Scopetani *et al.*, 2020). Novel separation techniques like electrostatic separation has been reported to reduce the sample mass of different types. This method involves separation on the basis of the conduction properties of different samples after being subjected to electrostatic charges. Biogenic materials are discharged early, being conductive in nature, whereas non-conductive microplastics are discharged slowly. This method results in a maximum reduction of sample volume with a microplastic recovery rate of almost 99% (Felsing *et al.*, 2018).

6.3.1.2 Chemical Methods

Microplastics, due to their size, are often ingested or grown on by marine organisms. Hence, removal of such interfering organic matters like biotic tissues, etc., for proper detection and identification is of the utmost importance. Chemical digestion using acids, alkalis, and other chemicals without alteration of the structural integrity of the microplastic particles is paramount in nature.

6.3.1.2.1 Acid Digestion

Acids such as HCl and HNO_3 are often used for organic material degradation. However, HNO_3 is often preferred over HCl due to its rapid acid-digestion abilities. HCl, on the other hand, is inconsistent and inefficient in biogenic compound digestion and has been shown to alter the microplastic surfaces of PVC and other types of microplastics (Karami *et al.*, 2017).

6.3.1.2.2 Alkali Digestion

Microplastic particles present inside the tissues of marine flora and fauna are extracted by digesting the tissues, using strong alkali such as NaOH and KOH. Studies on alkaline digestion using 10 M KOH solution was found to be highly time-consuming, since digestion of total organic matter was reported at a time interval of 2–3 weeks. However, this process is beneficial, since most microplastic types, except cellulose acetate-based plastic materials, were found to be KOH resistant after biogenic organic matter digestion (Kühn *et al.*, 2017). Similar studies with NaOH have degradation of microplastics along with organic matter, thereby making it unfeasible for application.

6.3.1.2.3 Other Forms of Digestion

Oxidation is another efficient method of removal of organic matter. H_2O_2 is one such oxidizing agent, observed to digest organic matter with minimal alterations on plastic polymers. The digestion process is slow and concentration dependent. Reduction in size and thinning of microplastic particles are commonly observed as side-effects of using H_2O_2.

Another efficient method of digestion is enzymatic or biological digestion, which, unlike chemical digestion, does not alter or degrade microplastics. Cellulase, chitinase, lipase, protease, and proteinase-K are commonly used enzymes used for microplastic separation from biogenic tissues. Studies using proteinase-K with $CaCl_2$, followed by oxidation with H_2O_2, have shown recovery of 97%, but the plastic surface accumulates a layer of calcium deposition (Karlsson *et al.*, 2017). Multistep detergent treatment with sodium dodecyl sulphate (SDS), followed by treatment with a mixture of three enzymes (protease, chitinase, and cellulose), along with H_2O_2, have been reported to efficiently remove 98% of organic matter, with a recovery rate of 83% (Löder *et al.*, 2017). However, enzymatic processes are time-consuming and need constant monitoring, since these are highly pH- and temperature-sensitive.

6.3.2   Identification and Detection of Microplastics

The extracted particles obtained from the separation process often contains non-plastic microplastic lookalikes, along with true microplastics. Identification steps confirm the presence of the microplastics by analyzing the chemical composition using various detection techniques like SEM-EDS, FTIR-ATR, NIR, Raman spectroscopy, and NMR spectroscopy.

6.3.2.1 SEM

SEM, or Scanning Electron Microscopy, involves the use of a powerful microscope, which emits a high-emission electron beam to analyze the topography and morphology of the microplastic sample. EDS (Electron Dispersive Spectroscopy) analysis detects the characteristic peaks of different elements found on the microplastic surface. SEM-EDS is extensively used for the screening of plastic and non-plastic pellets, which are not commonly differentiable using visual detection. Detection of strong carbon peaks, along with weak elemental peaks, are associated with plastic pellets, whereas the carbon peak is absent in cases of non-plastic pellets (Tirkey and Upadhyay, 2021). High resolution images of microplastic surfaces using SEM also confirms the traces of chemical and physical weathering by the detection of grooves and other fractures across the surface.

6.3.2.2 FTIR

Fourier-transform infrared spectroscopy, a non-destructive approach, is considered to be direct and extremely reliable. When the sample is subjected to infrared light, the radiation is absorbed, depending upon its molecular structure, and the transmission or reflection is measured. FTIR accurately identifies the polymer type and also the type of physiochemical weathering, along with oxidation intensity (Tirkey and Upadhyay, 2021). Attenuated total reflection (ATR), Focal Plane Array (FPA), and micro-FTIR are three variations of optimized FTIR technologies. Since FTIR can only detect microplastic upto $20\,\mu$ m, micro-FTIR can be used to identify and analyze samples less than $10\,\mu$ m. Micro-FTIR have been effectively used to differentiate between synthetic and semi-synthetic fibers. For the analysis of large microplastic, ATR-FTIR is used. Sample preparation is not necessary in this process (Tirkey and Upadhyay, 2021). Disadvantages with respect to ATR involves damage to fragile and weathered microplastic samples due to the probe's high pressure. FPA-FTIR can analyze microplastic particles less than $20\,\mu$ m, by scanning microplastic residues with high degree of lateral resolution. This method provides images with high resolution for better results with minimal analytical bias.

6.3.2.3 Other Analytical Techniques

NIR penetrates deeper into the microplastic materials, in which, like FTIR, there is no requirement of sample preparation. In this process, bulk samples can also be analyzed. Based on the C-H, C-O, and N-H bands, the plastic materials are identified (Tirkey and Upadhyay, 2021). Raman spectroscopy is a vibrational spectroscopy that works on the basis of inelastic scattering of light. The vibrational spectra obtained acts as a chemical fingerprint, thereby identifying the particles in the sample. Raman spectroscopy is capable of detecting microplastics of sizes such as $1\,\mu$ m, along with its chemical characteristics as well as its structures (Tirkey and Upadhyay, 2021).

6.4 DEGRADATION OF MICROPLASTIC

Environmental microplastic trash breaks down mechanically, chemically, and biologically. The structure, additives, chemical composition, and environmental factors like temperature, moisture content in air, depositional model (such as soil, water, or terrestrial sand, as opposed to marine), along with depositional environment, all affect how quickly a polymer degrades. The latter component is crucial in determining the extent of abrasion happening mechanically that takes place in different environments, like landfills to beaches, and how sunlight exposures are given to microplastics, immersed in the water column or in benthic division. Even while microplastics are capable of being broken down into nano plastics, oligomers, and monomers via mechanical, chemical, and biological processes, it is more typical for each of the three weathering mechanisms to work together to cause the degradation of the microplastic. Three types of degradation are outlined in this chapter, and instances of how microplastics have degraded naturally and in synthetic settings are given (Corcoran, P.L., 2022, pp. 531–542).

Microorganisms can remove microplastics by adapting to the habitat where they are present. Microorganisms react to the stress in numerous ways, including rate of growth, rate of metabolism, and new macromolecules synthesis for cellular defense (Cheng, H.-C. *et al.*, 2011, pp. 1–3). Since these enzymes illustrate its significant role in controlling the processing of cells, these responses to stresses are strongly tied to the activity of enzymes. Enzymes have a function in the breakdown of anthropogenic contaminants, such as microplastics, in addition to being engaged in cell regulation and function. For instance, a microorganism's degrading enzyme can precisely target the polymer structure of microplastic and break it down into its monomer, which will then be utilized as a carbon source in the microorganism's energy generation cycle (Othman, A.R. *et al.*, 2021, pp. 3057–3073).

There are two main mechanisms that make up the mechanism. It was suggested that enzyme surface modification methods via enzyme hydrolases (carboxylesterases, lipases, proteases, cutinases) are in charge of altering microplastic polymer surfaces susceptible to the process of degradation. Othman, A.R. *et al.* (2021, pp. 3057–3073) conducted a thorough analysis of the problem and asserted that certain microplastic hydrolases exclusively react to the microplastic's surface. This kind of enzyme is a surface-modifying enzyme (Austin, H.P. *et al.*, 2018, p. 115) (refer to Figure 6.4). As a result, the previously mentioned enzymes elevate the hydrophilic nature of the microplastic surface while sparing its constituent parts from degradation. The elemental spectroscopy chemical analysis (ESCA) has demonstrated and clarified the mechanism between the enzyme and surface of the microplastic (Kawai, F. *et al.*, 2019, pp. 4253–4268).

Although there has been little advancement, photocatalytic breakdown of microplastics is one of the main processes to eliminate microplastic photocatalysts engaged in microplastic breakdown through photocatalysis. Characteristics of the microplastics, the characteristics of the photocatalyst, the light source, the conditions of the solution, and environmental variables are some of the photocatalytic processes and factors impacting degradation. Non-metal and metal dopants are used to modify the surface of photocatalysts. To create multifunctional composites, the photocatalysts are doped and combined with safe, non-toxic, as well as ecologically acceptable polymers. Superoxide ions (O_2-) and hydroxyl radicals ($OH-$) are involved in degradation of (micro)plastics using the photocatalytic method, which results in the breakdown of the polymer chain and some intermediate production. ROS generation plays a vital role in photocatalytic microplastic degradation. A number of researches have verified the photocatalytic effectiveness of microplastics like PE, PS (chlorine- free

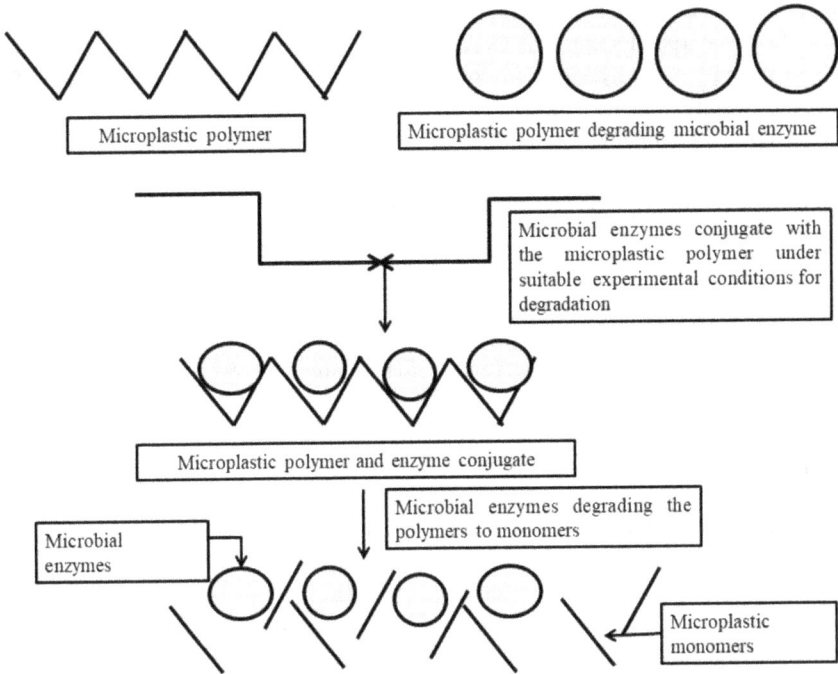

FIGURE 6.4 Schematic diagram on microplastics degradation using microbial enzymes.

polymers), or PVC. Multiple processes are associated with the photocatalytic breakdown of microplastics (Kim, S. *et al.*, 2005, pp. 24260–24267). MPs experience a photo-aging process when exposed to natural light. Direct photon absorption by the macromolecule creates an excited state, which leads to chain breakage, branching cross-linking, and oxidation processes.

TiO_2 is frequently employed as a model photocatalyst due to its potent oxidizing power for contaminants (organic). Although only a few studies have shown the usage of TiO_2/plastic composite materials to develop photodegradable plastic such as high- and low-density polyethylene (PE) films, polypropylene (PP), and polystyrene (PS), it is clear that extensive research is required to determine TiO_2's use of microplastics degradation. Low-density PE was more effectively removed using TiO_2 nanotubes when exposed to visible light, but photodegradable PS was more effectively removed using copper phthalocyanine, polypyrrole, and multiwalled carbon nanotubes (Nabi, I. *et al.*, 2020, p. 101326).

All across the world, one of the most-used polymers, PVC (polyvinyl chloride), is one of the worst for the environment because it produces harmful byproducts, including dioxins and polycyclic aromatic hydrocarbons. Chemical recovery techniques are now seen to be a potential strategy for recovering and reusing waste PVC. Recycling using a chemical method is the practice of partially transforming plastic waste to more valuable and smaller goods that may be utilized as chemical feedstock or fuel. The value that products from pyrolysis made from polymer wastes add is, however, not very high. Additionally, the oil products have a high chlorine content, which may cause the discharge of some poisonous or dangerous compounds.

The strong carbon hybrids demonstrated a remarkable MPs-degradation performance by catalyzing the production of reactive radicals from peroxymonosulfate. The exceptional stability of the carbocatalysts in the HT environment was ensured by the spiral design and extremely graphitic degree. Microplastics are extracted from the cosmetics diluted in de-ionized water, then treated with carbon hybrid nanoparticles under suitable conditions.

6.5   VALUE-ADDED PRODUCTS FROM MICROPLASTIC

Microplastics, which are in the range of 1–4 mm and primarily comprised of high-density polyethylene granules (mass ratio of more than 85%), can be utilized as carrier particles in fluidized bed reactors for the removal of carbon and nitrogen in septic wastewater. To recycle the secondary microplastics, the main sludge was collected, filtered through sieves with a 700-mesh opening, and then washed with tap water. The recycled microplastic products were then produced as composite particles (CP) with leftover clay (He, X. *et al.*, 2019, pp. 151, 107300). For the catalytic-pyrolysis steam reformation of microplastics waste that has been mixed in organic solvent like phenol, production of hydrogen and valued fuels can be done by generating dual functional Ni-Pt nanocatalysts mounted on Al_2O_3 and TiO_2.

6.6   CONCLUSION

Microplastics, both in their primary and secondary forms, are increasing in the marine environment by multiple folds. Various economic activities, coupled with improper waste management practices, contribute to this growing microplastic pollution. The particles, often from drainage, enters the marine environment, and ultimately, becomes a part of the marine food web. Its consumption by the marine fauna and subsequent biomagnification not only affects the fauna directly but also affects human health, being a part of the food chain. Problems in circulatory and digestive systems as well as muscle tissues of the marine fauna are most commonly observed across all species, affecting their enzymatic activities, locomotory functions, malnutrition, repressed reproduction, and even premature death. In humans, the circulatory and endocrine systems as well as the muscle cells are often severely affected. To understand the impacts, along with the important features of microplastics, different procedures of separation and identification and analysis have been employed. Sieving, filtration, density separation, etc., are some common physical techniques, along with various chemical digestion processes, which have reported significant separation and extraction of microplastics from the ocean. Analytical methods such as SEM-EDS, FTIR, FTIR-ATR, NMR, Raman spectroscopy, etc., have proved to be efficient tools in efficient identification as well as effective qualitative and quantitative analysis of various types of MPs present in the sample. Microplastic is largely degraded in an enzymatic way. The other methods may be photocatalysis and a microbial way of degradation. Studies are going on regarding the degradation of microplastics. There are a few value-added products like carrier particles, which are used in a fluidized bed reactor for waste water treatment, as well as nano composites. There is no doubt that the marine microplastic issue has a massive impact on the ecological and environmental health; therefore, identification of these particles in the ocean becomes a priority. Needless to say, only through proper identification of the microplastics can their successful removal from the ocean be performed. This process not only requires a more efficient and economically feasible solution but also, crucially, needs better waste-management and precautionary practices, all directed toward the goal of a cleaner and greener future.

REFERENCES

Alberghini, L. *et al.* (2023) 'Microplastics in fish and fishery products and risks for human health: A review', *International Journal of Environmental Research and Public Health*, 20(1), p. 789. Available at: Https://doi.org/10.3390/ijerph20010789.

Allen, S. *et al.* (2019) 'Atmospheric transport and deposition of microplastics in a remote mountain catchment', *Nature Geoscience*, 12(5), pp. 339–344. Available at: https://doi.org/10.1038/s41561-019-0335-5.

An, L. *et al.* (2020) 'Sources of microplastic in the environment', in *Handbook of Environmental Chemistry*. Springer Science and Business Media Deutschland GmbH, pp. 143–159. Available at: https://doi.org/10.1007/698_2020_449.

Anderson, P.J. *et al.* (2017) 'Microplastic contamination in Lake Winnipeg, Canada', *Environmental Pollution*, 225, pp. 223–231. Available at: https://doi.org/10.1016/j.envpol.2017.02.072.

Andrade, C., and Ovando, F. (2017) 'First record of microplastics in stomach content of the southern king crab Lithodes santolla (Anomura: lithodidae), Nassau bay, Cape Horn, Chile', *Anales del Instituto de la Patagonia*, 45(3), pp. 59–65. Available at: https://doi.org/10.4067/S0718-686X2017000300059.

Andrady, A.L. (2017) 'The plastic in microplastics: A review', *Marine Pollution Bulletin*, 119(1), pp. 12–22.

Arthur, C., Baker, J.E., and Bamford, H.A. (2009) 'Proceedings of the international research workshop on the occurrence, effects, and fate of microplastic marine debris', September 9–11, 2008, University of Washington Tacoma, Tacoma, WA, USA.

Austin, H.P., Allen, M.D., Donohoe, B.S., Rorrer, N.A., Kearns, F.L., Silveira, R.L., Pollard, B.C., Dominick, G., Duman, R., El Omari, K., and Mykhaylyk, V. (2018). 'Characterization and engineering of a plastic-degrading aromatic polyesterase', *Proceedings of the National Academy of Sciences*, 115(19), pp. E4350–E4357.

Ballent, A. *et al.* (2016) *Sources and Sinks of Microplastics in Canadian Lake Ontario Nearshore, Tributary and Beach Sediments*. Available at: https://ir.lib.uwo.ca/earthpubhttps://ir.lib.uwo.ca/earthpub/14.

Barnes, D.K.A. *et al.* (2009) 'Accumulation and fragmentation of plastic debris in global environments', *Philosophical Transactions of the Royal Society B: Biological Sciences*, 364(1526), pp. 1985–1998. Available at: https://doi.org/10.1098/rstb.2008.0205.

Batel, A. *et al.* (2016) 'Transfer of benzo[a]pyrene from microplastics to Artemia nauplii and further to zebrafish via a trophic food web experiment: CYP1A induction and visual tracking of persistent organic pollutants', *Environmental Toxicology and Chemistry*, 35(7), pp. 1656–1666. Available at: https://doi.org/10.1002/etc.3361.

Baulch, S., and Perry, C. (2014) 'Evaluating the impacts of marine debris on cetaceans', *Marine Pollution Bulletin*, 80(1), pp. 210–221. Available at: https://doi.org/10.1016/j.marpolbul.2013.12.050.

Bergmann, M., Gutow, L., and Klages, M. (2015) *Marine Anthropogenic Litter*. Springer Nature, Cham, p. 447.

Berntsen, P. *et al.* (2010) 'Biomechanical effects of environmental and engineered particles on human airway smooth muscle cells', *Journal of the Royal Society Interface*, 7(Suppl 3), pp. S331–S340. Available at: https://doi.org/10.1098/rsif.2010.0068.focus.

Besseling, E. *et al.* (2013) 'Effects of microplastic on fitness and PCB bioaccumulation by the lugworm *Arenicola marina* (L.)', *Environmental Science & Technology*, 47(1), pp. 593–600. Available at: https://doi.org/10.1021/es302763x.

Besseling, E. *et al.* (2015) 'Microplastic in a macro filter feeder: Humpback whale Megaptera novaeangliae', *Marine Pollution Bulletin*, 95(1), pp. 248–252. Available at: https://doi.org/10.1016/j.marpolbul.2015.04.007.

Bhattacharya, P. (2016) *A Review on the Impacts of Microplastic Beads Used in Cosmetics.* Available at: www.mcmed.us/journal/abs.

Bouwmeester, H., Hollman, P.C.H., and Peters, R.J.B. (2015) 'Potential health impact of environmentally released micro- and nanoplastics in the human food production chain: Experiences from nanotoxicology', *Environmental Science & Technology*, 49(15), pp. 8932–8947. Available at: https://doi.org/10.1021/acs.est.5b01090.

Canesi, L. *et al.* (2012) 'Bivalve molluscs as a unique target group for nanoparticle toxicity', *Marine Environmental Research*, 76, pp. 16–21. Available at: https://doi.org/10.1016/j.marenvres.2011.06.005.

Cedervall, T., Hansson, L.A., Lard, M., Frohm, B., and Linse, S. (2012) 'Food chain transport of nanoparticles affects behaviour and fat metabolism in fish', *PloS one*, 7(2), p. e32254.

Cheng, H.C., Qi, R.Z., Paudel, H., and Zhu, H.J. (2011) 'Regulation and function of protein kinases and phosphatases', *Enzyme Research*, 2011. Available at: https://doi.org/10.4061/2011/794089.

Cole, G., and Sherrington, C. (2016) *Study to Quantify Pellet Emissions in the UK.* Eunomia, Bristol, UK.

Cole, M. *et al.* (2015) 'The impact of polystyrene microplastics on feeding, function and fecundity in the marine copepod *Calanus helgolandicus*', *Environmental Science & Technology*, 49(2), pp. 1130–1137. Available at: https://doi.org/10.1021/es504525u.

Corcoran, P.L. (2022). 'Degradation of microplastics in the environment'. In Rocha-Santos, T., Costa, M.F., Mouneyrac, C. (eds) *Handbook of Microplastics in the Environment.* Cham: Springer International Publishing, pp. 531–542. Available at: https://doi.org/10.1007/978-3-030-39041-9_10

Crawford, C.B., and Quinn, B., 2016. *Microplastic Pollutants.* Elsevier, Amsterdam.

Davarpanah, E., and Guilhermino, L. (2019) 'Are gold nanoparticles and microplastics mixtures more toxic to the marine microalgae Tetraselmis chuii than the substances individually?' *Ecotoxicology and Environmental Safety*, 181, pp. 60–68. Available at: https://doi.org/10.1016/j.ecoenv.2019.05.078.

De Falco, F. *et al.* (2018) 'Evaluation of microplastic release caused by textile washing processes of synthetic fabrics', *Environmental Pollution*, 236, pp. 916–925. Available at: https://doi.org/10.1016/j.envpol.2017.10.057.

De Witte, B., Devriese, L., Bekaert, K., Hoffman, S., Vandermeersch, G., Cooreman, K., and Robbens, J. (2014) 'Quality assessment of the blue mussel (Mytilus edulis): Comparison between commercial and wild types', *Marine Pollution Bulletin*, 85(1), pp. 146–155.

Desforges, J.P.W., Galbraith, M., and Ross, P.S. (2015) 'Ingestion of microplastics by zooplankton in the Northeast Pacific Ocean', *Archives of Environmental Contamination and Toxicology*, 69, pp. 320–330.

Desforges, J.P.W. *et al.* (2014) 'Widespread distribution of microplastics in subsurface seawater in the NE pacific ocean', *Marine Pollution Bulletin*, 79(1–2), pp. 94–99. Available at: https://doi.org/10.1016/j.marpolbul.2013.12.035.

Devriese, L.I. *et al.* (2015) 'Microplastic contamination in brown shrimp (Crangon crangon, Linnaeus 1758) from coastal waters of the Southern North Sea and Channel area', *Marine Pollution Bulletin*, 98(1), pp. 179–187. Available at: https://doi.org/10.1016/j.marpolbul.2015.06.051.

Ding, Y. *et al.* (2021) 'The abundance and characteristics of atmospheric microplastic deposition in the northwestern South China Sea in the fall', *Atmospheric Environment*, 253. Available at: https://doi.org/10.1016/j.atmosenv.2021.118389.

Farrell, P., and Nelson, K. (2013) 'Trophic level transfer of microplastic: Mytilus edulis (L.) to Carcinus maenas (L.)', *Environmental Pollution*, 177, pp. 1–3. Available at: https://doi.org/10.1016/j.envpol.2013.01.046.

Felsing, S., Kochleus, C., Buchinger, S., Brennholt, N., Stock, F., and Reifferscheid, G. (2018) 'A new approach in separating microplastics from environmental samples based on their electrostatic behavior', *Environmental Pollution*, 234, pp. 20–28.

Foekema, E.M. *et al.* (2013) 'Plastic in north sea fish', *Environmental Science & Technology*, 47(15), pp. 8818–8824. Available at: https://doi.org/10.1021/es400931b.

Fossi, M.C. *et al.* (2014) 'Large filter feeding marine organisms as indicators of microplastic in the pelagic environment: The case studies of the Mediterranean basking shark (Cetorhinus maximus) and fin whale (Balaenoptera physalus)', *Marine Environmental Research*, 100, pp. 17–24. Available at: https://doi.org/10.1016/j.marenvres.2014.02.002.

Fries, E., Dekiff, J.H., Willmeyer, J., Nuelle, M.T., Ebert, M., and Remy, D. (2013) 'Identification of polymer types and additives in marine microplastic particles using pyrolysis-GC/MS and scanning electron microscopy', *Environmental Science: Processes & Impacts*, 15(10), pp. 1949–1956.

Gambardella, C. *et al.* (2017) 'Effects of polystyrene microbeads in marine planktonic crustaceans', *Ecotoxicology and Environmental Safety*, 145, pp. 250–257. Available at: https://doi.org/10.1016/j.ecoenv.2017.07.036.

Gasperi, J., Wright, S.L., Dris, R., Collard, F., Mandin, C., Guerrouache, M., Langlois, V., Kelly, F.J., and Tassin, B. (2018) 'Microplastics in air: Are we breathing it in?' *Current Opinion in Environmental Science & Health*, 1, pp. 1–5.

Hale, R.C. *et al.* (2020) 'A global perspective on microplastics', *Journal of Geophysical Research: Oceans*. Blackwell Publishing Ltd. Available at: https://doi.org/10.1029/2018JC014719.

Hämer, J. *et al.* (2014) 'Fate of microplastics in the marine isopod *Idotea emarginata*', *Environmental Science & Technology*, 48(22), pp. 13451–13458. Available at: https://doi.org/10.1021/es501385y.

He, X., Li, H., and Zhu, J. (2019). 'A value-added insight of reusing microplastic waste: Carrier particle in fluidized bed bioreactor for simultaneous carbon and nitrogen removal from septic wastewater', *Biochemical Engineering Journal*, 151, p. 107300.

Hidalgo-Ruz, V., Gutow, L., Thompson, R.C., and Thiel, M. (2012) 'Microplastics in the marine environment: A review of the methods used for identification and quantification', *Environmental Science & Technology*, 46(6), pp. 3060–3075.

Hirai, H. *et al.* (2011) 'Organic micropollutants in marine plastics debris from the open ocean and remote and urban beaches', *Marine Pollution Bulletin*, 62(8), pp. 1683–1692. Available at: https://doi.org/10.1016/j.marpolbul.2011.06.004.

Hudcová, H., Vymazal, J., and Rozkošný, M. (2019) 'Present restrictions of sewage sludge application in agriculture within the European Union', *Soil and Water Research. Czech Academy of Agricultural Sciences*, pp. 104–120. Available at: https://doi.org/10.17221/36/2018-SWR.

Karami, A., Golieskardi, A., Choo, C.K., Romano, N., Ho, Y.B., and Salamatinia, B. (2017) 'A high-performance protocol for extraction of microplastics in fish', *Science of the Total Environment*, 578, pp. 485–494.

Karlsson, T.M., Vethaak, A.D., Almroth, B.C., Ariese, F., van Velzen, M., Hasselöv, M., and Leslie, H.A. (2017) 'Screening for microplastics in sediment, water, marine invertebrates and fish: Method development and microplastic accumulation', *Marine Pollution Bulletin*, 122(1–2), pp. 403–408.

Kawai, F., Kawabata, T., and Oda, M. (2019) 'Current knowledge on enzymatic PET degradation and its possible application to waste stream management and other fields', *Applied Microbiology and Biotechnology*, 103, pp. 4253–4268.

Kim, S., Hwang, S.J., and Choi, W. (2005) 'Visible light active platinum-ion-doped TiO$_2$ photocatalyst', *The Journal of Physical Chemistry B*, 109(51), pp. 24260–24267.

Kim, S.W., and An, Y.-J. (2019) 'Soil microplastics inhibit the movement of springtail species', *Environment International*, 126, pp. 699–706. Available at: https://doi.org/10.1016/j.envint.2019.02.067.

Koongolla, J.B. *et al.* (2020) 'Occurrence of microplastics in gastrointestinal tracts and gills of fish from Beibu Gulf, South China Sea', *Environmental Pollution*, 258, p. 113734. Available at: https://doi.org/10.1016/j.envpol.2019.113734.

Kosuth, M., Mason, S.A., and Wattenberg, E.V. (2018). 'Anthropogenic contamination of tap water, beer, and sea salt', *PLoS One*, 13(4), p. e0194970.

Kühn, S., Van Werven, B., Van Oyen, A., Meijboom, A., Rebolledo, E.L.B., and Van Franeker, J.A. (2017) 'The use of potassium hydroxide (KOH) solution as a suitable approach to isolate plastics ingested by marine organisms', *Marine Pollution Bulletin*, 115(1–2), pp. 86–90.

Lares, M., Ncibi, M.C., Sillanpää, M., and Sillanpää, M. (2019) 'Intercomparison study on commonly used methods to determine microplastics in wastewater and sludge samples', *Environmental Science and Pollution Research*, 26(12), pp. 12109–12122.

Lind, P.M., and Lind, L. (2011) 'Circulating levels of bisphenol A and phthalates are related to carotid atherosclerosis in the elderly', *Atherosclerosis*, 218(1), pp. 207–213. Available at: https://doi.org/10.1016/j.atherosclerosis.2011.05.001.

Lithner, D., Larsson, Å., and Dave, G. (2011) 'Environmental and health hazard ranking and assessment of plastic polymers based on chemical composition', *Science of the Total Environment*, 409(18), pp. 3309–3324. Available at: https://doi.org/10.1016/j.scitotenv.2011.04.038.

Löder, M.G.J. *et al.* (2017) 'Enzymatic purification of microplastics in environmental samples', *Environmental Science & Technology*, 51(24), pp. 14283–14292. Available at: https://doi.org/10.1021/acs.est.7b03055.

Lusher, A.L. *et al.* (2014) 'Microplastic pollution in the Northeast Atlantic Ocean: Validated and opportunistic sampling', *Marine Pollution Bulletin*, 88(1–2), pp. 325–333. Available at: https://doi.org/10.1016/j.marpolbul.2014.08.023.

Marsden, P., Koelmans, A.A., Bourdon-Lacombe, J., Gouin, T., D'Anglada, L., Cunliffe, D., Jarvis, P., Fawell, J., and De France, J. (2019) *Microplastics in Drinking Water*. World Health Organization, Geneva.

Martins, A., and Guilhermino, L. (2018) 'Transgenerational effects and recovery of microplastics exposure in model populations of the freshwater cladoceran Daphnia magna Straus', *Science of the Total Environment*, 631–632, pp. 421–428. Available at: https://doi.org/10.1016/j.scitotenv.2018.03.054.

Masura, J., *et al.* (2015) *Laboratory Methods for the Analysis of Microplastics in the Marine Environment: Recommendations for Quantifying Synthetic Particles in Waters and Sediments.* NOAA Marine Debris Division, Silver Spring, MD, 31 p (NOAA Technical Memorandum NOS-OR&R-48). Available at: http://dx.doi.org/10.25607/OBP-604.

Mintenig, S.M., Int-Veen, I., Löder, M., and Gerdts, G. (2014) 'Mikroplastik in ausgewählten Kläranlagen des Oldenburgisch-Ostfriesischen Wasserverbandes (OOWV) in Niedersachsen', Sample Analysis Using Micro-FTIR Spectroscopy.

Mintenig, S.M. *et al.* (2019) 'Low numbers of microplastics detected in drinking water from ground water sources', *Science of the Total Environment*, 648, pp. 631–635. Available at: https://doi.org/10.1016/j.scitotenv.2018.08.178.

Moore, C.J. (2008) 'Synthetic polymers in the marine environment: A rapidly increasing, long-term threat', *Environmental Research*, 108(2), pp. 131–139.

Nabi, I., Li, K., Cheng, H., Wang, T., Liu, Y., Ajmal, S., Yang, Y., Feng, Y., and Zhang, L. (2020) 'Complete photocatalytic mineralization of microplastic on TiO$_2$ nanoparticle film', *Iscience*, 23(7).

Napper, I.E. *et al.* (2015) 'Characterisation, quantity and sorptive properties of microplastics extracted from cosmetics', *Marine Pollution Bulletin*, 99(1), pp. 178–185. Available at: https://doi.org/10.1016/j.marpolbul.2015.07.029.

Nguyen, B., Claveau-Mallet, D., Hernandez, L.M., Xu, E.G., Farner, J.M., and Tufenkji, N. (2019) 'Separation and analysis of microplastics and nanoplastics in complex environmental samples', *Accounts of Chemical Research*, 52(4), pp. 858–866.

Novotny, T.E., and Slaughter, E. (2014) 'Tobacco product waste: An environmental approach to reduce tobacco consumption', *Current Environmental Health Reports*. Springer, pp. 208–216. Available at: https://doi.org/10.1007/s40572-014-0016-x.

Obbard, R.W. *et al.* (2014) 'Global warming releases microplastic legacy frozen in Arctic Sea ice', *Earth's Future*, 2(6), pp. 315–320. Available at: https://doi.org/10.1002/2014ef000240.

Othman, A.R., Hasan, H.A., Muhamad, M.H., Ismail, N.I., and Abdullah, S.R.S. (2021). 'Microbial degradation of microplastics by enzymatic processes: A review', *Environmental Chemistry Letters*, 19, pp. 3057–3073.

Possatto, F.E. *et al.* (2011) 'Plastic debris ingestion by marine catfish: An unexpected fisheries impact', *Marine Pollution Bulletin*, 62(5), pp. 1098–1102. Available at: https://doi.org/10.1016/j.marpolbul.2011.01.036.

Romeo, T. *et al.* (2015) 'First evidence of presence of plastic debris in stomach of large pelagic fish in the Mediterranean Sea', *Marine Pollution Bulletin*, 95(1), pp. 358–361. Available at: https://doi.org/10.1016/j.marpolbul.2015.04.048.

Ryan, P.G. (2013) 'A simple technique for counting marine debris at sea reveals steep litter gradients between the Straits of Malacca and the Bay of Bengal', *Marine Pollution Bulletin*, 69(1–2), pp. 128–136. Available at: https://doi.org/10.1016/j.marpolbul.2013.01.016.

Schymanski, D., Goldbeck, C., Humpf, H.U., and Fürst, P. (2018) 'Analysis of microplastics in water by micro-Raman spectroscopy: Release of plastic particles from different packaging into mineral water', *Water Research*, 129, pp. 154–162.

Scopetani, C., Chelazzi, D., Mikola, J., Leiniö, V., Heikkinen, R., Cincinelli, A., and Pellinen, J. (2020) 'Olive oil-based method for the extraction, quantification and identification of microplastics in soil and compost samples', *Science of the Total Environment*, 733, p. 139338.

Selvam, S. *et al.* (2020) 'Microplastic presence in commercial marine sea salts: A baseline study along Tuticorin Coastal salt pan stations, Gulf of Mannar, South India', *Marine Pollution Bulletin*, 150, p. 110675. Available at: https://doi.org/10.1016/j.marpolbul.2019.110675.

Shaw, D.G., and Day, R.H. (1994) 'Colour-and form-dependent loss of plastic micro-debris from the North Pacific Ocean', *Marine Pollution Bulletin*, 28(1), pp. 39–43.

Shim, W.J., Hong, S.H., and Eo, S.E. (2017) 'Identification methods in microplastic analysis: A review', *Analytical Methods*, 9(9), pp. 1384–1391.

Stock, F., Kochleus, C., Bänsch-Baltruschat, B., Brennholt, N., and Reifferscheid, G. (2019) 'Sampling techniques and preparation methods for microplastic analyses in the aquatic environment–A review', *TrAC Trends in Analytical Chemistry*, 113, pp. 84–92.

Tirkey, A., and Upadhyay, L.S.B. (2021) 'Microplastics: An overview on separation, identification and characterization of microplastics', *Marine Pollution Bulletin*, 170, p. 112604.

Tiwari, M., Rathod, T.D., Ajmal, P.Y., Bhangare, R.C., and Sahu, S.K. (2019) 'Distribution and characterization of microplastics in beach sand from three different Indian coastal environments', *Marine Pollution Bulletin*, 140, pp. 262–273.

Tosetto, L., Williamson, J.E., and Brown, C. (2017) 'Trophic transfer of microplastics does not affect fish personality', *Animal Behaviour*, 123, pp. 159–167. Available at: https://doi.org/10.1016/j.anbehav.2016.10.035.

Van Cauwenberghe, L. *et al.* (2013) 'Microplastic pollution in deep-sea sediments', *Environmental Pollution*, 182, pp. 495–499. Available at: https://doi.org/10.1016/j.envpol.2013.08.013.

Von Moos, N., Burkhardt-Holm, P., and Köhler, A. (2012) 'Uptake and effects of microplastics on cells and tissue of the blue mussel *Mytilus edulis* l. After an experimental exposure', *Environmental Science & Technology*, 46(20), pp. 11327–11335. Available at: https://doi.org/10.1021/es302332w.

Waldschläger, K. *et al.* (2020) 'The way of microplastic through the environment—Application of the source-pathway-receptor model (review)', *Science of the Total Environment*. Elsevier B.V. Available at: https://doi.org/10.1016/j.scitotenv.2020.136584.

Wang, Z.M., Wagner, J., Ghosal, S., Bedi, G., and Wall, S. (2017) 'SEM/EDS and optical microscopy analyses of microplastics in ocean trawl and fish guts', *Science of the Total Environment*, 603, pp. 616–626.

Wick, P. *et al.* (2010) 'Barrier capacity of human placenta for nanosized materials', *Environmental Health Perspectives*, 118(3), pp. 432–436. Available at: https://doi.org/10.1289/ehp.0901200.

Wright, S.L., Thompson, R.C., and Galloway, T.S. (2013) 'The physical impacts of microplastics on marine organisms: A review', *Environmental Pollution*, 178, pp. 483–492.

www.grida.no. (n.d.) *Global Distribution of Microplastics | GRID-Arendal*. [online] Available at: www.grida.no/resources/13339.

Zubris, K.A.V., and Richards, B.K. (2005) 'Synthetic fibers as an indicator of land application of sludge', *Environmental Pollution*, 138(2), pp. 201–211. Available at: https://doi.org/10.1016/j.envpol.2005.04.013.

7 Edible Cutlery
A Tenable Solution to the Plastic Menace, Bolstering the Global Economy

Namratha B. and Santosh L. Gaonkar

7.1 INTRODUCTION: THE SAGA OF EDIBLE CUTLERY

The rapidly growing hospitality industry creates demand for a huge number of professionals each year. Many people are drawn to jobs that are challenging, exciting and, above all, provide opportunities for genuine satisfaction (Gisslen 2018).

Cutlery is one of the simplest but most useful things. These are the devices developed and used around the world to consume food. The pioneer of the application of cutlery still remains a question. Spoons are considered one of the oldest utensils created by living creatures. They made them from natural elements such as wood, animal bones and shells.

The first documented evidence of a spoon was present in England in 1259. In those days, spoons were used not only for taking food but also as a symbol of wealth and power in rituals. At the turn of the 18th century, even forks and knives were introduced as cutlery. Silver was the most popular metal for preparing cutlery, as it barely reacted with food, until the introduction of stainless steel. Stainless steel has become the metal of choice for most cutlery because it is easy to clean, non-reactive and durable. Aluminum was used for most of the utensils in the kitchen. It is a good conductor and is lightweight, making it easy to handle cutlery. However, aluminum is a relatively soft metal and should be handled carefully and not be banged around. It cannot be used with acidic foods, as this tends to cause chemical reactions. It may also discolor light-colored foods such as sauces. The finest heat conductor of all is copper, which was once a common material for cutlery. But it is costly and needs a lot of maintenance and is also very heavy. Moreover, copper chemically combines with a wide variety of foods to create hazardous chemicals. Only a few upscale dining establishments still utilize it today.

Plastic was first utilized for tableware and kitchen utensils after World War II, when there was a scarcity for metals. Companies began producing plastic cutlery in the 1960s as a less expensive substitute for conventional tableware.

The advent of plastic has greatly reduced the price of cutlery and greatly simplified its availability. Various types and sizes of cups, plates, spoons, forks and knives have been introduced for people to choose from. Today, stainless steel cutlery costs much

DOI: 10.1201/9781003449133-7

more than plastic and food. These single-use plastic utensils are made to be used once and then thrown away, negating the need for washing and conserving precious resources like water and electricity. Demand at fast food cafeterias has surged as a result of the availability of plastic cutlery. Because of this, the plastic utensils were damaging the ecology and leaching into the environment. Bisphenol A and polyvinyl chloride are main constituents of the plastics that are inherently cancer-causing and enter the human body through the food chain as microplastics. Most plastic utensils and cutlery give off a chemical, known as styrene, that causes a variety of illnesses. It is therefore imperative to take action and address this issue before it is too late.

The United States, India and Japan were the major purchasers of plastic cutlery. The majority of single-use plastic cutlery was manufactured in much disorganized, small units where hygiene rules were not followed. Users of plastic cutlery are exposed to chemical contamination, sometimes even making this cutlery sticky; that is, plastic cutlery must be repeatedly served to customers without disposal, exposing them to bacterial infection. It certainly has researchers thinking about alternatives to plastic cutlery. In the current technology era, new technologies to streamline various chores are introduced very frequently.

7.2   IMPACT OF PLASTICS

Over the course of their lifetime, plastics produce around 3.8% of the global greenhouse gas emissions. The effects of climate change and global warming, which are already wreaking havoc on our planet, are exacerbated by these emissions. We need to be concerned about more than just plastic trash and greenhouse gas emissions; we also need to be concerned about the environmental issues brought on by the extraction and production processes. Just the process of extracting and making plastic consumes a lot of energy, water, chemicals and other non-renewable resources. Utensils made of plastic typically contain dangerous substances like BPA (bisphenol A). These toxins may find their way into our food, which could have negative effects on our health. When technical cleanup methods are doubtful and natural mineralization processes take a long time, plastic pollution accumulating in environmental areas is considered 'difficult to reverse'. Large, hydrophobic polymer molecules are not readily biodegradable. This stability is what makes plastics useful, but this property turns out to be a curse when we look at the cumulative accumulation of products in the environment. Plastic waste is conspicuous everywhere, and countless used tires and shopping bags are lying far and wide. This plastic waste reduces soil fertility, clogs drainage systems and leads to unpleasant odors and the spread of infectious diseases.

Even grazing livestock can, over time, swallow plastic bags, which can be a health hazard. When municipal plastic waste is incinerated; air pollutants such as CO, CO_2, NO_x, particulate matter and tetrachlorodibenzo-p-dioxin are released into the atmosphere. Workers in manufacturing are victims of poisoning from monomers, plasticizers and other additives (Das and Das 2019). Labor in the PVC industry suffers from angiocarcinoma, hair loss, liver damage and reproductive problems. Changes in the carbon and nitrogen cycles, altered soil, sediment, and aquatic ecosystems, concurrent biological effects on endangered species, ecotoxicity and related societal repercussions are only a few of the impending effects of irreversible plastic pollution. When the amount of plastic pollution entering a quarter outpaces the rate of natural removal processes and cleanup efforts, the accumulation of plastics in the environment occurs.

The ocean floor is the main accumulation sector for plastic pollution and contains the maximum concentration of microplastic particles in the environment (Tekman *et al.* 2020). The low thermal conductivity of plastic materials is thought to be beneficial in certain applications such as insulation, but when distributed in aquatic ecosystem, these plastics contribute to global warming. They displace the same amount of water, restricting the flow of heat from the sun to the aquatic environment, causing sea-level rise and the release of energy directly into the surroundings. Plastics are also intentionally introduced into agricultural soils through plastic mulching with polyethylene and sludge-derived biosolids containing plastic residues and the application of polymeric stabilizers against soil erosion (Blasing and Amelung 2018). The amount of plastic in the world's agricultural soil is probably more than that on the surface of the ocean, based only on estimates of sewage sludge inputs. To better understand and manage the peril of plastic pollution in the environment, focus should be on environmental processes such as the accumulation of small, weathered particles, associated chemicals and the heteroaggregates with natural, organic carbon. Discovery-driven research aimed at identifying the currently anonymous effects of plastic weathering on biogeochemical cycles and animal health is also required (Iroegbu *et al.* 2020).

Rapid increases in population density and economic growth depend entirely on the depletion of fossil fuels. This is forcing today's generation to explore alternative energy sources. Liquid fuel production is a fantastic alternative, as the calorific value of plastic is comparable to that of motor fuel (40 MJ/kg) (Panda *et al.* 2010). Plastic waste is treated with different control strategies, depending on its origin, quality and characteristics (Buekens and Huang 1998). Plastics can be classified using various techniques such as IR spectroscopy and X-ray fluorescence. However, these methods may not be commercially affordable. Several entrepreneurs have come up with the idea of processing mixed plastic-waste into a substitute for wood or concrete in the manufacture of boat decks, benches, etc. Incineration of plastic waste to generate energy is theoretically possible because it reduces the burden of CO_2 emissions (Scott 2000). Additionally, plastic waste can be incinerated along with other municipal waste. The high calorific value of plastics can increase the heating of solid waste. However, water-to-energy technology is facing setbacks due to distrust from people in developed countries. This is due to the lack of clear explanations for the release of greenhouse gases and toxic pollutants such as dioxins and furans when plastic is incinerated. An oil refinery in China's Hunan province will be the first facility to process about 30,000 tons of plastic waste to produce 20,000 tons of gasoline and diesel. Then, in 1996, microbiologist Paul Baskis began economically breaking down plastic waste into lighter, cleaner oils. However, in 2005, Professor Alka Zadgaonkar established the world's first continuous process in India to convert plastic waste such as carrier bags, PVC pipes, waste bottles and broken buckets into liquid hydrocarbon fuels. Pyrolyzing plastics without generating toxins still requires some effort. In addition, efficient catalytic systems are envisioned that can facilitate large-scale conversion.

TABLE 7.1

Plastic Usage across Various Industries as Reported in "Innovations in Plastics: The Potential and Possibilities"

Industry	Plastic Usage (million tonnes)
Packaging	11.5
Building and construction	2.6
Electricals and electronics	2.6
Agriculture	1.8
Automotive	1.8
Households	0.1

India's plastic intake and waste production have been enhanced radically over the last five years. As statistics show, India's plastic consumption increased from 13.7 million tonnes in 2016–17 to 19.8 million tonnes in 2019–20, representing a compound annual growth rate (CARG) of 9.7%. Meanwhile, waste generation increased from 1.6 million tonnes to 3.4 million tonnes from 2016–17 to 2019–20, with a CARG of 20.7%. This is clear from a report by the Marico Innovation Foundation. The report also notes that, of the 3.4 million tonnes of waste generated, only 30% is recycled, the rest going to landfills. Among all states, Maharashtra, Gujrat and Tamil Nadu generate the most plastic waste, accounting for 38% of total production. Table 7.1 shows the use of plastics in various industries.

Indian entrepreneur Mariwala said, "Government has taken the first step towards single-use plastics but needs to play a more active role like introduction of landfill tax". He adds that there should be incentives for citizens to landfill or sort their waste. Waste sorting is done manually, which ultimately leads to poor-quality recycling.

The precautionary principle is currently applied in relation to some uses of plastics, especially where they are considered unnecessary or substitutable—or where there is an unacceptable risk of release to the environment. Plastic bags are now taxed or completely eliminated in many countries around the world (Horton 2022).

At composting facilities, garbage is collected in plastic bags and disposable plastic tableware, so it seems difficult to manage plastic. Dioxins are produced when chlorinated plastics are burned (Thornton 2001). Due to their high surface-area-to-weight ratio, plastic bags are very mobile. They are hydrophobic and less dense than water, so they can swim to reach water bodies (Thompson *et al.* 2004). Floating plastics are subject to biofilm formation and are less susceptible to decomposition by UV light (Gregory 2003). Degraded plastics can lose certain properties, such as physical strength and integrity, without removing the polymer structure (Greene 2014). Such degradable plastics are made photosensitive by introducing transition metal carbonyl and carbon monoxide groups into the polymer. Compostable plastics require a certain amount of moisture and oxygen for microbes to absorb (Song *et al.* 2009). This partially decomposed plastic has been sent to composting plants and sewage plants, where it has been consumed by microorganisms. However, it could not mineralize into carbon dioxide and water. Subsequently, starch-intercalated polymer chains were used in degraded plastics to improve UV sensitivity. However, such plastics lose their mechanical and physical properties faster than standard plastics. Conventional plastics rely on petroleum products (Andrady and Neal 2009). Degradable plastics, on the other hand, are made from renewable resources that reduce the need for oil reserves (Suriyamongkol *et al.* 2009). Additionally, these renewable plastics turn into micro-waste with a larger surface area. It provides adsorption sites for organic pollutants consumed by organisms that feed on land and sea filters. As such, partially decomposed plastic still poses a hazard to the environment. Conventional plastics rely on petroleum products. Degradable plastics, on the other hand, are made from renewable resources that reduce the need for oil reserves. Additionally, these renewable plastics turn into micro-waste with a larger surface area (Fendall and Sewell 2009). It provides adsorption sites for organic pollutants consumed by organisms that feed on land and sea filters (Rochman *et al.* 2013). As such, partially decomposed plastic still poses a hazard to the environment (Breslin and Swanson 1993).

7.3   DEMAND FOR EDIBLE CUTLERY: A GLOBAL PERSPECTIVE

According to the literature, edible cutlery was not a new concept but was introduced in the 1440s (Natarajan *et al.* 2019) Bread bowls were first introduced in 1427 to impress the British Duke. Japan, Taiwan, Poland, Belgium, France, South Africa, the United States and India are just a few of the countries producing edible cutlery. Recently, research into edible tableware using 3D printers has become popular.

Malafi and colleagues (1994) published a study of cutlery made from biodegradable resins and compared its performance with cutlery made from non-degradable polystyrene. The survey was conducted on 243 sailors on board three U.S. Navy ships. From a Navy perspective, the dumping of plastic waste into the ocean is a concern (Pruter 1987; Zarfl *et al.* 2001; Moore 2008; O'Brine and Thompson 2010; Law 2017). In addition, storing used cutlery on board is problematic and affects the health of the crew. In contrast, biodegradable cutlery can be thrown overboard and decompose. Sailors are captives and form a more controlled group than the average civilian. Sailors were once civilians and would become civilians off board, so they could be trendsetters in the civilian market. Observations were made when sailors were tricked into eating with both types of cutlery. The results show that polystyrene cutlery is tougher than biodegradable cutlery. The color of the polystyrene cutlery was more soothing to their eyes. However, the majority reported no difference in palatability of the foods eaten with the two types of cutlery.

The Defence Food Research Laboratory (Mysore, India) has been working on the development of edible cutlery technology. As reported by Indian newspaper *The Deccan Herald* in 2017, this indigenous cutlery is patent pending. The ingredients of this cutlery are not disclosed, but it is reported to have negligible calories. Since it is a product for military personnel, the cutlery is made to be as light as possible and reduce the burden. The Orto Cafe in Japan has edible tablewear created by Japanese designer Nobuhiko Arikawa. This tablewear includes chopsticks prepared from hardtack, the biscuit dough, baked in a bakery. These are tested to last for months, together, under dry conditions. Along the same lines, a small family business called Marushige in Hekinan (a Japanese city) makes edible plates with shrimp and potato starch that taste like sweet potatoes and onions. There is also an attempt by a Yokohama-based company to make small cups out of edible seaweed. Keio University in Tokyo is known for making edible toys for children in day care centres. The South African company founded by Georgina de Kock has been producing crispy wheat husks since 2011. It has a shelf life of 15 months, and its crispiness is restored when stored in the oven. Its ingredients are completely vegan and contain no artificial colors or yeast. Jerzy Wysocki, from Poland, makes microwave-safe cutlery out of wheat bran. A U.S. company called Loliware has launched a gluten-free, gelatin-free, non-genetically modified disposable edible cup made from agar, sweeteners and colors derived from fruits and vegetables.

Entrepreneurs are agents of change. They invent new solutions for old problems. In the case of ICMRI (The International Crops Research Institute for the Semi-Arid Tropic) groundwater researcher Narayana Peesapati, this has happened. The big idea came to this Indian while moving from Ahmedabad to Hyderabad. On the plane, he saw a fellow passenger eating with jowar (sorghum) chips. So 'jowar' inspired him to create organic spoons to replace plastic cutlery. This chase resulted in the development of an amazing product called edible cutlery. His scheme was so distinctive that it gained a spirited advantage and great response in the international market. During his research, he learned that related attempts had been made before, but either the people were unsuccessful or the ideas failed to take off. Cutlery had to meet certain parameters such as robustness, durability and shape retention. It was very difficult to meet all the specifications, but his foresight and diligence enabled Peesapati to find the right combination of components that made for a preferred edition of vegan and delicious cutlery. Machines and molds also had to be thoughtfully designed during the process.

7.4   PRODUCTION AND BRANDING

Half of all calories devoured by people come from maize, rice and wheat. Their inventories would wane as climate alters due to whimsical precipitation and climate extremes. Thus, there is an urgent requirement for growing hardier species to secure our needs. Millets being climate-resilient, growing well in drier circumstances, requiring very little water are the most excellent alternatives for the current worldwide issue (Kumar et al. 2018). Lands that cannot keep up anything else can still have millets growing on them (Ashoka et al. 2020). They have a concise cycle and can be created between major trim seasons, and they, additionally, enhance the soil fertility. As per the report published by Indian Institute of Millets Research, Hyderabad, millets contain 7–12 % protein, 2–5 % fat, 65–75 % carbohydrates and 15–20 % dietary fiber. Thus, millets can be termed 'powerhouses of nutrition' (Anitha et al. 2020). Small millets are more nutritious compared to fine cereals, as they contain a higher amount of protein, fat and fiber content. According to nourishment specialists, millets are not only naturally gluten-free but also significantly higher in iron and calcium than processed wheat and rice. This makes them a fantastic option for people attempting to manage insulin resistance or reduce blood sugar (Ren et al. 2018; Vedamanickam et al. 2020). Cereal-based food products supplemented with millets

FIGURE 7.1 Variety of available millets.

are gaining momentum in the market due to nutritional and economic rewards. It is high time that the standard society should begin to comprehend and appreciate the long-lost benefits of millets (Figure 7.1).

Millet sorghum was chosen as the base constituent, as it meets all the specifications required by Peesapati. The company's first cutlery was a spoon made from rice, sorghum and wheat flour—vegan, preservative-free, trans fat-free, dairy-free and naturally biodegradable. As they came in various sizes, they could be used for both eating and serving food. These cutleries come in a range of flavors such as cinnamon, ginger, cumin, sugary and spicy, so they were a treat in themselves. The product was dehydrated and hardened by high temperature sweltering. It could withstand hot drinks to cold ice cream for about 15–20 minutes without breaking. Cutlery would be eaten by animals if thrown away without being eaten and would rot in three days if no one ate it. When stored in cool, dry conditions, edible cutlery would last up to three years. This is because the main raw material was sorghum, which grows on dry land and consumes much less water than other crops. Thus, Peesapati solved the problem of low water table by using millet.

Peesapati was a scientist, not a cook, so he had to learn the art of baking by visiting different bakeries. This led him to name his manufacturing division Bakeys Private Limited in 2010.

FIGURE 7.2 Edible "Enchi Crunchi" cups advertisement.

As an eco-friendly initiative, a Mangaluru (a city in Karnataka, India) start-up has introduced an edible cup/cookie cup that can be used to drink hot and cold beverages under the brand name "Enchi Crunchi" (Figure 7.2).

The cups are made of ingredients including rice flour, ragi flour, refined flour, cornflour and vegetable oils with added flavors. This team suggests that it is a three-layer (batter) cup and undergoes multiple steaming processes. There are also edible

Tea cup

Muffin

Stacking cups

Spoon

Chat plate

Straw

Coffee cup

Cups

Ice cream cup

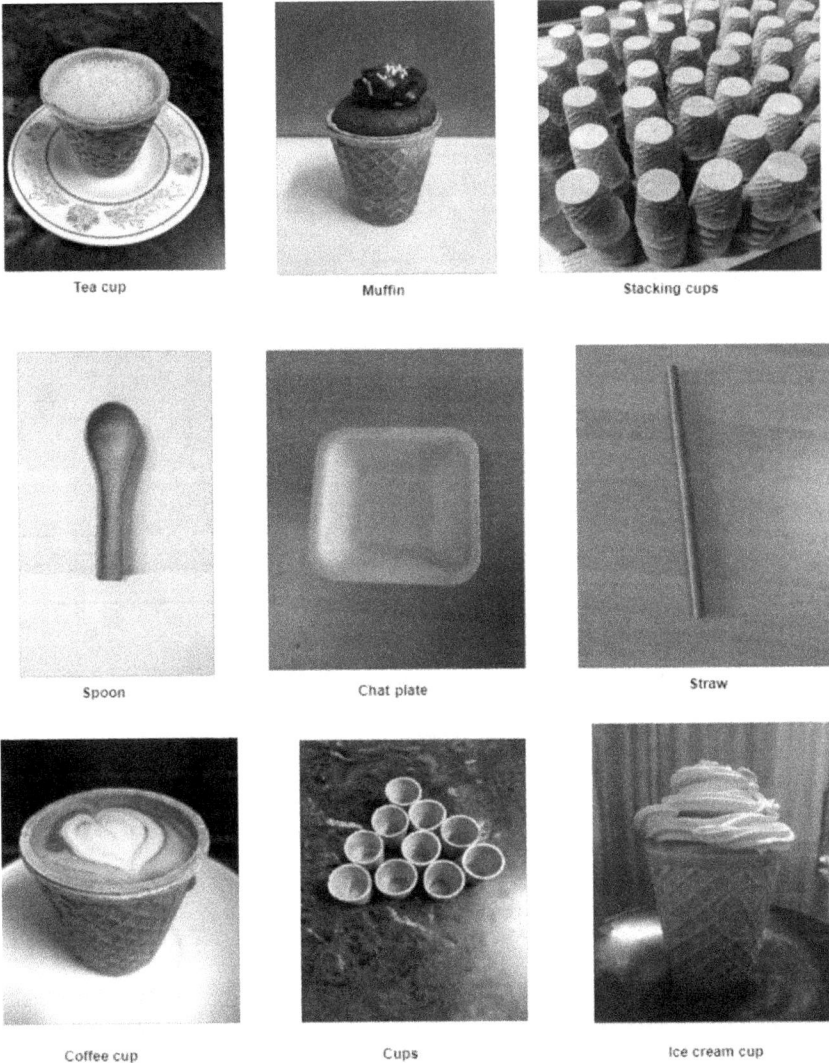

FIGURE 7.3 Edible cutlery by "Enchi Crunchi".

spoons, straws, 'chat' plates and ice cream bowls on the market (Figure 7.3). The "Enchi Crunchi" cups are Rs. 10 each and come in 90 ml and 120 ml capacities with different flavors—vanilla, chocolate and cardamom for hot drinks and additional strawberry flavor for cold drinks. The hot beverage will last for about 20 minutes. Edible cups are stronger than waffles and ice cream cones due to the different ingredients. This initiative is seen as a step toward a plastic-free Mangaluru.

Various plant and tree parts with Ayurvedic significance can also be incorporated into edible cutlery to enhance its properties. Shabaana and colleagues (2021) report using *Moringa oleifera* along with finger millet, foxtail millet, wheat and rice

1	• Washing and drying the raw materials
2	• Addition of binding agents
3	• Pulverisation into flour
4	• Weighing of the flour
5	• Dough preparation
6	• Rolling into sheet
7	• Shaping by moulds
8	• Drying in hot air oven (65 °C)
9	• Edible cutlery

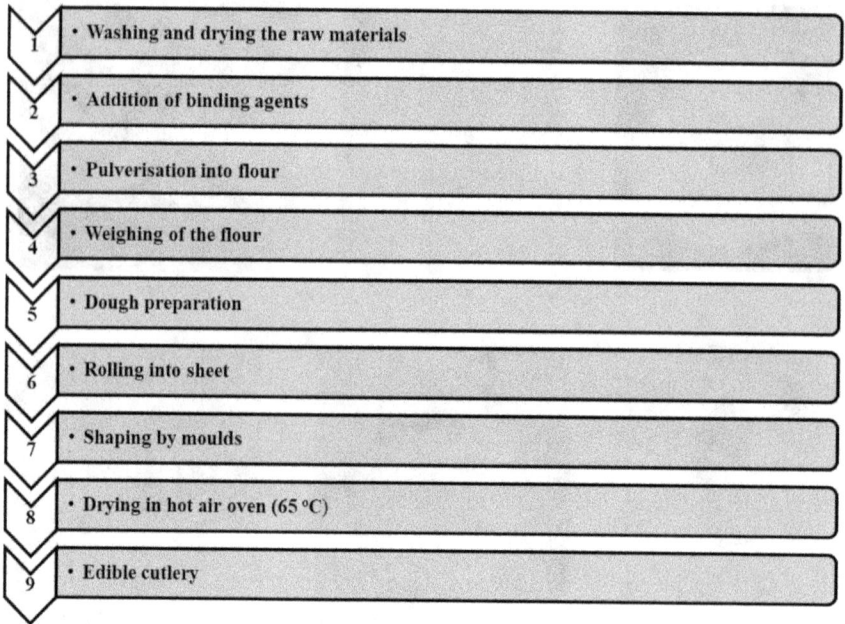

FIGURE 7.4 Flow chart for the preparation of edible cutlery using Moringa oleifera.

to make edible cutlery (Figure 7.4). *Moringa oleifera* is a rich source of nutrients (Lakshmipriya *et al.* 2016). It is a food fortificant in making bread, biscuits and soup (Adewumi and Oyeyinka 2018). They carried out nine trials between 60° C and 100° C with an increase in temperature of 5° C in each trial. Composition ratios were varied according to the final product from each experiment.

7.5   QUALITY ANALYSIS OF VARIOUS FORMULATIONS

Edible cutlery has great potential. They come in a variety of flavors that can be paired with certain types of foods. For instance, the flavor of the ice cream can be matched with the ice cream spoon.

Final product proximity testing such as moisture (Zambrano *et al.* 2019) and ash, and chemical analysis for fat (Marangoni *et al.* 2020), protein and fiber should be performed against AOAC (Association of Official Analytical Collaboration) standards to verify product compliance. Water absorption percentage, biodegradability testing, sensory evaluation and statistical analysis are closely monitored before a product is placed on the market.

7.6   CONCLUSION

The digital and physical parts of the business model are out of balance, lacking awareness and collaboration, resulting in asset heaviness, complexity and risk. The production of such edible cutlery results in a lot of wastage. Therefore, many entrepreneurs do not come to experiment. There are many innovative startups in the circular economy. But funding for these startups is too low. The success of edible cutlery depends not only on its degradability but also on its ability to meet consumer standards and demands. The edible cutlery available today is durable and does not wear out easily for consuming all types of food, from hot soups to *bhel puri*, salads to ice cream. Many manufacturers should start mass production to lower prices and make such edible cutlery more accessible to consumers. This report should serve as a framework for moving forward and making changes on the ground.

Millet has traditionally been considered the grain of the poor. Therefore, transitions must occur from 'upper strata'. Plant breeders need to develop high-yielding, short-lived millet cultivars with improved palatability that is particularly suitable for organic farming. Edible cutlery can create demand for a variety of millets, further leading to demand for growing such crops, potentially benefiting farmers across the globe. Innovation can reduce ineffectiveness, increase supply, diminish price-choice dilemmas and enable economies of scale that boost customer adoption. Meticulous and targeted regulation is needed to limit the production and use of virgin plastics and persuade innovation toward safer and more competitive materials. Broader social strategies should include avoiding the unnecessary use of plastic and promoting actions to minimize plastic waste.

REFERENCES

Adewumi, T.O., and Oyeyinka, S.A., 2018. Moringa oleifera as a food fortificant: Recent trends and prospects. *Journal of Saudi Society of Agricultural Sciences*, 17 (2), 127–136.

Andrady, A.L., and Neal, M.A., 2009. Applications and societal benefits of plastics. *Philosophical Transactions of the Royal Society B: Biological Sciences*, 364 (1526), 1977–1984.

Anitha, S., Govindaraj, M., and Kane-Potaka, J., 2020. Balanced amino acid and higher micronutrients in millets complements legumes for improved human dietary nutrition. *Cereal Chemistry*, 97 (1), 74–84.

Ashoka, P., Gangaiah, B., and Sunitha, N., 2020. Millets-foods of twenty first century. *International Journal of Current Microbiology and Applied Sciences*, 9 (12), 2404–2410.

Blasing, M., and Amelung, W., 2018. Plastics in soil: Analytical methods and possible sources. *Science of Total Environment*, 612, 422–435.

Breslin, V.T., and Swanson, R.L., 1993. Deterioration of starch-plastic composites in the environment. *Journal of the Air and Waste Management Association*, 43 (3), 325–335.

Buekens, A.G., and Huang, H., 1998. Catalytic plastics cracking for recovery of gasoline range hydrocarbons from municipal plastic wastes. *Resources, Conservation and Recycling*, 23 (3), 163–181.

Das, A.K., and Das, M., 2019. *Environmental Chemistry with Green Chemistry*. New Delhi, India: Books & Allied (P) Ltd.

Fendall, L.S., and Sewell, M.A., 2009. Contributing to marine pollution by washing your face: Microplastics in facial cleansers. *Marine Pollution Bulletin*, 58 (8), 1225–1228.

Gisslen, W., 2018. *Professional Cooking*. Hoboken, NJ: John Wiley & Sons, Inc.

Greene, K.L., and Tonjes, D.J., 2014. Degradable plastics and their potential for affecting solid waste systems. *Waste Management and The Environment VII*, 180, 91–102.

Gregory, M.R., and Andrady, A.L., 2003. Plastics in the marine environment. *In:* Andrady, A.L., ed. *Plastics and the Environment*. Hoboken, NJ: Wiley Interscience, 379–402.

Horton, A.A., 2022. Plastic pollution: When do we know enough? *Journal of Hazardous Material*, 422, 1–5.

Iroegbu, A.O.C., Sadiku, R.E., Ray, S.S., and Hamam, Y., 2020. Plastics in municipal drinking water and wastewater treatment plant effluents: Challenges and opportunities for South Africa-a review. *Environmental Science and Pollution Research*, 27, 12953–12966.

Kumar, A., Tomer, V., Kaur, A., Kumar, V., and Gupta, K., 2018. Millets: A solution to agrarian and nutritional challenges. *Agriculture & Food Security*, 7 (31), 1–15.

Lakshmipriya, G., Doriyaa, K., and Kumar, D.S., 2016. Moringa oleifera: A review on nutritive importance and its medicinal application. *Food Science and Human Wellness*, 5 (2), 49–56.

Law, K.L., 2017. Plastics in the marine environment. *Annual Review of Marine Science*, 9 (1), 205–229.

Malafi, T.N., Devine, M.A., and Lesher, L.L., 1994. A user evaluation of biodegradable cutlery. *Journal of Environmental Polymer Degradation*, 2(4), 219–223.

Marangoni, A.G., van Duynhoven, J.P.M., Acevedo, N.C., Nicholsona, R.A., and Patel, A.R., 2020. Advances in our understanding of the structure and functionality of edible fats and fat mimetics. *Soft Matter*, 16, 289–306.

Moore, C.J., 2008. Synthetic polymers in the marine environment: A rapidly increasing, long-term threat. *Environmental Research*, 108 (2), 131–139.

Natarajan, N., Vasudevan, M., Velusamy, V.V., and Selvara, M., 2019. Eco-friendly and edible waste cutlery for sustainable environment. *International Journal of Engineering and Advanced Technology*, 9 (1S4), 615–624.

O'Brine, T., and Thompson, R.C., 2010. Degradation of plastic carrier bags in the marine environment. *Marine Pollution Bulletin*, 60 (12), 2279–2283.

Panda, A.K., Singh, R.K., and Mishra, D.K., 2010. Thermolysis of waste plastics to liquid fuel A suitable method for plastic waste management and manufacture of value added products-A world prospective. *Renewable and Sustainable Energy Reviews*, 14, 233–248.

Pruter, A.T., 1987. Sources, quantities and distribution of persistent plastics in the marine environment. *Marine Pollution Bulletin*, 18 (6), 305–310.

Ren, X., Yin, R., Hou, D., Xue, Y., Zhang, M., Diao, X., Zhang, Y., Wu, J., Hu, J., Hu, X., and Shen, Q., 2018. The glucose-lowering effect of foxtail millet in subjects with impaired glucose tolerance: A self-controlled clinical trial. *Nutrients*, 10 (10), 1509.

Rochman, C.M., Browne, M.A., Halpern, B.S., Hentschel, B.T., Hoh, E., Karapanagioti, H.K., Rios-Mendoza, L.M., Takada, H.S., and Thompson, R.C., 2013. Classify plastic waste as hazardous. *Nature*, 494 (7436), 169–171.

Scott, G., 2000. Green polymers. *Polymer Degradation and Stability*, 68, 1–7.

Shabaana, M., Firdouse, T.F., and Prabha, P.H., 2021. Development and quality evaluation of eco-Friendly *Moringa Oleifera* leave powder incorporated edible cutlery. *International Journal of Advances in Engineering and Management*, 3 (3), 160–166.

Song, J.H., Murphy, R.J., Narayan, R., and Davies, G.B.H., 2009. Biodegradable and compostable alternatives to conventional plastics. *Philosophical Transactions of the Royal Society B: Biological Sciences*, 364 (1526), 2127–2139.

Suriyamongkol, P., Weselake, R., Narine, S., Moloney, M., and Shah, S., 2009. Biotechnological approaches for the production of polyhydroxyalkanoates in microorganisms and plants—A review. *Biotechnology Advances*, 25 (2), 148–175.

Tekman, M.B., Wekerle, C., Lorenz, C., Primpke, S., Hasemann, C., Gerdts, G., and Bergmann, M., 2020. Tying up loose ends of microplastic pollution in the arctic: Distribution from the sea surface through the water column to deep-sea sediments at the HAUSGARTEN observatory. *Environmental Science and Technology*, 54, 4079–4090.

Thompson, R.C., Olsen, Y., Mitchell, R.P., Davis, A., Rowland, S.J., John, A.W.G., McGonigle, D., and Russell, A.E., 2004. Lost at sea: Where is all the plastic? *Science*, 304 (5672), 838.

Thornton, J., 2001. *Pandora's Poison*. Cambridge, MA: MIT Press.

Vedamanickam, R., Anandan, P., Bupesh, G., and Vasanth, S., 2020. Study of millet and non-millet diet on diabetics and associated metabolic syndrome. *Biomedicine*, 40, 55–58.

Zambrano, M.V., Dutta, B., Mercer, D.G., MacLean, H.L., and Touchie, M.F., 2019. Assessment of moisture content measurement methods of dried food products in small-scale operations in developing countries: A review. *Trends in Food Science & Technology*, 88, 484–496.

Zarfl, C., Fleet, D., Fries, E., Galgani, F., Gerdts, G., Hanke, G., and Matthies, M., 2001. Microplastics in oceans. *Marine Pollution Bulletin*, 62 (8), 1589–1591.

8 Bioengineering Solutions for Microplastic Pollution
Leveraging Microbial Assistance

Sakina Bombaywala, Vinay Pratap and Ashootosh Mandpe

8.1 INTRODUCTION

Traditionally, plastic pollution is regarded as an irreversible threat with a low rate of degradation. Therefore, some initiatives can be taken to decrease the loss of microplastics (MPs) in the environment during manufacturing, use, and discarding. Due to the complexity of environmental matrices, it is difficult to separate microplastics in isolation (Wagner and Lambert, 2018). In the last decade, considerable progress has been made to produce biodegradable plastics, chiefly from renewable materials of natural origin, to reduce the dependence on synthetic plastics, although the microbe-assisted degradation of the plastics produces MPs (Anand et al., 2023). Physical separation methods like filtration and sedimentation can remove MPs but are not capable of complete remediation, and some residuals are detected in the sludge fraction of wastewater treatment plants (WWTPs).

Since the 1970s, several studies have reported microbe-assisted degradation of plastics (Hidalgo-Ruz et al., 2012). In recent times, synthetic organic polymers (polyamide and polyesters) and biodegradable plastics have been developed that can be easily degraded by microbes. But a large fraction of plastics are highly recalcitrant, have complex molecular structures, and remain in heterogeneous environmental matrices (Wierckx et al., 2019). Also, potential microbes remain within the confinements of WWTPs, making it even more challenging to remediate MPs (Kitamoto et al., 2011). Therefore, a holistic and multidisciplinary approach is required to address the problem of MP pollution. This involves isolating, screening, and characterizing potential microorganisms using advanced methods like genomics and *in situ* toxicity studies. For instance, recalcitrant polyolefins were used as a feedstock for microbial growth, leading to biodegradation and biotransformation before chemical depolymerization or glycolysis. Before feeding, a thorough assessment of MPs is required, which includes the study of crystalline structure and degradation efficiency by microbial activities. Biosorption and bioaccumulation of MPs involve the

DOI: 10.1201/9781003449133-8

use of both living and non-living components. Aquatic eukaryotes have been recently studied for immobilization and degradation of MPs (Miri et al., 2022). However, the application of various microbial species for the bioremediation of MPs can be explored in the future.

Microbial-assisted degradation benefits from the fact that microorganisms can easily adapt to any changes in environmental conditions by acquiring necessary metabolic functions, including the potential to degrade harmful contaminants (Brooks et al., 2011).

Microbes can be used to degrade MPs in an eco-friendly manner. Thus, the cleanup of the natural environment is supported without any adverse side effects (Restrepo-Flórez et al., 2014; Kumar Sen and Raut, 2015; Qi et al., 2017). At present, only a few potent microbes are isolated, and fewer microbial interactions and functional activities are unexplored for efficient MP removal (Shah et al., 2013). The chapter thoroughly summarizes the present state of knowledge regarding MP bioremediation and methods used to improve microbial metabolism, and functional activities for efficient degradation of MP are also discussed in detail.

8.2 BIODEGRADATION PRINCIPLE OF MICROPLASTICS

The microbial bioremediation of MPs involves three steps: degradation of polymers into particles of small molecular size; breakdown of particles into oligomers, dimers, and monomers; followed by complete mineralization into non-toxic byproducts and the biotransformation of intermediate molecules that can be used as a carbon source for increasing biomass production (Espinoza, 2019). Extracellular enzymes like manganese peroxidase, laccase, lipase, and esterase assist in making plastic polymers more hydrophilic. These enzymes help to replace functional groups on the exterior of MPs with either carbonyl or alcoholic groups to promote the attachment of microbes (Shahnawaz et al., 2019; Taniguchi et al., 2019). Hydrolase enzymes such as poly (3-hydroxybutyrate) depolymerases, esterase, cutinase, and lipase are able to break large polymer chains into small particles (Sol et al., 2020). The addition of sensitive chemical groups on polymer chains aids in rapid cleavage into small chains. Because of its large size (>25 mm), enzymes are not able to penetrate inside; thus, they only act on the surface of the polymers. Microbes assimilate and mineralize MPs by transporting monomers into their cytoplasm, followed by metabolic digestion (Zettler et al., 2013). Previous research aimed at understanding the functions of extracellular enzymes involved only the first step of MP degradation. Whereas intracellular enzymes are involved in metabolic digestion, they remain unknown. More investigations are required for a thorough understanding of enzyme characteristics and transport mechanisms for the assimilation of monomers. The monomers or intermediates with hydroxyl or carbonyl group are metabolized by intracellular enzymes of the tricarboxylic acid cycle and β-oxidation (Taniguchi et al., 2019). The byproducts of the complete mineralization of MPs are water, nitrogen, carbon dioxide, and methane (Zettler et al., 2013). Fleming et al. (2017) studied the complex relationship between microbes and polymers. The study showed that microbial attachment to the biofilm surface involves the following steps: biofouling; plasticizer degradation; breaking of the polymer backbone, hydrolysis; and finally, infiltration of microorganisms inside

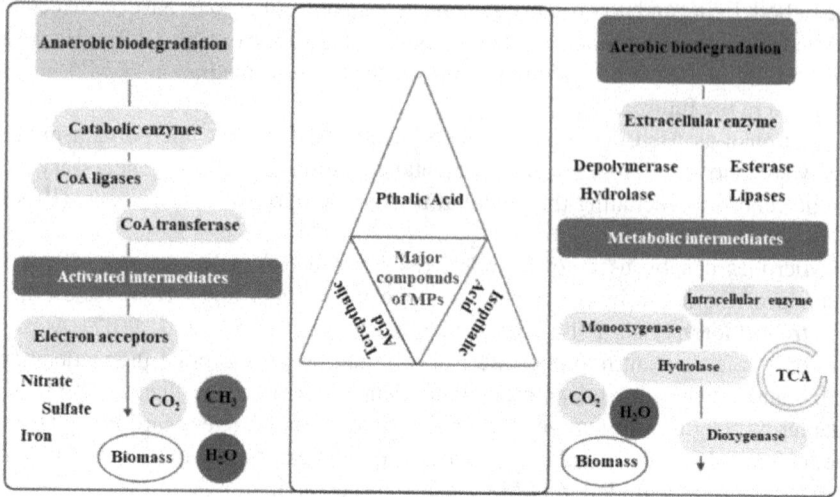

FIGURE 8.1 Processes of MP biodegradation through abiotic and biotic factors.

the polymer structure. Small particles have a high surface-volume ratio; therefore, they are more susceptible to biodegradation (Sivan, 2011). Biodegradation takes place at optimal conditions like pH, temperature, moisture, and salinity, with microorganisms possessing suitable encoding genes, enzymes, and metabolic pathways (Gong et al., 2012). The physical (structure) and chemical (functional groups) characteristics of MPs must also allow for easy adherence of microorganisms. Another factor affecting biodegradation is the degree of branching and polymerization of particular MP. This includes a proportion of amorphous and crystalline regions, crystal size, and lamellar thickness. For instance, polyhydroxyalkanoates (PHA) depolymerase can easily hydrolyze amorphous PHA polymer chains present on the surface, whereas crystalline structures are only eroded (Shabbir et al., 2020). The microbial and physiochemical processes associated with degradation of MPs are illustrated in Figure 8.1. Bioremediation approaches like composting, phytoextraction, composting, and enzymatic degradation are promising techniques for reducing MP contamination. However, the success is based on thorough understanding of microbial functions (Ru et al., 2020). More investigations are needed to identify and characterization of enzymes from biotic origins. Pre-treatment methods should also be explored for enhancing depolymerase activity of the microorganisms.

8.3 BIODEGRADATION OF MPS USING MICROBIAL APPROACHES

Microbial assisted degradation of MPs is an integrated approach that is affected by physiochemical conditions or environmental factors (Ammala et al., 2011). Since MPs are more resistant to microbial decomposition than other materials, transformation methods can promote microbial attack (Rujnić-Sokele and Pilipović, 2017). MPs offer a unique ecological habitat for microbes by providing an adhesion surface

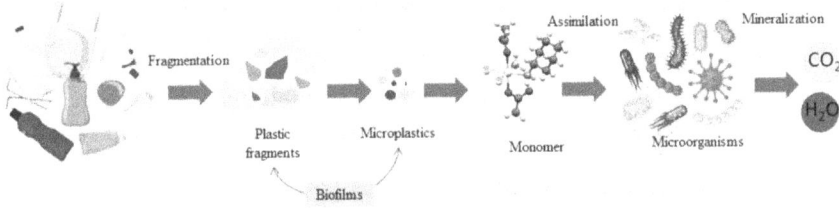

FIGURE 8.2 Aerobic and anaerobic pathways and associated enzymes in biodegradation of MPs.

for colony formation and as a carbon and energy source. Several microorganisms with the capability to degrade MPs have been isolated, screened, and characterized from various environmental niches. The microbes involved in MP biodegradation are categorized into single bacterial culture, consortia, and fungal and algal cultures. The effect of MPs on microbial growth, biodegradation rate, and factors affecting the degradation process are summarized in the following sections. Various aerobic and anaerobic pathways, along with accompanying enzymes, are presented in Figure 8.2.

8.3.1 BACTERIAL DEGRADATION

Bacteria are the most abundant of all living things on earth. Many bacterial species have the capability to break pollutants (Bakir et al., 2014). The increasing number of bacteria with the ability to utilize MPs has been isolated by enrichment techniques from wastewater, soil, sludge, or sediments. In vitro studies usually employ single bacterial cultures, and the investigations are focused on understanding the pathways involved and the effect of various physiological factors on the degradation process. Additionally, the complete process of MP biodegradation by specific bacteria and subsequent structural transformations of MPs are strictly examined (Janssen et al., 2002). In a recent study, bacteria with the capacity to utilize PP-MP were isolated from mangrove sediment (Auta et al., 2017). The study reported a 6.4% and 4% decrease in the weight of MPs by using *Rhodococcus sp.* 36 and *Bacillus sp.* 27, respectively, within 40 days. The formation of pores and surface irregularities were observed after the treatment with bacteria. Hence, the study demonstrated the ability of pure bacterial culture to adhere, inhabit, transform, and break down MPs. In the same study, *Bacillus gottheilii* and *Bacillus cereus* were investigated for degradation of different MPs like polyethylene terephthalate (PET), polyethylene (PE), polystyrene (PS), and polypropylene (PP). *Bacillus cereus* was able to degrade 1.6, 6.6, and 7.4% of PE, PET and PS, respectively, whereas, *Bacillus gottheilii* was able to reduce the weight of PE, PP, PET, and PS by 6.2, 3, 3.6, and 5.8%, respectively. Also, surface roughness with multiple grooves and cracks was observed on MPs after bacterial activity. In-depth analysis has revealed that bacteria are able to alter the functional group, appearance, chemical structure, and physical properties of MPs. For instance, after incubation of PE with the culture of *B. cereus*, carbonyl bonds present in PE were not detected in the FTIR spectrum. Moreover, additional absorption bands at

3419 and 3738 cm−1 appeared in the PE spectrum that can be ascribed to new O=H and N=H bonds on the PE structure. The study by Jeon and Kim (2013) highlighted the role of novel bacteria, *Stenotrophomonas maltophilia* LB 2–3, in the degradation of polylactic acid. The bacterial activity resulted in a reduction of tensile characteristics and molecular weight of polylactic acid. Uscátegui et al. (2016) reported 1–2% degradation of modified plastics like polyurethanes by *E.coli* in 72 hrs of incubation. The biodegradation of PP and PP films and bioriented polypropylene was investigated for a period of 11 months. Microbial activity was reported based on erosion patterns and cracking observed on the PP surface. *Pseudomonas aeruginosa* was shown to degrade up to 10% of PS-PLA nanocomposites (Shimpi et al., 2012). Mohan et al. (2016) investigated the debromination of high-impact polystyrene film by *Bacillus* strains. A 23% reduction in weight was reported after 30 days. Thus, it is indicated that the bacteria are better at degrading modified plastics. It can be attributed to the fact that modified plastics are more prone to biodegradation. Apart from direct isolation or enrichment methods for obtaining a pure bacterial culture from a polluted environment, several other methods can also be employed. A study reported waxworms could digest PE films. The ability was attributed to two bacterial species: *Enterobacter asburiae* YT1 and *Bacillus sp.* YP1, found in the gut of waxworms. In addition, the guts of insects and ticks have also been explored to isolate potent bacteria. One drawback of the method is that a small set of bacteria are isolated and screened, wherein most belonged to *Pseudomonas, Bacillus, Chelatococcus*, and *Lysinibacillus fusiformis*. Also, bacterial activity can degrade a small amount of MPs, causing weight loss of only up to 10–15%. The duration of MP degradation is significantly greater with bacteria, extending up to three months. Future studies are needed to improve selection of functional bacteria and optimizing experimental conditions for enhanced and rapid degradation of MPs. Bacterial strains with the ability to degrade MPs are detailed in Table 8.1.

TABLE 8.1
Degradation of Microplastics by Potent Bacterial Strains

Microbes	Type of Microplastics	Experimental Period (Days)	Gravimetric Weight Loss (%)	Reference
Paenibacillus sp. and *Bacillus sp.*	PP	60	14.7	(Park and Kim, 2019)
Engyodontium album	PP	365	0.5	(Taniguchi et al., 2019)
Bacillus sp. and *Bacillus simplex*	Low-density polyethylene (LDPE)	21	–	(Huerta Lwanga et al., 2018)
Bacillus sp.	PP	60	10.7	(Paço et al., 2017)
Exiguobacterium sp. strain YT2	PS	28	7.4+-0.4	(Yang et al., 2015)
Stenotrophomonas maltophilia LB 2–3	Polylactic acid (PLA)	40	6.4	(Jeon and Kim, 2013)

8.3.2 ALGAL DEGRADATION

Microalga or its enzymes and toxins can be employed for breaking down polymers (Manzi et al., 2022). Microalgal systems are advantageous, since they do not require expensive carbon sources and easily adapt to varied environmental conditions, compared to bacterial systems (Yan et al., 2016). Plastic surfaces in wastewater streams are found to be colonized by microalga, wherein adhesion or close proximity initiates degradation. Exopolysaccharides and lignolytic enzymes are released during microalgal growth, which is responsible for MP degradation. Moreover, the rate of microalgal growth increases by using polymers as carbon sources for the accumulation of cellular carbohydrates and proteins. In a recent study, scanning electron microscopy (SEM) analysis gave insights into the breakdown or degradation of low-density polyethylene sheets by algae (Sanniyasi et al., 2021). Several processes involved in the algal degradation of MP are penetration, hydrolysis, fouling, and corrosion (Chia et al., 2020). Kumar Sen and Raut (2015) demonstrated microbial colony formation and degradation of low-density polyethylene surface by *Oscillatoria subbrevis* and *Phormidium lucidum* without any pretreatment or additions of prooxidative agents. A combination of algae and bacteria, including *Chlamydomonas mexicana, Stephanodiscus hantzschii, Chlorella fusca var. vacuolated*, and *Chlorella vulgaris*, were used for the degradation of bisphenols, a polymer additive with estrogenic activity (Hirooka et al., 2005; Li et al., 2009; Ji et al., 2014). The formation of microbial biofilms is common on MP surfaces for effective degradation. Cyanobacterial species such as *Rivularia, Prochlorothrix, Microcystis, Pleurocapsa, Leptolyngbya Calothrix, Scytonema*, and *Synechococcus* are reported to form biofilm on the surface of MPs (Bryant et al., 2016; Debroas et al., 2017; Dussud et al., 2018; Muthukrishnan et al., 2019). In addition to cyanobacteria, diatoms were found in the biofilms that carry out photosynthesis (Zettler et al., 2013). With the advent of biotechnological tools, genetic modifications can be used to make genetically modified microalgal cells that can produce and excrete huge amounts of enzymes for utilizing MPs. Kim et al. (2020) genetically modified the microalgae *Chlamydomonas reinhardtii*, which was capable of degrading terephthalic acid and polyethylene terephthalate films. Likewise, genetically modified *P. tricornutum* was able to produce polyethylene terephthalate hydrolase for depolymerization of degradation of polyethylene terephthalate glycol and polyethylene terephthalate (Moog et al., 2019). Synthetic biology and genetic engineering have enhanced the degradation ability of microalgae, thus making it an efficient and sustainable solution for the biological treatment of polyethylene terephthalate. Thus, microalgae are effective MP degraders owing to their ease of culture and capability to use MPs as a carbon source.

8.3.3 FUNGAL DEGRADATION

Besides bacteria, fungi have also been reported to attach, grow, and degrade MP (Mitik-Dineva et al., 2009). Fungal growth reduces the hydrophobicity of MPs by the addition of functional groups such as carbonyl, carboxyl, and ester. A wide diversity of fungal strains is available, having a rapid growth rate. Further, fungi can assist in biotransformation and bioavailability (Chen et al., 2016), thereby changing the morphology as well as internal properties of MP. Recent studies have reported that fungal culture can utilize MP as a carbon source. In vitro studies have identified novel fungal strains showing enhanced utilization of MP. However, fungal strains are difficult to isolate by ectopic screening, and as such, only a few studies have reported the successful use of fungus for MP biodegradation. The study by Yamada-Onodera et al. (2001) showed that 0.5% PE in solid media was degraded by *Penicillium simplicissimum* in 500 hrs with UV irradiation.

Further analysis showed that only small particles of PE remained in the liquid growth medium of the fungus after three months. The study suggests UV pretreatment as an effective method to enhance biodegradation by fungi. Complete mineralization of thermo-oxidized, low-density polyethylene by *Aspergillus niger* and *Penicillium pinophilum* was reported by Volke-Sepúlveda et al. (2002). The study showed a 0.5–0.3% reduction in the polymer weight within three months of incubation with fungal strains. Specifically, the weight of the polymer was reduced by three crystalline lamellar units (0.4–1.8 Å) with a 3.2% increase in the content of small crystals, and a mean crystalline size of 8.4–14 Å was reported. In a similar study, serine hydrolase was found in two *Pestalotiopsis microspora* strains that promoted polyurethane degradation (Russell et al., 2011). Many fungal cultures were analyzed for their ability to degrade high-density polyethylene found in PE waste discarded near marine coasts.

Using in vitro screening techniques, Devi et al. (2015) were able to isolate *Aspergillus flavus* VRKPT2 and *Aspergillus tubingensis* VRKPT1 that had the ability to degrade 8.51% and 6.02% of high-density polyethylene. In a similar study, fungal strains—namely, white-rot fungi IZU154, *Trametes versicolor*, and *Phanerochaete chrysosporium*—were able to decompose nylon-6,6 by an oxidative attack. Further, the fungus-mediated degradation of MPs can be enhanced by using genomic and proteomics methods. Potential fungal strains with the capability to degrade MP while showing resistance to corrosion and oxidation should be screened and characterized. The studies will lay the foundation for designing novel methods by using fungus to reduce the impact of MPs on the environment (Paço et al., 2017). Studies at the pilot scale should be done to address the issue of real-world applications. Therefore, at present, fungal-based MP degradation is an active area of research.

8.3.4 Enzymatic Degradation

Enzymes are being studied for polymer degradation; however, some of the factors that affect catalytic reactions are surface topology, water absorbency, crystallinity, reaction temperature, and orientation of the polymer. Enzymatic hydrolysis of PET has been investigated for use in the production of nanoparticles, fibers, and films (Welch et al., 2009). Enzymes have also been employed for surface modifications, including changes in functional groups attached to the surface of polymers and degradation of monomers (Pellis et al., 2016). The surface modification makes plastic polymers hydrophilic by increasing wettability, dyeability, anti-pilling behavior, and finishing fastness (Kawai et al., 2019). The study by Barth et al. (2015) reported lipases and cutinases from *Candida Antarctica* and *Aspergillus oryzae*, respectively, for use in enhancing the hydrophilic properties of plastic fabrics. The rate of MP degradation also depends on the surface-to-volume ratio; that is, a small polymer size or a high ratio indicates rapid degradation by enzymatic activities. Enzymatic hydrolysis is a heterogeneous process that is comparable to cellulase action on cellulose. Many bio-based plastics degradation have been reported that involved the use of enzymes. For example, several studies have demonstrated the depolymerization of polyhydroxybutyrate (PHB) by employing PHB depolymerase (Roohi et al., 2018).

Consequently, dimers and monomers of biobased plastics can be mineralized into water and carbon dioxide. Most bio-based plastics like PLA, polyethylenechlorinates (PEC), polyglycolide (PGA), and polycaprolactone (PCL) are completely degraded into the water, nitrogen, methane, carbon dioxide, and hydrogen, although PLA should be decomposed under specific environmental conditions to prevent the formation of harmful intermediate and byproducts that causes enzyme inhibition (Sankhla et al., 2020). Several microbes with a crucial role in plastics degradation are reported to produce depolymerases, including *Klebsiella sp., Pseudomonas sp., Bacillus sp., Azotobacter sp.*, and *Escherichia sp.* (Jayasekara et al., 2005). Important factors affecting the degradation of monomers and dimers of plastics are crystallinity, surface topology, hydrophobicity, and molecular size. Plastic degradation is carried out in aqueous solutions due to a decrease in the transition temperature (Tg) of plastics in water. The hydrogen bonds between MPs chains are broken, thereby causing randomization and flexibility while increasing bioaccessibility for enzymes (Kawai et al., 2014). Plastics can have variations in crystallinity based on the nature of their applications. For instance, PET that is employed in the manufacturing of textiles has up to 40% crystallinity. At the same time, polyvinyl chloride (PVC) is amorphous in nature, with a less crystalline structure, and is used in pipe and tires. More crystallinity corresponds to lesser flexibility of polymer chains, thereby limiting accessibility to enzymes. For instance, polyvinyl alcohol, PVC, and PEC are readily hydrolyzed by proteases, cutinases, and carboxylesterases. However, no hydrolases are reported for degradation of PET present in films and textiles (Horton et al., 2017). It is increasingly necessary to find a bottleneck for using the full potential of enzymes for the effective and rapid degradation of plastics. Functional enzymes involved in MP degradation are detailed in Table 8.2.

8.3.5 BIOFILMS FOR MICROPLASTIC DEGRADATION

The MPs present in aquatic ecosystems are in proximity to organic matter, inorganic substances, and microbial communities (Parrish and Fahrenfeld, 2019). Therefore, MPs can act as a substrate for the adherence of various types of microbes, including algae, fungi, bacteria, protest, and viruses (Oberbeckmann and Labrenz, 2020). Microbial colonization causes biofilm formation, which is a complex system comprised of cellular secretions, inorganic and organic matter, etc. (Fleming et al., 2017). The exterior of MPs has different chemical compositions and physical forms, such as high or low density and smooth or rough surfaces. The biofilm formation can cause changes in the physical appearance and chemical composition of MPs by secreting degrading/modifying enzymes, degrading additives, releasing harmful byproducts, and masking surface characteristics (Miao et al., 2019). Thus, biofilms are able to modify the chemical and physical properties of MPs leading to their degradation/

TABLE 8.2

Enzymes Involved in MP Degradation

Microbe	Type of Microplastics	Enzyme	Reference
Comamonas acidovorans	Polyurethane (PUR)	Esterases	(Nakajima-Kambe et al., 1997)
Paenibacillus amylolyticus	PLA	Protease/Esterase	(Teeraphatpornchai et al., 2003)
Candida rugosa Pseudomonas chlororaphis	PUR	Putative polyurethanes	(Russell et al., 2011)
Pestalotiopsis microspore	PUR	Serine hydrolases	(Russell et al., 2011)
Bacillus subtilis	PUR	Estreases	(Shah et al., 2013)
Pseudomonas aeruginosa E7	PE	Rubredoxin, Rubredoxin reductase, and Alkane monooxygenase	(Jeon and Kim, 2013)
Ideonella sakaiensis	PE, PET	Glycoside hydrolases	(Yoshida et al., 2016)
Bacillus licheniformis, B. subtilis and Thermobifidafusca	PET oligomers	Carboxylesterases	(Wei and Zimmermann, 2017)

decomposition. Lobelle and Cunliffe (2011) demonstrated the effect of biofilm on PE-MPs. Approximately one week's time was required for the appearance of biofilm, and it continued to grow for up to three weeks. The abundance of heterotrophic bacteria present in the biofilm increased from 1.4×10^4 cells cm^{-2} in one week to 1.2×10^5 cells cm^{-2} within three weeks. After three weeks, the biofilm becomes increasingly hydrophilic, apparent from the sinking of MPs below the seawater-air interface.

Moreover, the weight of PE-MPs decreased, owing to decomposition or degradation by the microbial activity of the biofilm. Another study by Hossain et al. (2019) demonstrated the degradation of MPs in microcosms amended with bacterial cultures, including *Acinetobacter calcoaceticus*, *Escherichia coli*, and *Burkholderia cepacia*. The eroded PP-MP disks had more bacterial colonization, compared to the non-eroded surface, thereby suggesting the major role of microbes in altering the physio-chemical characteristics of MPs. The effects of biofilms on MPs and other substances are different. Miao et al. (2019) studied the biofilm activity of MP (PP and PE) and natural materials like wood and cobblestone, using DNA sequence data. The difference was analyzed on the basis of microbial compositions along with other metrics like species diversity, richness, and evenness of the two biofilms. As expected, biofilm on natural materials had a lower number of microbial communities related to MP degradation. Biofilm on MPs was selective for the growth of *Pirellulaceae*, *Cyclobacteriaceae*, *Roseococcus*, and *Phycisphaerales*. Biodegradation mechanisms of MPs by a microbial community of biofilms are more complicated than those of single bacteria or fungus. The MP degradation takes place in four stages: the first stage involves the attachment of microbes to the MP surface, followed by changes in surface properties like hydrophilicity and appearance. The second stage involves enzymatic hydrolysis of MPs and the associated release of monomers and additives. This is followed by the third stage, in which the mechanical stability of MPs decreases as more enzymes and oxidative radicals attack MPs. In the final stage, microbial filaments and water penetrated the embrittled structure of MP, resulting in the utilization and further degradation of MPs by the microbial population (Yuan et al., 2020). The study suggests that MPs act as a different environmental niche that absorbs nutrients and promotes microbial growth and functional activities. It also helps in the immobilization of microbes on support media and serves as a carbon and energy source for the proliferation of microbes (Galloway et al., 2017). However, a systematic study of MP degradation by biofilm is required to determine complex interactions between microbial communities and MP surfaces. Until now, interactions have been assessed in relation to changes in weight, chemical composition, and physical appearance of MPs. It is essential to characterize enzymes and encode genes involved, together with analysis of degradation byproducts. Controlled experiments, byproduct tracking, interaction analysis, and degradation behavior study are needed for an in-depth understanding of biofilm-based degradation.

8.4 FACTORS AFFECTING MICROBIAL DEGRADATION OF MPS

Multiple factors affect the microbe-assisted degradation of MPs. They are divided into factors associated with microbial growth, MP properties, and environmental factors. The physicochemical properties of MPs are molecular weight, density, fraction of crystalline structure, substituent functional groups, and 3D structure (Shah et al., 2013). The surface area of MPs also affects degradation susceptibility. For instance, MPs with large surface areas are more susceptible to degradation than those with smaller surface areas. Additives and plasticizers also assist in the biodegradation of MPs. Environmental factors that affect MP biodegradation are light (Vis/UV), humidity, and concentration of stress-causing molecules like detergents, antibiotics, and heavy metals. This can affect microbial proliferation and functional activity.

Further, the oxidative environment results in damage and aging of MPs that increase microbial biomass (Krueger et al., 2015). Sudhakar et al. (2008) studied the effect of UV light, stressors, density, and additives on the biodegradation of MPs. The addition of ethanol increased the biodegradation rate of TO-LDPE by 0.64% and 0.5% when incubated with *A. niger* and *P. pinophilum* (Volke-Sepúlveda et al., 2002). Changes in interaction between TO-LDPE and fungal culture after amendment with ethanol suggested that additives can promote the biodegradation of MPs. Satlewal et al. (2008) developed consortia of native microbes with the ability to degrade 21.7% of LDPE and 22.41% of high-density polyethylene (HDPE). The study demonstrated the effect of density on MPs degradation. In another study, Jeon and Kim (2013) reported a significant decrease in the tensile strength and molecular weight of PLA-MP after exposure to UV irradiation and *S. maltophilia LB 2–3* culture in liquid media.

Similar results were obtained for PLA-MP degradation in compost materials (Lau et al., 2009). After exposure to UV light, the formation of solid-white brittle was observed, which was associated with microbial assimilation by partial degradation of PLA structure. The chromophoric groups present in MPs are thought to absorb UV light that triggers photo-oxidation reactions. Additives like light and heat stabilizers, flame retardants, antioxidants, and plasticizers have significant effects on MP biodegradation. A 2–4% increase in degradation of MPs has been reported after the addition of photosensitizers (Gu, 2003).

Moreover, the biodegradation of MPs is significantly influenced by many environmental factors and interrelationships, among the factors. Therefore, a comprehensive assessment is needed to understand interactions among the factors and determine optimum conditions for MP biodegradation. This will also help in promoting MP biodegradation under changing environmental conditions.

8.5   ADVANCED METHODS FOR ENHANCED DEGRADATION

It is still not agreed upon which biotechnological method is most suitable for the biodegradation of MPs. The combination of several methods or optimization of methods is needed to degrade various MPs under various environmental habitats. The bacterial culture of *Ideonella sakaiensis 201-F6* is reported to mineralize polyethylene terephthalate to terephthalic and ethylene glycol by the action of polyethylene terephthalate hydrolase. In a different study, competent bacterial cells were genetically modified to carry gene encoding for PTE hydrolase from *Ideonella sakaiensis 201-F6*. In a recent study by Moog et al. (2019), genetically modified *Phaeodactylum tricornutum, a* photosynthetic microalgae, was used for PTE hydrolase and was capable of degrading significant amounts of PTE to non-hazardous monomers. Liu et al. (2021) use genetic engineering to capture PVC in the biofilm. For this, *wsp* genes in *Pseudomonas aeruginosa* were deleted, resulting in an increase in the formation of exopolymeric substances with consequent accumulation of PVC in the biofilm. Furthermore, the yhjH gene was added, which was positively induced by the arabinose-induced promoter. Induction of yhjH resulted in a decline in levels of a cyclic dimer of guanosine-monophosphate, thereby loosening the structure of the biofilm for the release of adhered MPs. The systematic capture and release enabled the creation of an effective MP scavenger system for bioremediation. With the advent of genetic engineering and the ease of manipulating the genetic material of microorganisms, the development of cost-effective and sustainable bioremediation methods has paced. Several methods involve withering recombinant DNA technology, genetic modification, and gene cloning to enhance the biodegradation capability of the microbes (Kumar et al., 2020). However, only limited studies have demonstrated the application of genetically modified microbes for MP biodegradation. It is important to identify genes responsible for degrading or metabolizing MPs and the suitability of expression of target genes in *E.coli*. Genetic engineering has been used to either alter the specificity and affinity of enzymes or to construct novel degradation pathways. The processes used are site-directed mutagenesis, polymerase chain reaction, and antisense ribonucleic acid (RNA) technology. Antisense ribonucleic acid involves the synthesis of antisense RNA that can regulate gene expression in host cells. In contrast, the process of site-directed mutagenesis involves the addition of a mutation in the sequence of the target gene to enhance the catalytic activity of the encoded enzyme. In the study by (Lameh et al., 2022), in-silico, site-directed mutagenesis of the carboxylesterase gene in *Archaeoglobus fulgidus* was performed. It leads to the production of BTA-hydrolase involved in the degradation of PET. In a similar study, manganese-dependent peroxidase was secreted by the modified strains of *S. cerevisiae BY 4741* and *E.coli*. Likewise, laccase was produced by modified strains of *E.coli BL21* and *P. chrysosporium* (Sharma et al., 2018; Paço et al., 2019). Genetically modified strains have been used for enhanced biodegradation of PET. The PET degradation has also been demonstrated by cutinases. The enzyme is responsible for breaking polyester bonds in PET and works optimally in the temperature range of 70–75 °C. Cutinase produced by the engineered strains of yeast assists in PET degradation while preventing bond formation with sugar molecules at strategic positions (Shirke et al., 2018). It is reported that cutinase was able to decrease biodegradation time from 41.8 hrs required by the wild strain to 6.2 hrs by the modified strain (Islam et al., 2019). The consortia of marine microorganisms also showed an increased ability to degrade MPs (Tsiota et al., 2018). Despite the ability of genetically engineered strains for increased biodegradation potential, they have failed to show successful results in field applications.

8.5.1 METAGENOMICS FOR EXPLORING POTENTIALS OF UNCULTURED MICROBIAL COMMUNITY

Bioinformatics has been used as an effective tool for improving MP biodegradation. Various databases like The Environmental Contaminant Biotransformation Pathway Resource, The University of Minnesota Biocatalysis/Biodegradation Database, MetaCyc, and BioCyc databases are curated to include information on enzymes, genes, and metabolic pathways involved in MP biodegradation (Karp et al., 2019; Caspi et al., 2020). Several computational methods, along with databases, can help identify and study enzymes and pathways associated with plastic degradation. This will aid in designing novel methods for enhanced degradation (Ali et al., 2021). However, experimental validation is essential to make full use of bioinformatics approaches. Also, extensive investigation is needed to fill the knowledge gap regarding degrading microbes, key metabolic processes, and enzymes. Therefore, a combination of bioinformatics, metabolic engineering, molecular biology, genetics, and systems biology may be helpful in finding sustainable options for MP biodegradation.

8.5.2 GENE EDITING TOOLS FOR ENHANCED MP DEGRADATION

Genetic engineering has helped researchers to manipulate the genetic pathways of indigenous microbes for increasing degradation potential or to introduce MP degrading genes in cloned microbes. The gene editing tools are used for the expression of target genes in microbes, plants, and animals (Paço et al., 2019; Anand et al., 2023). Various gene editing tools like transcription activator-like effector nucleases (TALENS), chemical transformation, electroporation, zinc finger proteins, and clustered, regularly interspaced palindromic repeats (CRISPR-Cas9) are used for improving or modifying the genetic makeup of microbes (Gaj et al., 2013). The first step in making genetic modification is to find target genes. Recombinant DNA is made by inserting the target gene into a suitable vector, followed by introduction in a host cell. The expression of target genes can also be altered by gene editing. It can involve either loss or gain of functional activities. Genome editing approaches can be used on specific genes to enhance MP biodegradation. The target genes encode esterase depolymerase for polyethylene terephthalate hydrolase, dehalogenase, and laccase that are involved in MP biodegradation. Shao et al. (2019) identified three distinct CRISPR sequences in the genome of *Streptomyces albogriseolus* LBX-2, which can be effectively engineered for oxygenase genes associated with polyethylene degradation. It is important to have knowledge of enzymes and pathways involved in polymer degradation in order to effectively use genetic engineering tools (Oesterle et al., 2017). Until now, no microorganisms have been successfully modified that showed enhanced MP biodegradation.

8.6 CONCLUSION

MPs are pollutants of emerging concern that can persist for a long time, ranging up to ten years, in the environment. Microbe-assisted degradation of MPs is a cost-effective and environment-friendly approach for remediation of polluted environments. Several microbes with the ability to utilize MPs have been isolated and characterized. The monomers, dimers, and oligomers are obtained as byproducts of biodegradation that can be reused as raw materials. A greater number of aerobic microbes have been explored for their degradation ability, compared to anaerobic microbes. Moreover, microorganisms that are effective against high-molecular-weight plastics and various types of MPs are less known. It is also important to classify materials and polymers that are used in the manufacture of plastics in order to develop better biodegradation strategies. Meta-omics approaches can be applied to explore uncultured microorganisms and to identify functional enzymes. The use of enzymes for the transformation or complete degradation of plastics is challenging. For this, in situ methods can be employed to understand the mechanism of enzymatic degradation in real time. A detailed summarization of MP degradation studies under similar conditions and then optimization of various degradation conditions like pH, temperature, substrate concentration, and inoculum size is needed to enhance the rate of degradation further. Various techniques, such as enzyme technology, meta-omics, and nanotechnology, can be integrated to develop better biotechnological solutions for MP pollution.

REFERENCES

Ali, S.S., Elsamahy, T., Koutra, E., Kornaros, M., El-Sheekh, M., Abdelkarim, E.A., Zhu, D., Sun, J., 2021. Degradation of conventional plastic wastes in the environment: A review on current status of knowledge and future perspectives of disposal. Sci. Total Environ. 771, 144719. https://doi.org/10.1016/j.scitotenv.2020.144719.

Ammala, A., Bateman, S., Dean, K., Petinakis, E., Sangwan, P., Wong, S., Yuan, Q., Yu, L., Patrick, C., Leong, K.H., 2011. An overview of degradable and biodegradable polyolefins. Prog. Polym. Sci. 36, 1015–1049. https://doi.org/10.1016/j.progpolymsci.2010.12.002.

Anand, U., Dey, S., Bontempi, E., Ducoli, S., Vethaak, A.D., Dey, A., Federici, S., 2023. Biotechnological methods to remove microplastics: A review. Environ. Chem. Lett. 21, 1787–1810. https://doi.org/10.1007/s10311-022-01552-4.

Auta, H.S., Emenike, C.U., Fauziah, S.H., 2017. Screening of Bacillus strains isolated from mangrove ecosystems in Peninsular Malaysia for microplastic degradation. Environ. Pollut. 231, 1552–1559. https://doi.org/10.1016/j.envpol.2017.09.043.

Bakir, A., Rowland, S.J., Thompson, R.C., 2014. Enhanced desorption of persistent organic pollutants from microplastics under simulated physiological conditions. Environ. Pollut. 185, 16–23. https://doi.org/10.1016/j.envpol.2013.10.007.

Barth, M., Oeser, T., Wei, R., Then, J., Schmidt, J., Zimmermann, W., 2015. Effect of hydrolysis products on the enzymatic degradation of polyethylene terephthalate nanoparticles by a polyester hydrolase from Thermobifida fusca. Biochem. Eng. J. 93, 222–228. https://doi.org/10.1016/j.bej.2014.10.012.

Brooks, A.N., Turkarslan, S., Beer, K.D., Yin Lo, F., Baliga, N.S., 2011. Adaptation of cells to new environments. WIREs Syst. Biol. Med. 3, 544–561. https://doi.org/10.1002/wsbm.136.

Bryant, J.A., Clemente, T.M., Viviani, D.A., Fong, A.A., Thomas, K.A., Kemp, P., Karl, D.M., White, A.E., DeLong, E.F., 2016. Diversity and activity of communities inhabiting plastic debris in the North pacific gyre. mSystems 1, e00024–16. https://doi.org/10.1128/mSystems.00024-16.

Caspi, R., Billington, R., Keseler, I.M., Kothari, A., Krummenacker, M., Midford, P.E., Ong, W.K., Paley, S., Subhraveti, P., Karp, P.D., 2020. The MetaCyc database of metabolic pathways and enzymes—a 2019 update. Nucleic Acids Res. 48, D445–D453. https://doi.org/10.1093/nar/gkz862.

Chen, Y., Stemple, B., Kumar, M., Wei, N., 2016. Cell surface display fungal laccase as a renewable biocatalyst for degradation of persistent micropollutants bisphenol A and sulfamethoxazole. Environ. Sci. Technol. 50, 8799–8808. https://doi.org/10.1021/acs.est.6b01641.

Chia, W.Y., Ying Tang, D.Y., Khoo, K.S., Kay Lup, A.N., Chew, K.W., 2020. Nature's fight against plastic pollution: Algae for plastic biodegradation and bioplastics production. Environ. Sci. Ecotechnol. 4, 100065. https://doi.org/10.1016/j.ese.2020.100065.

Debroas, D., Mone, A., Ter Halle, A., 2017. Plastics in the North Atlantic garbage patch: A boat-microbe for hitchhikers and plastic degraders. Sci. Total Environ. 599–600, 1222–1232. https://doi.org/10.1016/j.scitotenv.2017.05.059.

Devi, S.R., Rajesh Kannan, V., Nivas, D., Kannan, K., Chandru, S., Robert Antony, A., 2015. Biodegradation of HDPE by Aspergillus spp. from marine ecosystem of Gulf of Mannar, India. Mar. Pollut. Bull. 96, 32–40. https://doi.org/10.1016/j.marpolbul.2015.05.050.

Dussud, C., Meistertzheim, A.L., Conan, P., Pujo-Pay, M., George, M., Fabre, P., Coudane, J., Higgs, P., Elineau, A., Pedrotti, M.L., Gorsky, G., Ghiglione, J.F., 2018. Evidence of niche partitioning among bacteria living on plastics, organic particles and surrounding seawaters. Environ. Pollut. 236, 807–816. https://doi.org/10.1016/j.envpol.2017.12.027.

Espinoza, R.M.B., 2019. *Microplastics in Wastewater Treatment Systems and Receiving Waters* [Doctoral dissertation University of Glasgow] Enlighten Publications Record. https://doi.org/10.5525/gla.thesis.76781.

Fleming, D., Chahin, L., Rumbaugh, K., 2017. Glycoside hydrolases degrade polymicrobial bacterial biofilms in wounds. Antimicrob. Agents Chemother. 61, e01998–16. https://doi.org/10.1128/AAC.01998-16.

Gaj, T., Gersbach, C.A., Barbas, C.F., 2013. ZFN, TALEN, and CRISPR/Cas-based methods for genome engineering. Trends Biotechnol. 31, 397–405. https://doi.org/10.1016/j.tibtech.2013.04.004.

Galloway, T.S., Cole, M., Lewis, C., 2017. Interactions of microplastic debris throughout the marine ecosystem. Nat. Ecol. Evol. 1, 0116. https://doi.org/10.1038/s41559-017-0116.

Gong, J., Duan, N., Zhao, X., 2012. Evolutionary engineering of Phaffia rhodozyma for astaxanthin-overproducing strain. Front. Chem. Sci. Eng. 6, 174–178. https://doi.org/10.1007/s11705-012-1276-3.

Gu, J.-D., 2003. Microbiological deterioration and degradation of synthetic polymeric materials: Recent research advances. Int. Biodeterior. Biodegrad. 52, 69–91. https://doi.org/10.1016/S0964-8305(02)00177-4.

Hidalgo-Ruz, V., Gutow, L., Thompson, R.C., Thiel, M., 2012. Microplastics in the marine environment: A review of the methods used for identification and quantification. Environ. Sci. Technol. 46, 3060–3075. https://doi.org/10.1021/es2031505.

Hirooka, T., Nagase, H., Uchida, K., Hiroshige, Y., Ehara, Y., Nishikawa, J., Nishihara, T., Miyamoto, K., Hirata, Z., 2005. Biodegradation of bisphenol A and disappearance of its estrogenic activity by the green alga chlorella Fusca var. Vacuolata. Environ. Toxicol. Chem. 24, 1896. https://doi.org/10.1897/04-259R.1.

Horton, A.A., Walton, A., Spurgeon, D.J., Lahive, E., Svendsen, C., 2017. Microplastics in freshwater and terrestrial environments: Evaluating the current understanding to identify the knowledge gaps and future research priorities. Sci. Total Environ. 586, 127–141. https://doi.org/10.1016/j.scitotenv.2017.01.190.

Hossain, M.R., Jiang, M., Wei, Q., Leff, L.G., 2019. Microplastic surface properties affect bacterial colonization in freshwater. J. Basic Microbiol. 59, 54–61. https://doi.org/10.1002/jobm.201800174.

Huerta Lwanga, E., Thapa, B., Yang, X., Gertsen, H., Salánki, T., Geissen, V., Garbeva, P., 2018. Decay of low-density polyethylene by bacteria extracted from earthworm's guts: A potential for soil restoration. Sci. Total Environ. 624, 753–757. https://doi.org/10.1016/j.scitotenv.2017.12.144.

Islam, S., Apitius, L., Jakob, F., Schwaneberg, U., 2019. Targeting microplastic particles in the void of diluted suspensions. Environ. Int. 123, 428–435. https://doi.org/10.1016/j.envint.2018.12.029.

Janssen, P.H., Yates, P.S., Grinton, B.E., Taylor, P.M., Sait, M., 2002. Improved culturability of soil bacteria and isolation in pure culture of novel members of the divisions *Acidobacteria*, *Actinobacteria*, *Proteobacteria*, and *Verrucomicrobia*. Appl. Environ. Microbiol. 68, 2391–2396. https://doi.org/10.1128/AEM.68.5.2391-2396.2002.

Jayasekara, R., Harding, I., Bowater, I., Lonergan, G., 2005. Biodegradability of a selected range of polymers and polymer blends and standard methods for assessment of biodegradation. J. Polym. Environ. 13, 231–251. https://doi.org/10.1007/s10924-005-4758-2.

Jeon, H.J., Kim, M.N., 2013. Biodegradation of poly(l-lactide) (PLA) exposed to UV irradiation by a mesophilic bacterium. Int. Biodeterior. Biodegrad. 85, 289–293. https://doi.org/10.1016/j.ibiod.2013.08.013.

Ji, M.-K., Kabra, A.N., Choi, J., Hwang, J.-H., Kim, J.R., Abou-Shanab, R.A.I., Oh, Y.-K., Jeon, B.-H., 2014. Biodegradation of bisphenol A by the freshwater microalgae Chlamydomonas mexicana and Chlorella vulgaris. Ecol. Eng. 73, 260–269. https://doi.org/10.1016/j.ecoleng.2014.09.070.

Karp, P.D., Billington, R., Caspi, R., Fulcher, C.A., Latendresse, M., Kothari, A., Keseler, I.M., Krummenacker, M., Midford, P.E., Ong, Q., Ong, W.K., Paley, S.M., Subhraveti, P., 2019. The BioCyc collection of microbial genomes and metabolic pathways. Brief. Bioinform. 20, 1085–1093. https://doi.org/10.1093/bib/bbx085.

Kawai, F., Kawabata, T., Oda, M., 2019. Current knowledge on enzymatic PET degradation and its possible application to waste stream management and other fields. Appl. Microbiol. Biotechnol. 103, 4253–4268. https://doi.org/10.1007/s00253-019-09717-y.

Kawai, F., Oda, M., Tamashiro, T., Waku, T., Tanaka, N., Yamamoto, M., Mizushima, H., Miyakawa, T., Tanokura, M., 2014. A novel Ca^{2+}-activated, thermostabilized polyesterase capable of hydrolyzing polyethylene terephthalate from Saccharomonospora viridis AHK190. Appl. Microbiol. Biotechnol. 98, 10053–10064. https://doi.org/10.1007/s00253-014-5860-y.

Kim, J.W., Park, S.-B., Tran, Q.-G., Cho, D.-H., Choi, D.-Y., Lee, Y.J., Kim, H.-S., 2020. Functional expression of polyethylene terephthalate-degrading enzyme (PETase) in green microalgae. Microb. Cell Factories 19, 97. https://doi.org/10.1186/s12934-020-01355-8.

Kitamoto, H.K., Shinozaki, Y., Cao, X., Morita, T., Konishi, M., Tago, K., Kajiwara, H., Koitabashi, M., Yoshida, S., Watanabe, T., Sameshima-Yamashita, Y., Nakajima-Kambe, T., Tsushima, S., 2011. Phyllosphere yeasts rapidly break down biodegradable plastics. AMB Express 1, 44. https://doi.org/10.1186/2191-0855-1-44.

Krueger, M.C., Harms, H., Schlosser, D., 2015. Prospects for microbiological solutions to environmental pollution with plastics. Appl. Microbiol. Biotechnol. 99, 8857–8874. https://doi.org/10.1007/s00253-015-6879-4.

Kumar, M., Xiong, X., He, M., Tsang, D.C.W., Gupta, J., Khan, E., Harrad, S., Hou, D., Ok, Y.S., Bolan, N.S., 2020. Microplastics as pollutants in agricultural soils. Environ. Pollut. 265, 114980. https://doi.org/10.1016/j.envpol.2020.114980.

Kumar Sen, S., Raut, S., 2015. Microbial degradation of low density polyethylene (LDPE): A review. J. Environ. Chem. Eng. 3, 462–473. https://doi.org/10.1016/j.jece.2015.01.003.

Lameh, F., Baseer, A.Q., Ashiru, A.G., 2022. Retracted: Comparative molecular docking and molecular-dynamic simulation of wild-type- and mutant carboxylesterase with BTA-hydrolase for enhanced binding to plastic. Eng. Life Sci. 22, 13–29. https://doi.org/10.1002/elsc.202100083.

Lau, A.K., Cheuk, W.W., Lo, K.V., 2009. Degradation of greenhouse twines derived from natural fibers and biodegradable polymer during composting. J. Environ. Manage. 90, 668–671. https://doi.org/10.1016/j.jenvman.2008.03.001.

Li, R., Chen, G.-Z., Tam, N.F.Y., Luan, T.-G., Shin, P.K.S., Cheung, S.G., Liu, Y., 2009. Toxicity of bisphenol A and its bioaccumulation and removal by a marine microalga Stephanodiscus hantzschii. Ecotoxicol. Environ. Saf. 72, 321–328. https://doi.org/10.1016/j.ecoenv.2008.05.012.

Liu, S.Y., Leung, M.M.-L., Fang, J.K.-H., Chua, S.L., 2021. Engineering a microbial 'trap and release' mechanism for microplastics removal. Chem. Eng. J. 404, 127079. https://doi.org/10.1016/j.cej.2020.127079.

Lobelle, D., Cunliffe, M., 2011. Early microbial biofilm formation on marine plastic debris. Mar. Pollut. Bull. 62, 197–200. https://doi.org/10.1016/j.marpolbul.2010.10.013.

Manzi, H.P., Abou-Shanab, R.A.I., Jeon, B.-H., Wang, J., Salama, E.-S., 2022. Algae: A frontline photosynthetic organism in the microplastic catastrophe. Trends Plant Sci. 27, 1159–1172. https://doi.org/10.1016/j.tplants.2022.06.005.

Miao, L., Wang, P., Hou, J., Yao, Y., Liu, Z., Liu, S., Li, T., 2019. Distinct community structure and microbial functions of biofilms colonizing microplastics. Sci. Total Environ. 650, 2395–2402. https://doi.org/10.1016/j.scitotenv.2018.09.378.

Miri, S., Saini, R., Davoodi, S.M., Pulicharla, R., Brar, S.K., Magdouli, S., 2022. Biodegradation of microplastics: Better late than never. Chemosphere 286, 131670. https://doi.org/10.1016/j.chemosphere.2021.131670.

Mitik-Dineva, N., Wang, J., Truong, V.K., Stoddart, P., Malherbe, F., Crawford, R.J., Ivanova, E.P., 2009. Escherichia coli, pseudomonas aeruginosa, and staphylococcus aureus attachment patterns on glass surfaces with nanoscale roughness. Curr. Microbiol. 58, 268–273. https://doi.org/10.1007/s00284-008-9320-8.

Moog, D., Schmitt, J., Senger, J., Zarzycki, J., Rexer, K.-H., Linne, U., Erb, T., Maier, U.G., 2019. Using a marine microalga as a chassis for polyethylene terephthalate (PET) degradation. Microb. Cell Factories 18, 171. https://doi.org/10.1186/s12934-019-1220-z.

Muthukrishnan, T., Al Khaburi, M., Abed, R.M.M., 2019. Fouling microbial communities on plastics compared with wood and steel: Are they substrate- or location-specific? Microb. Ecol. 78, 361–374. https://doi.org/10.1007/s00248-018-1303-0.

Nakajima-Kambe, T., Onuma, F., Akutsu, Y., Nakahara, T., 1997. Determination of the polyester polyurethane breakdown products and distribution of the polyurethane degrading enzyme of Comamonas acidovorans strain TB-35. J. Ferment. Bioeng. 83, 456–460. https://doi.org/10.1016/S0922-338X(97)83000-0.

Oberbeckmann, S., Labrenz, M., 2020. Marine microbial assemblages on microplastics: Diversity, adaptation, and role in degradation. Annu. Rev. Mar. Sci. 12, 209–232. https://doi.org/10.1146/annurev-marine-010419-010633.

Oesterle, S., Wuethrich, I., Panke, S., 2017. Toward genome-based metabolic engineering in bacteria, in: Advances in Applied Microbiology. Elsevier, pp. 49–82. https://doi.org/10.1016/bs.aambs.2017.07.001.

Paço, A., Duarte, K., Da Costa, J.P., Santos, P.S.M., Pereira, R., Pereira, M.E., Freitas, A.C., Duarte, A.C., Rocha-Santos, T.A.P., 2017. Biodegradation of polyethylene microplastics by the marine fungus Zalerion maritimum. Sci. Total Environ. 586, 10–15. https://doi.org/10.1016/j.scitotenv.2017.02.017.

Paço, A., Jacinto, J., Da Costa, J.P., Santos, P.S.M., Vitorino, R., Duarte, A.C., Rocha-Santos, T., 2019. Biotechnological tools for the effective management of plastics in the environment. Crit. Rev. Environ. Sci. Technol. 49, 410–441. https://doi.org/10.1080/10643389.2018.1548862.

Park, S.Y., Kim, C.G., 2019. Biodegradation of micro-polyethylene particles by bacterial colonization of a mixed microbial consortium isolated from a landfill site. Chemosphere 222, 527–533. https://doi.org/10.1016/j.chemosphere.2019.01.159.

Parrish, K., Fahrenfeld, N.L., 2019. Microplastic biofilm in fresh- and wastewater as a function of microparticle type and size class. Environ. Sci. Water Res. Technol. 5, 495–505. https://doi.org/10.1039/C8EW00712H.

Pellis, A., Haernvall, K., Pichler, C.M., Ghazaryan, G., Breinbauer, R., Guebitz, G.M., 2016. Enzymatic hydrolysis of poly(ethylene furanoate). J. Biotechnol. 235, 47–53. https://doi.org/10.1016/j.jbiotec.2016.02.006.

Qi, X., Ren, Y., Wang, X., 2017. New advances in the biodegradation of Poly(lactic) acid. Int. Biodeterior. Biodegrad. 117, 215–223. https://doi.org/10.1016/j.ibiod.2017.01.010.

Restrepo-Flórez, J.-M., Bassi, A., Thompson, M.R., 2014. Microbial degradation and deterioration of polyethylene—A review. Int. Biodeterior. Biodegrad. 88, 83–90. https://doi.org/10.1016/j.ibiod.2013.12.014.

Roohi, B.K., Zaheer, M.R., Kuddus, M., 2018. PHB (poly-β-hydroxybutyrate) and its enzymatic degradation. Polym. Adv. Technol. 29, 30–40. https://doi.org/10.1002/pat.4126.

Ru, J., Huo, Y., Yang, Y., 2020. Microbial degradation and valorization of plastic wastes. Front. Microbiol. 11, 442. https://doi.org/10.3389/fmicb.2020.00442.

Rujnić-Sokele, M., Pilipović, A., 2017. Challenges and opportunities of biodegradable plastics: A mini review. Waste Manag. Res. J. Sustain. Circ. Econ. 35, 132–140. https://doi.org/10.1177/0734242X16683272.

Russell, J.R., Huang, J., Anand, P., Kucera, K., Sandoval, A.G., Dantzler, K.W., Hickman, D., Jee, J., Kimovec, F.M., Koppstein, D., Marks, D.H., Mittermiller, P.A., Núñez, S.J., Santiago, M., Townes, M.A., Vishnevetsky, M., Williams, N.E., Vargas, M.P.N., Boulanger, L.-A., Bascom-Slack, C., Strobel, S.A., 2011. Biodegradation of Polyester Polyurethane by Endophytic Fungi. Appl. Environ. Microbiol. 77, 6076–6084. https://doi.org/10.1128/AEM.00521-11.

Sankhla, I.S., Sharma, G., Tak, A., 2020. Fungal degradation of bioplastics: An overview, in: New and Future Developments in Microbial Biotechnology and Bioengineering. Elsevier, pp. 35–47. https://doi.org/10.1016/B978-0-12-821007-9.00004-8.

Sanniyasi, E., Gopal, R.K., Gunasekar, D.K., Raj, P.P., 2021. Biodegradation of low-density polyethylene (LDPE) sheet by microalga, Uronema africanum Borge. Sci. Rep. 11, 17233. https://doi.org/10.1038/s41598-021-96315-6.

Satlewal, A., Soni, R., Zaidi, M., Shouche, Y., Goel, R., 2008. Comparative biodegradation of HDPE and LDPE using an indigenously developed microbial consortium. J. Microbiol. Biotechnol. 18, 477–482.

Shabbir, S., Faheem, M., Ali, N., Kerr, P.G., Wang, L.-F., Kuppusamy, S., Li, Y., 2020. Periphytic biofilm: An innovative approach for biodegradation of microplastics. Sci. Total Environ. 717, 137064. https://doi.org/10.1016/j.scitotenv.2020.137064.

Shah, Z., Krumholz, L., Aktas, D.F., Hasan, F., Khattak, M., Shah, A.A., 2013. Degradation of polyester polyurethane by a newly isolated soil bacterium, Bacillus subtilis strain MZA-75. Biodegradation 24, 865–877. https://doi.org/10.1007/s10532-013-9634-5.

Shahnawaz, M., Sangale, M.K., Ade, A.B., 2019. Bioremediation Technology for Plastic Waste. Springer Singapore, Singapore. https://doi.org/10.1007/978-981-13-7492-0.

Shao, H., Chen, M., Fei, X., Zhang, R., Zhong, Y., Ni, W., Tao, X., He, X., Zhang, E., Yong, B., Tan, X., 2019. Complete genome sequence and characterization of a polyethylene biodegradation strain, streptomyces Albogriseolus LBX-2. Microorganisms 7, 379. https://doi.org/10.3390/microorganisms7100379.

Sharma, B., Dangi, A.K., Shukla, P., 2018. Contemporary enzyme based technologies for bioremediation: A review. J. Environ. Manage. 210, 10–22. https://doi.org/10.1016/j.jenvman.2017.12.075.

Shimpi, N., Borane, M., Mishra, S., Kadam, M., 2012. Biodegradation of polystyrene (PS)-poly(lactic acid) (PLA) nanocomposites using Pseudomonas aeruginosa. Macromol. Res. 20, 181–187. https://doi.org/10.1007/s13233-012-0026-1.

Shirke, A.N., White, C., Englaender, J.A., Zwarycz, A., Butterfoss, G.L., Linhardt, R.J., Gross, R.A., 2018. Stabilizing leaf and branch compost Cutinase (LCC) with glycosylation: Mechanism and effect on PET hydrolysis. Biochemistry 57, 1190–1200. https://doi.org/10.1021/acs.biochem.7b01189.

Sivan, A., 2011. New perspectives in plastic biodegradation. Curr. Opin. Biotechnol. 22, 422–426. https://doi.org/10.1016/j.copbio.2011.01.013.

Sol, D., Laca, A., Laca, A., Díaz, M., 2020. Approaching the environmental problem of microplastics: Importance of WWTP treatments. Sci. Total Environ. 740, 140016. https://doi.org/10.1016/j.scitotenv.2020.140016.

Sudhakar, M., Doble, M., Murthy, P.S., Venkatesan, R., 2008. Marine microbe-mediated biodegradation of low- and high-density polyethylenes. Int. Biodeterior. Biodegrad. 61, 203–213. https://doi.org/10.1016/j.ibiod.2007.07.011.

Taniguchi, I., Yoshida, S., Hiraga, K., Miyamoto, K., Kimura, Y., Oda, K., 2019. Biodegradation of PET: Current status and application aspects. ACS Catal. 9, 4089–4105. https://doi.org/10.1021/acscatal.8b05171.

Teeraphatpornchai, T., Nakajima-Kambe, T., Shigeno-Akutsu, Y., Nakayama, M., Nomura, N., Nakahara, T., Uchiyama, H., 2003. Isolation and characterization of a bacterium that degrades various polyester-based biodegradable plastics. Biotechnol. Lett. 25, 23–28. https://doi.org/10.1023/A:1021713711160.

Tsiota, P., Karkanorachaki, K., Syranidou, E., Franchini, M., Kalogerakis, N., 2018. Microbial degradation of HDPE secondary microplastics: Preliminary results, in: Cocca, M., Di Pace, E., Errico, M.E., Gentile, G., Montarsolo, A., Mossotti, R. (Eds.), Proceedings of the International Conference on Microplastic Pollution in the Mediterranean Sea, Springer Water. Springer International Publishing, Cham, pp. 181–188. https://doi.org/10.1007/978-3-319-71279-6_24.

Uscátegui, Y.L., Arévalo, F.R., Díaz, L.E., Cobo, M.I., Valero, M.F., 2016. Microbial degradation, cytotoxicity and antibacterial activity of polyurethanes based on modified castor oil and polycaprolactone. J. Biomater. Sci. Polym. Ed. 27, 1860–1879. https://doi.org/10.1080/09205063.2016.1239948.

Volke-Sepúlveda, T., Saucedo-Castañeda, G., Gutiérrez-Rojas, M., Manzur, A., Favela-Torres, E., 2002. Thermally treated low density polyethylene biodegradation by Penicillium pinophilum and Aspergillus niger: Low Density Polyethylene Biodegradation. J. Appl. Polym. Sci. 83, 305–314. https://doi.org/10.1002/app.2245.

Wagner, M., Lambert, S. (Eds.), 2018. Freshwater Microplastics: Emerging Environmental Contaminants? The Handbook of Environmental Chemistry. Springer International Publishing, Cham. https://doi.org/10.1007/978-3-319-61615-5.

Wei, R., Zimmermann, W., 2017. Microbial enzymes for the recycling of recalcitrant petroleum-based plastics: How far are we? Microb. Biotechnol. 10, 1308–1322. https://doi.org/10.1111/1751-7915.12710.

Welch, M.J., Hawker, C.J., Wooley, K.L., 2009. The advantages of nanoparticles for PET. J. Nucl. Med. 50, 1743–1746. https://doi.org/10.2967/jnumed.109.061846.

Wierckx, N., Narancic, T., Eberlein, C., Wei, R., Drzyzga, O., Magnin, A., Ballerstedt, H., Kenny, S.T., Pollet, E., Avérous, L., O'Connor, K.E., Zimmermann, W., Heipieper, H.J., Prieto, A., Jiménez, J., Blank, L.M., 2019. Plastic biodegradation: Challenges and opportunities, in: Steffan, R.J. (Ed.), Consequences of Microbial Interactions with Hydrocarbons, Oils, and Lipids: Biodegradation and Bioremediation. Springer International Publishing, Cham, pp. 333–361. https://doi.org/10.1007/978-3-319-50433-9_23.

Yamada-Onodera, K., Mukumoto, H., Katsuyaya, Y., Saiganji, A., Tani, Y., 2001. Degradation of polyethylene by a fungus, Penicillium simplicissimum YK. Polym. Degrad. Stab. 72, 323–327. https://doi.org/10.1016/S0141-3910(01)00027-1.

Yan, N., Fan, C., Chen, Y., Hu, Z., 2016. The potential for microalgae as bioreactors to produce pharmaceuticals. Int. J. Mol. Sci. 17, 962. https://doi.org/10.3390/ijms17060962.

Yang, Y., Yang, J., Wu, W.-M., Zhao, J., Song, Y., Gao, L., Yang, R., Jiang, L., 2015. Biodegradation and mineralization of polystyrene by plastic-eating mealworms: Part 2. Role of gut microorganisms. Environ. Sci. Technol. 49, 12087–12093. https://doi.org/10.1021/acs.est.5b02663.

Yoshida, S., Hiraga, K., Takehana, T., Taniguchi, I., Yamaji, H., Maeda, Y., Toyohara, K., Miyamoto, K., Kimura, Y., Oda, K., 2016. A bacterium that degrades and assimilates poly(ethylene terephthalate). Science 351, 1196–1199. https://doi.org/10.1126/science.aad6359.

Yuan, J., Ma, J., Sun, Y., Zhou, T., Zhao, Y., Yu, F., 2020. Microbial degradation and other environmental aspects of microplastics/plastics. Sci. Total Environ. 715, 136968. https://doi.org/10.1016/j.scitotenv.2020.136968.

Zettler, E.R., Mincer, T.J., Amaral-Zettler, L.A., 2013. Life in the "plastisphere": Microbial communities on plastic marine debris. Environ. Sci. Technol. 47, 7137–7146. https://doi.org/10.1021/es401288x.

9 Scenario of Microplastics Waste in the Ecosystem

Strategies Toward Monitoring and Management

Sri Bharti, V.P. Sharma, Rachana Kumar and Papita Das

9.1 INTRODUCTION

Microplastics (MPs) are microscopic plastic particles that are less than 5mm in size. They are a growing concern in the environment because they do not biodegrade and may persist in the ecosystem for hundreds of years. Microplastics can be produced by a number of processes, including the breakdown of larger plastic items, synthetic fibers in clothing, and even personal care products like facial scrubs and toothpaste. The impact of microplastics on the ecosystem is not fully understood, but research has shown that they can have severe adverse effects on the environment. Microplastics can be ingested by marine life, such as fish and shellfish, and can accumulate in their tissues over time. This can lead to various health problems, such as inflammation, reduced reproduction, and even death (1, 2).

Microplastics can also have a negative impact on the food chain. When smaller organisms consume microplastics, they can be passed on to larger organisms that feed on them, eventually making their way up the food chain to humans. This could have potential health implications for humans who consume seafood. The presence of microplastics in the environment has also been linked to the release of harmful chemicals. Plastic particles can absorb and accumulate chemicals like pesticides and persistent organic pollutants (POPs), which can then be released into the environment when they are ingested by marine life. The disposal of microplastics is a growing concern, and it is important to take steps to reduce their release into the environment (3).

The use of disposable plastics should be reduced, waste management and recycling practices should be improved, and plastic replacements should be investigated as potential solutions. A significant environmental problem is the presence of microplastics in the environment. Although more research is needed to fully understand the consequences of microplastics on the environment, it is clear that they have a negative impact on ecosystems and human health (4, 5).

DOI: 10.1201/9781003449133-9

9.1.1   ENVIRONMENTAL IMPACTS

Microplastics have been found to have significant environmental impacts, including harming marine life and contaminating the food chain. Microplastics can also accumulate toxins, which can be harmful to the health of both animals and humans (6).

9.1.2   HUMAN HEALTH

There is growing concern about the prospective health impacts of microplastics on the health of living beings. According to research, microplastics can enter the human body through ingestion, inhalation, or through the skin. While the long-term health effects are not fully understood, there is evidence to suggest that microplastics can cause inflammation, oxidative stress, and even disrupt hormone systems in the body (7).

9.1.3   MITIGATION

Efforts to mitigate the impacts of microplastics are ongoing, including reducing plastic waste, improving waste management systems, and implementing regulations to ban or limit the use of microplastics in various products. Consumers can also play a role in reducing microplastics by making sustainable choices, such as using reusable products and avoiding products containing microbeads (8).

9.1.4   NATIONAL AND INTERNATIONAL SCENARIO

India is one of the largest consumers of plastics in the world, and as a result, microplastic pollution has become a significant environmental concern in the country.

9.1.5   REGULATION AND POLICIES

In order to minimize the usage of plastic, governments of varied countries have initiated efforts through several initiatives—viz. ban on single use plastics such as bags, straws, cutlery, etc. The policy intends to encourage the use of degradable substitutes to reduce marine litter by 30% in coming decade. However, there have been a number of obstacles to the successful implementation and enforcement of this ban (9).

9.1.6   RESEARCH AND MONITORING

In India, numerous research projects have been conducted to gauge the extent of the country's microplastic contamination. In recent studies, the contamination of commercial salt with microplastics has emerged as a significant concern, posing challenges for both sampling and quantification. Additionally, an R&D study conducted by IIT Mumbai on the contamination of Indian sea salts with microplastics, along with a potential prevention strategy, has sparked increased interest and curiosity within the scientific community. Additionally, the Central Pollution Control Board (CPCB) has been instructed by the National Green Tribunal (NGT), an environmental judicial body, to begin a thorough investigation into the prevalence of microplastics in rivers and lakes across India (3, 10).

9.2 PUBLIC AWARENESS AND EDUCATION

There is growing awareness among the Indian public about the impact of plastic pollution on the environment, including the issue of microplastics. Several initiatives and campaigns have been launched by the government, NGOs, and private organizations to raise awareness about the issue and promote sustainable alternatives to single-use plastics (11).

Many countries have recognized the need to address the issue of microplastics in their respective territories. Some countries have taken steps to ban microbeads in personal care products, which are a major source of microplastics in waterways. In the United Kingdom, a ban on microbeads in rinse-off personal care products came into effect in 2018. However, there is still a lack of comprehensive legislation regulating the use and disposal of microplastics in many countries. In some cases, microplastics are not even classified as a hazardous waste (12, 13).

In 2021–2030, the decade of Ocean Science for Sustainable Development aims to promote ocean science-related research, innovation, and capacity building. Furthermore, the European Union (EU) has put into effect laws aimed at reducing the amount of microplastics in particular items and enhancing waste-management procedures (14). The EU Single-Use Plastics Directive, which took effect in 2021, forbids the use of some single-use plastic products while requiring EU Member States to reduce the usage of others. Even if there has been some improvement in the fight against microplastics, much more has to be done on a national and worldwide basis. Governments, businesses, and consumers must work together to reduce the production and use of plastic goods and to manage and dispose of plastic waste in an efficient manner (15).

9.3 MICROPLASTICS IN DIFFERENT MATRICES

Milk: Microplastics have been found in milk from various animal species, including cows, goats, and humans. The sources of microplastics in milk are believed to be from plastic packaging, processing equipment, and feed. Studies have shown that microplastics can enter the milk during various stages of milk production, including milking, transportation, and processing. Levels of microplastics in milk are generally low; the long-term health effects of consuming microplastics are not fully understood. Studies have suggested that microplastics may accumulate in the human body and potentially cause inflammation and other health problems. Efforts are underway to reduce the presence of microplastics in milk, including improving waste management systems and implementing regulations to limit the use of plastic materials in animal feed and dairy-processing equipment (16).

Soil: Microplastics have been found in soil in various parts of the world, with sources including synthetic textiles, car tires, and plastic debris. Microplastics can enter soil through various pathways, including atmospheric deposition, sewage sludge application, and plastic mulch usage in agriculture. Studies have shown that microplastics can have negative impacts on soil organisms and soil health. They can physically damage soil organisms, reduce soil microbial activity, and alter nutrient cycling. The concentration of microplastics in soil can vary depending on the location, with urban areas generally having higher levels of microplastics due to the higher density of human activities and associated plastic waste. Efforts are underway to reduce the presence of microplastics in soil, including improving waste management practices, reducing plastic consumption, and developing sustainable agricultural practices that minimize plastic usage (17).

Blood: Recent research indicates that microplastics can enter the human body through ingestion or inhalation and are detectable in human blood. The sources of microplastics in the human body are believed to be from food, water, and air. A study published in *Environment International* discovered and measured plastic particle pollution in human blood, providing evidence that exposure to plastic particles can be absorbed into the bloodstream of individuals. This suggests that some of the plastic particles we come into contact with can enter our body and that the organs responsible for eliminating these particles, such as the liver, kidney, or biliary tract, are slower than the rate at which the particles are absorbed into the blood. Additionally, it reveals that some plastic particles can transfer to and accumulate in organs. It is yet unknown whether plastic particles are carried by certain cell types or are present in the plasma (and how much these cells may be responsible for transporting plastic particles from the mucosa to the bloodstream) (18).

The discovery of microplastics (MPs) in placenta tissue calls for a reassessment of our understanding of the immunological mechanisms that allow self-tolerance. MPs can enter the bloodstream and reach the placenta from the respiratory system or gastrointestinal tract (GIT) via M cells-mediated endocytosis or paracellular transport. The most likely transport mechanism for MPs involves particle uptake and translocation, with subsequent translocation to other organs depending on factors such as hydrophobicity, surface charge, surface fictionalization, protein corona, and particle size. Once MPs reach the maternal surface of the placenta, they can penetrate deep into the tissue through both active and passive transport mechanisms (19). Efforts are underway to better understand the sources and impacts of microplastics in the human body and develop strategies to reduce their presence. This includes improving waste-management practices, reducing plastic consumption, and implementing regulations to limit the use of microplastics in products such as personal care items and cleaning products.

Water: Microplastics are a pollutant in water bodies, including oceans, rivers, lakes, and even groundwater. The sources of microplastics in water include plastic waste from various human activities, such as packaging, textiles, and personal care products as well as microplastic fibers shed from synthetic clothing during laundering. Microplastics have also been found in drinking water sources, including tap water (49.67 ± 17.49 items/L) and bottled water (72.32 ± 44.64 items/L), although the levels are generally low (20). Studies have shown that microplastics can have significant negative impacts on aquatic organisms, including fish, turtles, and birds, as well as on the overall health of marine ecosystems. Microplastics can physically harm aquatic organisms, cause inflammation, and disrupt feeding and reproduction (21).

Furthermore, microplastics can also be ingested by humans through consumption of contaminated seafood and drinking water. A recent study published in the journal *Food Control* examined 160 different fish species obtained from coastal waters in Nigeria to determine the presence of microplastics in their gastrointestinal tracts. The study also estimated the amount of microplastics that adults in the region may be consuming annually. The researchers found a total of 5,744 microplastics across all the fish species analyzed, with an average of 39.65 ± 5.67 items per individual fish. The most common type of microplastic found in the fish was microbeads, which were present in all the guts examined, followed by fragments, burnt film, thread, fibers, and pellets. The majority of the microplastics found were smaller than 1,000 micrometers, with the smallest being 85 micrometers. Overall, the study highlights the concerning prevalence of microplastics in fish and the potential for human consumption of these harmful particles. Efforts are underway to reduce the presence of microplastics in water, including improving waste-management practices, reducing plastic consumption, and implementing regulations to limit the use of microplastics in various products (22).

9.4 MICROPLASTICS: MODES OF MONITORING AND MANAGEMENT

Microplastics are tiny plastic particles that have become a growing concern due to their widespread presence in the environment and potential impact on human and ecosystem health. Here are some modes of monitoring and management for microplastics:

Monitoring: Understanding the distribution, concentration, and environmental impact of microplastics requires constant monitoring. Animal and plant-based foods, food additives, beverages, and plastic food packaging are just a few of the sources that expose people to microplastics. These particles may build up in living things' cells and tissues, providing dangers for long-term biological consequences and health problems to people. Strict supervision and regulations are required to protect the safety of food and control the use of plastic. It is suggested that study be expanded beyond seafood to take into account the prevalence of microplastics in various kinds of food. The varied forms, shapes, and sizes of plastic particles should also be taken into consideration while examining the toxicity mechanisms and routes in the human body (23).

There are various methods of monitoring microplastics, including surface water and sediment sampling, air sampling, and biological sampling. Researchers use techniques such as microscopy and spectroscopy to identify and quantify microplastics in different environmental compartments (24).

Source reduction: Focusing on addressing their sources and reducing their impact are essential if we are to address the environmental problem posed by microplastics. There are several potential measures to remediate microplastic sources, such as decreasing the production of plastic waste, enhancing waste management systems, implementing regulations on microplastic use in consumer goods, creating public awareness campaigns, and investing in innovative technologies. These actions aim to prevent plastic waste from degrading into microplastics and decrease the introduction of microplastics into the environment. Furthermore, promoting the adoption of biodegradable plastics, developing effective filtration systems for wastewater treatment plants, and exploring other innovative solutions can help mitigate the release of microplastics into the environment. Educating the public about the risks associated with microplastics and providing guidance on proper plastic waste disposal practices can also encourage communities to take responsibility and reduce their impact. Managing microplastics primarily involves minimizing their release into the environment, which can be achieved through strategies such as reducing plastic consumption, promoting recycling and circular economy practices, and combating plastic littering and dumping (24, 25).

Wastewater treatment: In wastewater treatment, a number of physiochemical techniques are used to remove microplastics from the environment.

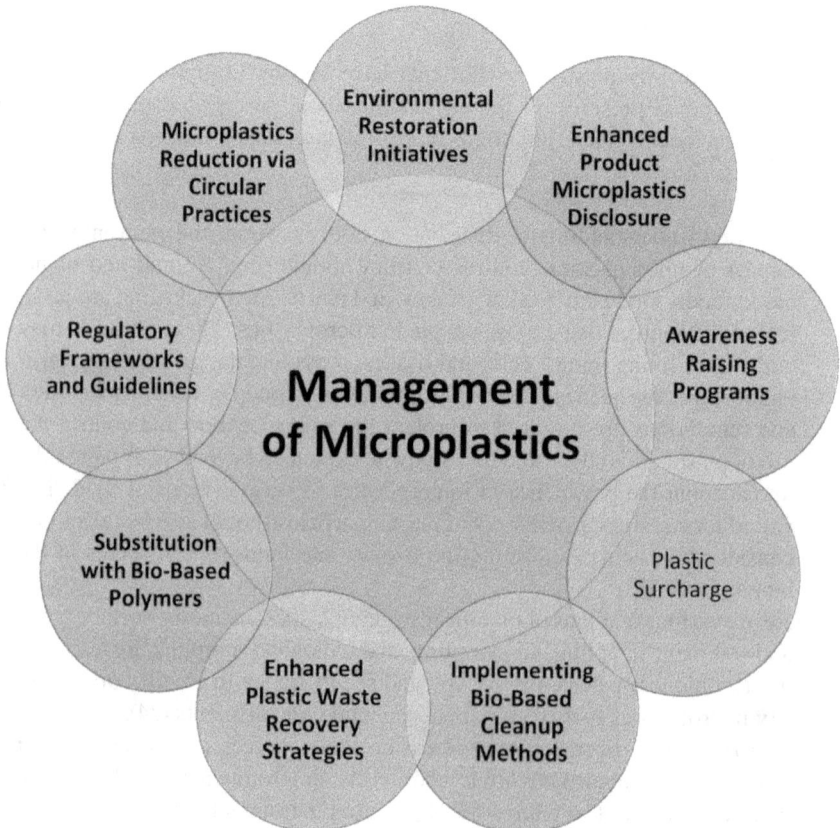

FIGURE 9.1 Management of microplastics.

These techniques use chemical and physical methods to lessen the amount of microplastics in influent water that enters treatment facilities. Although they still contribute to their accumulation, methods like filtering and membrane bioreactors are used to reduce the amount of microplastics in the environment. Other techniques for separating microplastics include electro-coagulation and agglomeration, however these often call for extra filtration steps. Techniques like FTIR and electron microscopy are frequently used to study the breakdown of microplastics because they shed light on potential structural alterations (26).

Biodegradable plastics (BDPs): There is growing interest in developing bio-degradable plastics as an alternative to conventional plastics that persist in the environment. Biodegradable plastics are designed to break down into natural substances in the environment, reducing the accumulation of microplastics. However, the application of has brought some potential problems to light. BDPs require specific conditions for biodegradation, which are difficult to achieve in natural environments. Without these conditions, BDPs can have the same longevity as conventional plastics (27). BDPs may can create biodegradable microplastics (BMPs) that can harm the environment. The impacts of BDPs on the environment and ecology, including the adsorption and release of harmful compounds, their function as vectors for microbes, epiphytes, and plants, and ecotoxicology, still need to be thoroughly investigated despite the significant study on BDP degradation and application.

Cleanup efforts: The unsustainable utilization and disposal of plastic products pose a significant danger to economies, ecosystems, and human well-being due to the generation of plastic pollution. Despite attempts at clean-up strategies, the sheer volume of plastic entering the environment continues to overwhelm these efforts. Hence, it is crucial to prioritize the reduction of plastic inputs into the environment through a comprehensive global approach involving various disciplines. Poorly managed waste is a major contributor to land-based plastic pollution, and this can be mitigated by improving the life cycle of plastics, particularly in production, consumption, and disposal, through the implementation of an Integrated Waste Management System. There are ten recommendations for stakeholders to decrease plastic pollution, which include regulating the production and consumption of plastics, promoting eco-design, increasing the demand for recycled plastics, minimizing plastic usage, utilizing renewable energy for recycling processes, implementing extended producer responsibility for waste management, enhancing waste collection systems, prioritizing recycling efforts, utilizing bio-based and biodegradable plastics, and improving the recyclability of electronic waste (e-waste) (28).

Overall, managing microplastics requires a multifaceted approach that involves reducing their release into the environment, removing them from wastewater, promoting the use of biodegradable plastics, and cleaning up existing pollution.

9.5 STEPS THAT CAN BE TAKEN TO MONITOR AND MANAGE MICROPLASTICS WASTE

There are several steps to improve waste collection systems to prevent plastic waste from entering natural ecosystems. This can involve educating the general people about the significance of appropriate waste management and infrastructure investment.

1. *Monitor sources of microplastics:* Identifying the sources of microplastics can help in reducing their production and entry into the environment. Sources of microplastics include synthetic fibers from textiles, microbeads from personal care products, and fragmented plastics from larger plastic items.
2. *Develop technologies to remove microplastics from the environment:* Scientists and engineers are developing various technologies to remove microplastics from the environment. These include filters and skimmers for water bodies and vacuum or magnetic techniques for removing microplastics from soil.
3. *Educate the public:* The usage of microplastics can be decreased and alternative solutions can be promoted by teaching the general population about their harmful effects on the surroundings and human wellness.
4. *Reduce plastic use:* Reducing plastic use can help in reducing the amount of plastic waste that enters the environment. This can be done by promoting the use of reusable bags, bottles, and containers and by avoiding single-use plastics.
5. *Encourage recycling:* Recycling plastics can help in reducing the amount of plastic waste that enters the environment. It is essential to remember that recycling alone will not suffice and that there must be a decrease in the amount of plastic produced in the initial phase.
6. *Support policies to reduce microplastics waste:* Governments can play a crucial role in reducing microplastics waste by implementing policies to ban microbeads in personal care products, promoting the use of biodegradable plastics, and imposing taxes on single-use plastics.
7. *Foster research:* In order to fully understand the causes and effects of microplastics in the environment, more research is required. This will help in developing more effective strategies for monitoring and managing microplastics waste.

FIGURE 9.2 Source pathways and receptors model.

9.6 MICROPLASTICS: CHALLENGES IN MONITORING

There are a number of difficulties involved with gauging microplastics, despite the fact that it is essential to do so in order to comprehend their dispersion and effects on the surroundings and human well-being. Here are some of the main challenges in monitoring microplastics:

Sampling: Collecting representative samples can be challenging, as microplastics are distributed heterogeneously in the environment and can easily be lost or degraded during collection, handling, and processing. Sampling methods and equipment must be carefully selected to ensure accurate and reliable data.

Detection: Microplastics are small and transparent, making them difficult to detect using traditional laboratory methods. Specialized techniques such as microscopy, spectroscopy, and chemical analysis are required for their identification and quantification.

Standardization: There is the absence of standardized microplastic monitoring techniques, which makes it difficult to compare data across studies and regions. Establishing standardized methods for sampling, processing, and analysis is critical for producing reliable and comparable data.

Identification: Different types of microplastics have different chemical and physical properties, which can affect their toxicity and behavior in the environment. Identifying and characterizing the types of microplastics present is important for understanding their impacts.

Cost: Microplastic monitoring requires specialized equipment and expertise, which can be expensive. The cost of monitoring can be a barrier for smaller organizations and countries with limited resources.

Overall, while monitoring microplastics is essential, addressing the challenges associated with sampling, detection, standardization, identification, and cost is critical to ensure accurate and reliable data that can inform effective management strategies.

9.7 MANAGEMENT PRACTICES FOR REMEDIATION OF MICROPLASTICS

Remediation of microplastics is a complex issue that requires the adoption of effective management practices. Here are some management practices that can be used for the remediation of microplastics:

Source Reduction: Preventing microplastics from entering the ecosystem in the initial instance is the best strategy to limit the quantity of the materials there. This can be achieved through source reduction practices, such as reducing the use of single-use plastic items, promoting recycling, and encouraging the use of eco-friendly materials.

Cleanup and Removal: Another management practice is the cleanup and removal of microplastics from the environment. This can be done through various methods, such as beach cleanups, sediment dredging, and the use of specialized microplastic removal equipment.

Monitoring and Assessment: Regular monitoring and assessment of microplastic pollution levels in various environments can help identify areas that require remediation. This can be done through water and sediment sampling as well as visual surveys.

Education and Awareness: The number of microplastics in the atmosphere can be decreased through educating the public regarding the harm that microplastics do to the surroundings and by raising awareness. This can be done through campaigns, educational programs, and workshops.

Policy and Regulation: Finally, policy and regulation can play a critical role in managing microplastic pollution. Governments can impose bans on single-use plastics, regulate the use of microplastic-containing products, and enforce penalties for polluters.

9.8 BIOACCUMULATION OF MICROPLASTICS

Bioaccumulation of microplastics involves the gradual buildup of tiny plastic particles in living organisms' tissues over time. This process can occur in various ecosystems, including freshwater, marine, and terrestrial environments and can have detrimental effects on both organism health and ecosystem functioning. The disintegration of bigger plastic objects, microbeads in personal care products, and the deterioration of synthetic textiles are some of the causes of microplastics, which are plastic particles smaller than 5 millimeters. Microplastics can be consumed by creatures after being discharged into the atmosphere. They also build up in their tissues. This bioaccumulation can lead to negative health effects in organisms, such as disruption of their endocrine systems, inflammation, and reduced reproductive success. Furthermore, the transfer of microplastics up the food chain can impact the health of other organisms, and ultimately, affect the functioning of ecosystems. Mitigating the bioaccumulation of microplastics involves reducing their release into the environment through source reduction measures, such as minimizing the use of single-use plastics and promoting the adoption of eco-friendly materials. Additionally, effective management practices, including regular monitoring and assessment of microplastic pollution levels, as well as the cleanup and removal of microplastics from the environment, can help alleviate the adverse impacts of microplastics on organisms and ecosystems (29, 30).

Size and Shape of Microplastics: Smaller microplastics (<100 μm) are more likely to be ingested by organisms and can accumulate in their tissues more easily. Additionally, the shape of microplastics can also impact their bioaccumulation potential.

Habitat and Diet of Organisms: The habitat and diet of an organism can also impact its exposure to microplastics. For example, organisms that live in areas with high levels of microplastic pollution, such as coastal regions, are more likely to ingest microplastics (31).

Hydrophobicity: Microplastics are often hydrophobic, which means that they do not dissolve in water and are more likely to attach to organic matter in the environment, such as sediments, algae, or plankton. This can increase the uptake of microplastics by organisms.

Chemical composition: Microplastics can adsorb and accumulate toxic chemicals, such as persistent organic pollutants (POPs), that are present in the surrounding water. This can increase the toxicity of microplastics to organisms and contribute to their bioaccumulation potential.

Tropic Level: Organisms higher up in the food chain, such as top predators, are more likely to accumulate higher levels of microplastics due to biomagnifications.

9.9 MICROPLASTICS: ADVERSE IMPLICATION FOR THE REPRODUCTIVE SYSTEM

A major issue is the harmful effects of microplastics on the reproductive system. Despite the fact that this field of study is still in its early stages, investigations have shown that microplastics may be hazardous to reproductive health. The ability of microplastics to alter endocrine systems is one potential cause of harm. Chemicals like phthalates and bisphenol A (BPA), which are known to disturb the endocrine system, can be found in or absorbed by microplastics. These substances can affect the development and functionality of reproductive organs, hormonal balance, and the generation of reproductive hormones. They can also disrupt normal reproductive processes and interfere with hormone signalling. Fertility problems, reproductive disorders, and abnormal child development could happen as a result.

The physical presence of microplastics in reproductive tissues is a further cause for concern. Numerous reproductive organs, including the ovaries, testicles, and placenta, have been found to contain microplastics. They may cause physical injury, inflammation, and cellular stress in these delicate tissues, which could impair reproductive function. Microplastics can also cause oxidative stress and start inflammatory processes within the body. Reproductive diseases, such as infertility, miscarriages, and unfavorable pregnancy outcomes, have been linked to these processes. It is important to acknowledge that further research is necessary to fully comprehend the magnitude of the damaging effects that microplastics have on reproductive health. Microplastics' size, shape, composition, and concentration, as well as their exposure time and manner, can all have an impact on their potential impact. By improving trash disposal procedures and promoting sustainable alternatives to disposable plastics, it is critical to reduce the release of microplastics into the environment in order to address possible risks. To further our understanding of the precise processes and long-term effects of microplastics on reproductive health, more research is required. This will make it possible to enact powerful rules and safety measures (32).

9.10 MICROPLASTICS: POTENTIAL HEALTH IMPACT OF CHEMICAL CONTAMINANTS ON CHILDREN

Injurious compounds such heavy metals, persistent organic pollutants (POPs), and endocrine-disrupting chemicals (EDCs) can be absorbed and accumulated by microplastics. Polychlorinated biphenyls (PCBs), phthalates, bisphenol A (BPA), and pesticides are a few examples of these contaminants. These substances have been linked to a number of detrimental health effects, including developmental and reproductive abnormalities, neurotoxicity, and disturbance of hormonal systems.

Children may be more susceptible to the possible health consequences of chemical pollutants from microplastics due to their developing bodies and organ systems. Through ingestion, inhalation, and skin contact, their actions, such as mouthing things and playing in settings with microplastics, may expose them. Preventive actions are advocated until study is conducted in order to comprehend the specific health consequences of microplastics and the compounds they are connected with. These include controlling plastic production, encouraging eco-friendly substitutes, and minimizing the introduction of microplastics into the environment. Further investigation is also required to fully evaluate the degree of exposure, potential toxicity, and long-term health effects of microplastics, especially among vulnerable populations like children.

Though further investigation is required to fully comprehend the effects of microplastics on children's health, it is critical to limit their exposure to chemical pollutants. By reducing the use of products containing microplastics, improving waste management techniques, and reducing the discharge of chemical pollutants into the environment, this can be accomplished. By avoiding plastic food and drink containers, choosing natural personal care items, and maintaining a healthy and balanced diet, people can also lessen their own exposure to these toxins (7).

9.11 BIODEGRADATION OF MICROPLASTICS

Biodegradation is the process by which microorganisms break down and metabolize organic compounds, including plastics, into simpler compounds. While plastics are generally considered to be non-biodegradable, recent studies have shown that some types of microplastics can biodegrade under certain conditions (33).

Microplastics, which are synthetic particles, pose a serious threat to oceans, water ecosystems, and land. According to their biodegradability, they can be categorized into three groups.: low, relatively low, and relatively high. The mechanisms for depolymerization and biodegradation of low biodegradable MPs like PP, PE, PS, and PVC are not well understood, but pyrolysis is commonly used to break them down. The mechanisms responsible for the depolymerization and biodegradation of low-biodegradable microplastics (MPs) such as PP, PE, PS, and PVC are not well understood. Pyrolysis is a commonly employed method to break them down. To preserve enzymes, whole-cell decay is recommended. In the case of highly degradable MPs like PET, surface-modifying enzymes and esterases have been successfully employed. Biodegradable MPs can degrade under certain conditions, and highly crystalline MPs can be broken down through sustained thermal hydrolysis in industrial composting plants. The first step to reducing microplastic pollution is to use natural alternatives and isolate MPs in environmental matrices. The efficacy of biotechnological techniques has to be confirmed, and field tests are needed to understand the rates and mechanisms underlying decomposition for various types of polymers utilising freshly recovered microbes (34). The biodegradation of microplastics can occur through different mechanisms, including enzymatic degradation, photo-degradation, and bio-fragmentation. Enzymatic degradation involves the breakdown of plastics by enzymes secreted by microorganisms, while photo-degradation involves the breakdown of plastics by exposure to sunlight (35). Numerous habitats, including soil, freshwater, and ocean settings, have been investigated for the breakdown of microplastics. In some cases, biodegradation can occur relatively quickly, with some studies reporting up to 90% degradation of microplastics within a few months under certain conditions (36).

Despite the fact that biodegradation appears to hold promise as a means of reducing plastic pollution, it is important to recognize that some polymers cannot biodegrade, and even those that can sometimes do so only partially. Furthermore, the prerequisites for biodegradation might not exist in many habitats, particularly in marine ones where the sluggish disintegration of plastic is caused by elements like low oxygen-levels and colder temperatures. Nanoplastics, which have unknown effects on the surroundings and human well-being, may also be produced through the biological breakdown of microplastics. The consumption of single-use plastics needs to be curbed, and waste management procedures must be improved, as critical first steps in decreasing the environmental impact of plastic trash, even though biodegradation, offers some promise (37).

9.12   MICROPLASTICS IN PERSONAL CARE PRODUCTS

Microplastics are minute plastic fragments that are smaller than 5 millimeters in size, commonly added to personal care products for their abrasive, thickening, and emulsifying properties. However, their presence in these products has raised concerns about environmental impact, particularly in terms of water pollution and the threat posed to marine and freshwater ecosystems. As these products are washed down the drain, microplastics enter water bodies, where they can be ingested by aquatic organisms, causing physical harm and disrupting their biological functions. Furthermore, these particles can carry toxic chemicals that may ultimately harm humans through the food chain. To tackle this issue, several countries, including the United States, Canada, and some European nations, have implemented bans or restrictions on microplastics in personal care products. Additionally, some companies have voluntarily ceased using microplastics and have opted for natural alternatives, like oatmeal and apricot seeds, as abrasives. Consumers also have a role to play by choosing products free from microplastics or opting for environmentally friendly alternatives that minimize pollution. It is crucial to address the environmental consequences associated with microplastics by implementing regulations and promoting sustainable practices. Through our choices as consumers, we can contribute to protecting the environment and our own health (13).

9.13   DETECTION OF MICROPLASTICS: MODERN-DAY TECHNIQUE

There are several modern-day techniques used to detect microplastics in different environments, including water, sediment, and organisms. Some of the commonly used techniques include:

Visual inspection: Microplastics (MPs) require closer examination under a microscope, especially smaller particles, while larger ones can be visually distinguished. Visual identification of MPs is based on consistent color, illumination and cellular functions are not present. Researchers include methods such as the hot-needle test and visual inspection to confirm the presence of plastic material and differentiate it from organic or inorganic particles. Criteria defined by Goswami et al. for visual identification and counting of MPs include the absence of cellular or biological features, presence of unsegmented fibers that do not resemble twisted ribbons, particles with homogeneous colors, and particles that melt when subjected to heat from a hot needle. MPs are categorized based on visual inspection considering their size, color, shape, and origin. Visual sorting is a simple way to count MPs, but it runs the danger of misclassifying non-plastic particles as plastic. Based on the size categories utilised for categorization, this could result in a substantial over- or underestimation of the amount of plastic that is actually available (38).

FTIR and Raman spectroscopy: Microplastics (MPs) polymer types are typically identified using the FTIR and Raman approach, which includes spectroscopy and microscopy. These techniques are the most common method for MPs analysis in any sample. With spectra ranging from 200 to 3500 cm1, Raman spectroscopy has been used to detect MPs in sand and biota throughout the east coast of India. This technique allows for the identification of MPs as small as 1 nm while also providing data on particle number, size distribution, and morphological characteristics. The advantages of Raman spectroscopy over FTIR include better resolution (1 m vs. 20 m), a wider range of spectrum coverage with unique fingerprint spectra, and less interference from water. Surface-enhanced Raman spectroscopy (SERS) is also one of the most used techniques for micro and nanoplastic identification. Raman spectroscopy does have drawbacks, though, in that weak Raman scattering signals require longer acquisition times in order to obtain acceptable signal-to-noise ratios. Weak signals may also have an impact on Raman microscopy, which is used to characterize MPs. By extending the measurement time and taking into account elements like fluorescence interference, color, biofouling, and degradation, this problem can be solved. It is important to keep in mind that the Raman spectrum of worn MPs may alter and that there isn't currently a Raman library accessible that is especially for altered MPs. It is crucial to create a specific spectrum database for weathering plastics and use it to find unidentified MPs in environmental specimens (39).

Pyrolysis-Gas Chromatography-Mass Spectrometry (Py-GC-MS): Py-GC-MS is a technique used to analyze the chemical composition of a sample by heating it to high temperatures, which breaks down the sample into simpler compounds that can be analyzed using gas chromatography and mass spectrometry. The method is used for the quantitative identification of chemical constituents of microplastics in food and other samples. The extracted sample is pyrolyzed at very high temperatures in double-shot pyrolysis GS/MS to remove undesirable co-extracted interfering compounds. This technique is highly useful for identification and quantification of microplastics in complex organic rich samples. Py-GC-MS is a highly effective technique known for its rapid measurements and good repeatability. It has been widely utilized for the identification of various polymers with specific compositions.

Liquid Chromatography techniques (UHPLC and LC-Tandem Mass spectrometry): Liquid chromatography (LC) is widely used analytical technique for analysis of different analytes and is also employed for detection if microplastics in environmental samples. LC techniques are quick with low detection limit and high repeatability. Depolymerization of polymers like polycarbonate (PC), polyethylene terephthalate (PET), and biodegradable polylactic acid, followed by analysis by LC-Tandem Mass spectrometry, is also reported (40).

Microscopic techniques: Microscopic techniques, such as Scanning Electron Microscopy (SEM) and Transmission Electron Microscopy (TEM), are employed for visualizing and identifying microplastics based on their physical characteristics, such as size, shape, and color. The morphology, aging, and genesis of microplastics (MPs) are investigated using a scanning electron microscope (SEM) and energy-dispersive X-ray spectrometer (EDS). This method offers both qualitative data regarding the chemical composition and high-resolution surface data. SEM/EDS has been used in India to characterize MPs that have been extracted from sediment, water, biota, and salt samples. SEM/EDS is a common technique for morphological and elemental characterization, but it is also time- and money-consuming. Selection bias may impact the chemical characterization since the ability of the observer determines how MPs are separated.

Nile Red staining Fluorescence microscopy: The Nile Red (NR) staining approach has demonstrated potential for locating MPs. The NR dye is adsorbed onto plastic surfaces using this technique, causing them to glow when exposed to blue light. Simple photography with an orange filter can be used to capture fluorescence emission, and picture analysis allows for the recognition and quantification of fluorescent particles. In India, MPs in soil and bivalve samples have been identified and measured using the NR staining fluorescence microscope technique. The outcomes are recorded with a research-grade camcorder mounted to a fluorescence microscope after the filter paper has been stained with NR dye at a concentration of 10 mg/ml (in acetone). Using a blue excitation range of 420–495 nm, the MPs are detected. Given that NR is a solvatochromic dye, the polarity of the solvent affects the fluorescence emission, which may allow MPs to be divided into several chemical groups according to their fluorescent shifts.

Atomic force microscope (AFM): The atomic force microscope (AFM) enables imaging at the nanoscale by utilizing specialized probes that can interact with objects in contact or non-contact modes. AFM investigates the patterns of abrasion and weathering observed on microplastics extracted from commercially available sea salts. The ability of microplastics to adsorb substances and the degree of bonding amongst polycyclic aromatic hydrocarbons and microplastics were both examined using AFM. Nonetheless, it is important to note that AFM has limitations. Firstly, it typically requires slow scanning rates to achieve high-quality images. Secondly, there is a possibility of introducing artifacts due to the interactions between the AFM tip and the sample as well as during the image processing stages.

Thermogravimetry (TGA): A thermo-analytical technique known as thermo-gravimetry examines the weight of a specimen as a function of time or temperature in an atmosphere (inert or air) that is predefined, while the temperature is set to range. To investigate the temperature profiles of MPs collected from the Narmada estuary, Sharma et al. used TGA in one investigation. Variations in the MP thermograms derived from estuary sediments suggested the existence of several polymer compositions, including polyamides, polyvinyl chloride, and other polymers. By keeping track of the mass loss of MPs during a heating program, TGA offers quantitative analysis. However, because of the complexity of the sample matrix, this method necessitates time-consuming cleaning and pre-concentration processes before analysis.

9.14 CONCLUSION

In conclusion, the scenario of microplastics waste in the ecosystem presents a significant challenge that requires urgent attention and effective strategies for monitoring and management. Microplastics, with their pervasive presence and detrimental impacts on wildlife and human health, demand comprehensive actions from individuals, industries, and governments alike. To tackle this issue, it is essential to implement robust monitoring systems that can accurately assess the extent of microplastic contamination across various ecosystems. Furthermore, the development and adoption of innovative waste management practices, including the reduction of plastic consumption, improved recycling infrastructure, and the promotion of alternative materials, are crucial steps toward mitigating the microplastics crisis. Additionally, raising awareness and educating the public about the harmful consequences of microplastic pollution can foster behavioral changes and encourage responsible consumption and disposal habits. By combining these strategies, we can strive toward a healthier ecosystem, ensuring the preservation of biodiversity and safeguarding the well-being of present and future generations. It is through collective efforts and a commitment to sustainable practices that we can successfully address the challenges posed by microplastics waste and create a cleaner and more resilient environment for all.

REFERENCES

1. Yaseen, A., Assad, I., Sofi, M. S., Hashmi, M. Z., & Bhat, S. U. (2022). A global review of microplastics in wastewater treatment plants: Understanding their occurrence, fate and impact. *Environmental Research*, 113258. https://doi.org/10.1016/j.envres.2022.113258.
2. Gao, Z., Chen, L., Cizdziel, J., & Huang, Y. (2023). Research progress on microplastics in wastewater treatment plants: A holistic review. *Journal of Environmental Management*, *325*, 116411. https://doi.org/10.1016/j.jenvman.2022.116411.

3. Vaid, M., Mehra, K., & Gupta, A. (2021). Microplastics as contaminants in Indian environment: A review. *Environmental Science and Pollution Research*. https://doi.org/10.1007/s11356-021-16827-6.

4. Mamun, A. A., Prasetya, T. A. E., Dewi, I. R., & Ahmad, M. (2023). Microplastics in human food chains: Food becoming a threat to health safety. *The Science of the Total Environment*, *858*(Pt 1), 159834. https://doi.org/10.1016/j.scitotenv.2022.159834.

5. Shruti, V. C., Kutralam-Muniasamy, G., Pérez-Guevara, F., Roy, P. D., & Elizalde-Martínez, I. (2023). First evidence of microplastic contamination in ready-to-use packaged food ice cubes. *Environmental Pollution*, *318*, 120905. https://doi.org/10.1016/j.envpol.2022.120905.

6. Bhuyan, Md. S. (2022). Effects of microplastics on fish and in human health. *Frontiers in Environmental Science*, *10*. https://doi.org/10.3389/fenvs.2022.827289.

7. Campanale, C., Massarelli, C., Savino, I., Locaputo, V., & Uricchio, V. F. (2020). A Detailed review study on potential effects of microplastics and additives of concern on human health. *International Journal of Environmental Research and Public Health*, *17*(4), 1212. https://doi.org/10.3390/ijerph17041212.

8. Pandey, P., Dhiman, M., Kansal, A., & Subudhi, S. P. (2023). Plastic waste management for sustainable environment: Techniques and approaches. *Waste Disposal & Sustainable Energy*. https://doi.org/10.1007/s42768-023-00134-6.

9. Alpizar, F., Carlsson, F., Lanza, G., Carney, B., Daniels, R. C., Jaime, M., Ho, T., Nie, Z., Salazar, C., Tibesigwa, B., & Wahdera, S. (2020). A framework for selecting and designing policies to reduce marine plastic pollution in developing countries. *Environmental Science & Policy*, *109*, 25–35. https://doi.org/10.1016/j.envsci.2020.04.007.

10. National Green Tribunal Resolved to Address Environmental Disputes in 2021. (2022, January 2). *The Economic Times*. https://economictimes.indiatimes.com/news/india/national-green-tribunal-resolved-to-address-environmental-disputes-in-2021/articleshow/88646205.cms?from=mdr.

11. Union Government is Generating Public Awareness and Undertaking Campaigns on Plastic Waste Management and Elimination of Single-Use Plastics in the Country. (n.d.). Pib.gov.in. Retrieved April 26, 2023, from https://pib.gov.in/PressReleaseIframePage.aspx?PRID=1897852.

12. *Microbead*. (2023, April 21). https://en.wikipedia.org/wiki/Microbead#:~:text=In%20the%20US%2C%20the%20Microbead.

13. Habib, R. Z., Aldhanhani, J. A. K., Ali, A. H., Ghebremedhin, F., Elkashlan, M., Mesfun, M., Kittaneh, W., Al Kindi, R., & Thiemann, T. (2022). Trends of microplastic abundance in personal care products in the united Arab emirates over the period of 3 years (2018–2020). *Environmental Science and Pollution Research*. https://doi.org/10.1007/s11356-022-21773-y.

14. UNESCO. (2019, June 20). *United Nations Decade of Ocean Science for Sustainable Development (2021–2030)*. UNESCO. https://en.unesco.org/ocean-decade.

15. *EU Restrictions on Certain Single-Use Plastics*. (n.d.). Environment.ec.europa.eu. https://environment.ec.europa.eu/topics/plastics/single-use-plastics/eu-restrictions-certain-single-use-plastics_en.

16. Da Costa Filho, P. A., Andrey, D., Eriksen, B., Peixoto, R. P., Carreres, B. M., Ambühl, M. E., Descarrega, J. B., Dubascoux, S., Zbinden, P., Panchaud, A., & Poitevin, E. (2021). Detection and characterization of small-sized microplastics (≥ 5 µm) in milk products. *Scientific Reports*, *11*(1). https://doi.org/10.1038/s41598-021-03458-7.

17. He, D., Luo, Y., Lu, S., Liu, M., Song, Y., & Lei, L. (2018). Microplastics in soils: Analytical methods, pollution characteristics and ecological risks. *TrAC Trends in Analytical Chemistry*, *109*, 163–172. https://doi.org/10.1016/j.trac.2018.10.006.

18. Leslie, H. A., J. M. van Velzen, M., Brandsma, S. H., Vethaak, D., Garcia-Vallejo, J. J., & Lamoree, M. H. (2022). Discovery and quantification of plastic particle

pollution in human blood. *Environment International, 163*(107199), 107199. https://doi. org/10.1016/j.envint.2022.107199.

19. Ragusa, A., Svelato, A., Santacroce, C., Catalano, P., Notarstefano, V., Carnevali, O., Papa, F., Rongioletti, M. C. A., Baiocco, F., Draghi, S., D'Amore, E., Rinaldo, D., Matta, M., & Giorgini, E. (2021). Plasticenta: First evidence of microplastics in human placenta. *Environment International, 146*(106274), 106274. https://doi.org/10.1016/j. envint.2020.106274.

20. Li, H., Zhu, L., Ma, M., Wu, H., An, L., & Yang, Z. (2023). Occurrence of microplastics in commercially sold bottled water. *Science of the Total Environment, 867*, 161553. https://doi.org/10.1016/j.scitotenv.2023.161553.

21. Alfaro-Núñez, A., Astorga, D., Cáceres-Farías, L., Bastidas, L., Soto Villegas, C., Macay, K., & Christensen, J. H. (2021). Microplastic pollution in seawater and marine organisms across the tropical eastern pacific and Galápagos. *Scientific Reports, 11*(1), 6424. https://doi.org/10.1038/s41598-021-85939-3.

22. Li, Q., Ma, C., Zhang, Q., & Shi, H. (2021). Microplastics in shellfish and implications for food safety. *Current Opinion in Food Science, 40*, 192–197. https://doi.org/10.1016/j. cofs.2021.04.017.

23. Mamun, A. A., Prasetya, T. A. E., Dewi, I. R., & Ahmad, M. (2023). Microplastics in human food chains: Food becoming a threat to health safety. *The Science of the Total Environment, 858*(Pt 1), 159834. https://doi.org/10.1016/j.scitotenv.2022.159834.

24. Huang, Z., Hu, B., & Wang, H. (2022). Analytical methods for microplastics in the environment: A review. *Environmental Chemistry Letters*. https://doi.org/10.1007/ s10311-022-01525-7.

25. Osman, A. I., Hosny, M., Eltaweil, A. S., Omar, S., Elgarahy, A. M., Farghali, M., Yap, P.-S., Wu, Y.-S., Nagandran, S., Batumalaie, K., Gopinath, S. C. B., John, O. D., Sekar, M., Saikia, T., Karunanithi, P., Hatta, M. H. M., & Akinyede, K. A. (2023). Microplastic sources, formation, toxicity and remediation: A review. *Environmental Chemistry Letters*. https://doi.org/10.1007/s10311-023-01593-3.

26. Mohammad R, A., & Qusay, A.-A. (2023). Eco-friendly microplastic removal through physical and chemical techniques: A review. *Annals of Advances in Chemistry, 7*(1). https://doi.org/10.29328/journal.aac.1001038.

27. Wang, C., Yu, J., Lu, Y., Hua, D., Wang, X., & Zou, X. (2021). Biodegradable microplastics (BMPs): A new cause for concern? *Environmental Science and Pollution Research, 28*(47), 66511–66518. https://doi.org/10.1007/s11356-021-16435-4.

28. Prata, J. C., Silva, A. L. P., da Costa, J. P., Mouneyrac, C., Walker, T. R., Duarte, A. C., & Rocha-Santos, T. (2019). Solutions and integrated strategies for the control and mitigation of plastic and microplastic pollution. *International Journal of Environmental Research and Public Health, 16*(13), 2411. https://doi.org/10.3390/ijerph16132411.

29. Miller, M. E., Hamann, M., & Kroon, F. J. (2020). Bioaccumulation and biomagnification of microplastics in marine organisms: A review and meta-analysis of current data. *PLoS One, 15*(10), e0240792. https://doi.org/10.1371/journal.pone.0240792.

30. Amelia, T. S. M., Khalik, W. M. A. W. M., Ong, M. C., Shao, Y. T., Pan, H.-J., & Bhubalan, K. (2021). Marine microplastics as vectors of major ocean pollutants and its hazards to the marine ecosystem and humans. *Progress in Earth and Planetary Science, 8*(1). https://doi.org/10.1186/s40645-020-00405-4.

31. Botterell, Z. L. R., Beaumont, N., Dorrington, T., Steinke, M., Thompson, R. C., & Lindeque, P. K. (2019). Bioavailability and effects of microplastics on marine zooplankton: A review. *Environmental Pollution, 245*(245), 98–110. https://doi.org/10.1016/j. envpol.2018.10.065.

32. Dubey, I., Khan, S., & Kushwaha, S. (2022). Developmental and reproductive toxic effects of exposure to microplastics: A review of associated signaling pathways. *Frontiers in Toxicology, 4*, 901798. https://doi.org/10.3389/ftox.2022.901798.

33. Zeenat, Elahi, A., Bukhari, D. A., Shamim, S., & Rehman, A. (2021). Plastics degradation by microbes: A sustainable approach. *Journal of King Saud University—Science*, *33*(6), 101538. https://doi.org/10.1016/j.jksus.2021.101538.
34. Miri, S., Saini, R., Davoodi, S. M., Pulicharla, R., Brar, S. K., & Magdouli, S. (2022). Biodegradation of microplastics: Better late than never. *Chemosphere*, *286*(Pt 1), 131670. https://doi.org/10.1016/j.chemosphere.2021.131670.
35. Kotova, I. B., Taktarova, Yu. V., Tsavkelova, E. A., Egorova, M. A., Bubnov, I. A., Malakhova, D. V., Shirinkina, L. I., Sokolova, T. G., & Bonch-Osmolovskaya, E. A. (2021). Microbial degradation of plastics and approaches to make it more efficient. *Microbiology*, *90*(6), 671–701. https://doi.org/10.1134/s0026261721060084.
36. Folino, A., Karageorgiou, A., Calabrò, P. S., & Komilis, D. (2020). Biodegradation of wasted bioplastics in natural and industrial environments: A review. *Sustainability*, *12*(15), 6030. https://doi.org/10.3390/su12156030.
37. Evode, N., Qamar, S. A., Bilal, M., Barceló, D., & Iqbal, H. M. N. (2021). Plastic waste and its management strategies for environmental sustainability. *Case Studies in Chemical and Environmental Engineering*, *4*(4), 100142. https://doi.org/10.1016/j.cscee.2021.100142.
38. Veerasingam, S., Ranjani, M., Venkatachalapathy, R., Bagaev, A., Mukhanov, V., Litvinyuk, D., Verzhevskaia, L., Guganathan, L., & Vethamony, P. (2020). Microplastics in different environmental compartments in India: Analytical methods, distribution, associated contaminants and research needs. *TrAC Trends in Analytical Chemistry*, *133*, 116071. https://doi.org/10.1016/j.trac.2020.116071.
39. Zhang, J., Peng, M., Lian, E., Xia, L., Asimakopoulos, A. G., Luo, S., & Wang, L. (2023). Identification of Poly(ethylene terephthalate) Nanoplastics in commercially bottled drinking water using surface-enhanced Raman spectroscopy. *Environmental Science & Technology*, *57*(22), 8365–8372.
40. Wang, L., Peng, Y., Xu, Y., Zhang, J., Zhang, T., Yan, M., & Sun, H. (2022). An in situ depolymerization and liquid chromatography–tandem mass spectrometry method for quantifying polylactic acid microplastics in environmental samples. *Environmental Science & Technology*, *56*(18), 13029–13035.

10 Risk and Remediation of Microplastic Pollution in Marine Life

Rajdeep Shaw, Dibyendu Khan, Madhushree Ghorui, MD Nazir, Punarbasu Chaudhuri and Rajib Bandopadhyay

10.1 INTRODUCTION

Plastic is a synthetic polymer comprised of many substances that improve manufacturing and performance at low cost. Due to its chemical resistance properties and availability, their use in various industries is increasing on a regular basis (da Costa et al. 2016). During the World War II, plastic manufacturing gained more popularity than other natural substances like wood, metal, ivory, stone, etc. About 4% of petroleum is used annually in plastic manufacturing, and 4% of petroleum also provides the energy during the plastic manufacturing process. After significant work, the plastic debris ends up in the environment, amassed over a long time-span (Shahul-Hamid et al. 2018). There are several plastic-producing countries, but China takes the first position in plastic manufacturing. Several items of microplastics (MPs) were reported from different regions of China, including Oujiang Estuary, about $680.0 \pm 284.6/m^3$; Jiaojiang Estuary, about $955.6/m^3$; Minjiang Estuary, about $1245.8/m^3$; Bohai Sea, about $0.33/m^3$; Yangtze Estuary, about $4137.3/m^3$; and South China Sea, about $131.5/m^3$ (Zhao et al. 2014, 2015). The abundance of MPs in the aquatic environment documented in Waimushan Beach, Taiwan is about $508/0.0125$ m^3 items (Kunz et al. 2016). MP items from different beach sediment in Iran are reported from: Bostanu, about 1258/kg; Gorsozan, 122/kg; Suru, 14/kg; and Khor-e-Azini, about 2/kg (Naji et al. 2017). MP items reported from India include Bay of Bengal (about a $20,000/km^2$ concentration from surface water) and Vembanad Lake (about a $252.80/m^2$ concentration from sediment) (Eriksen et al. 2018; Sruthy and Ramasamy 2017). MPs reported from USA beach sediment include: Cape Hatteras National Seashore, about 123/kg; Fort Pulaski National Monument, about 306/kg; Gulf Islands National Seashore, about 253/kg of items (Yu et al. 2018). MP concentration of about 0.0032 to 1.18 items/m³, mainly rendered by polyethylene and polypropylene, were reported from Ross Sea, Antarctica (Cincinelli et al. 2017). MPs were reported from an ice core in the Arctic Ocean of about 38–234 items/m³, composed of polypropylene, polystyrene, polyethylene, etc., which are mostly used for domestic and industrial purposes (Obbard et al. 2014). MP size ranges from 1 nm-20 cm. Accumulation of MPs in beach sediment is responsible for the enlargement of sand

DOI: 10.1201/9781003449133-10

particles, which, in turn, increase water permeability and decrease heat conductivity. Several pathogenic microorganisms are able to colonize plastic particles of beach sediment. MPs also serve as an egg-laying base of many insects or seasnails—e.g., *Homalopoma micans*, laying eggs five–48 in number per plastic pellet. Those floating plastic pellets are responsible for species invasiveness in the marine ecosystem. Inhalation of MPs by marine organisms causes indigestion and accumulation, then transfer from lower tropic levels to the higher tropic levels causes the deregulation of bodies' normal functions, including false satiation, reproductive complications, pathological stress, liver inflammation, oxidative stress, reduced growth rate, disturbed immune system, and granulocytoma formation in the marine as well as terrestrial biota (Shahul-Hamid et al. 2018). Accumulation of microplastic monomers in the liver cells promotes the virus attack (Hepatitis C virus and others) that is responsible for liver inflammation and hepatocellular carcinoma development in immune-compromised individuals (Shaw et al. 2023). Thus, MP monomers serve as a micro-toxin for living beings.

10.2 MICROPLASTICS DERIVATION IN THE AQUATIC ECOSYSTEM

MPs in the marine environment draw much attention by researchers due to endless use of this pollutant and its environmental hazards (Barboza and Gimenez 2015). MPs originates in the marine environment by various routes, like direct running off into the ocean or fragments of mesoplastic or microplastic, degraded into small fragments. Both land-based and sea-based sources are responsible for plastic debris accumulation in the ocean. In land systems, municipal drainage and sewage water are directly mixed with the river. Due to unsuitable management, plastic particles and other apparatuses directly fall into ocean water. About 80% of MPs are directly mixed into seawater from the terrestrial environment (Yang et al. 2021). According to International Union for Conservation of research (IUCN), synthetic textiles are the major source of MP contamination in the ocean world. Plastic debris includes large- and small-fragment particles that are manmade for industrial and consumer uses (Shaw and Day 1994). Exfoliants in the beauty products are made up of micron-sized plastic particles. Liquid hand cleanser and facial cleanser contain polyethylene MPs (Andrady 2011; Cesa et al. 2017). About 93% of polyethylene is used in cosmetics, along with nylon, polymethyl methacrylate, and polypropylene (Pereao et al. 2020). Thus, when these products are used, the plastic particles wash off and run off into the ocean (Fendall and Sewell 2009). The sea-based sources of MPs include fishing, shipping, etc. Due to leakage in shipping, a large amount of plastic material is mixed with water. Fishing equipment, if it falls into the sea or is damaged, can generate a very high quantity of plastic fragments (Yang et al. 2021). From pharmaceutical and medical industries, small plastic particles are transported in sewage due to their use in drug delivery. Weathering of large plastic particles in beach environments causes an accumulation on beach surfaces, the surface of ocean water, and in the deep ocean; for example, plastic particles in the sandy soil become dark-pigmented and more heated (~ 40°C) by the absorption of solar infrared radiation. In the carpeting and garment industries, small plastic particles are also mixed eventually into aquatic habitats (Pereao et al. 2020). The polar regions are geographically isolated from other countries; thus, it was thought that this region is free from MP contamination. However, a citizen science project has documented MPs from ocean sediment and surface water. The MPs' source in the Antarctic water includes wastewater released from a research station, several tourist vessels, fishing, etc.

The primary or secondary MP particles run off into the ocean and accumulate in the ocean sediment. MP fibers are mistakenly eaten by marine organisms and accumulate in the body, these organisms are then consumed by large marine organisms, and thus, the MPs particles enter into the food chain, which destroys the immune system, metabolic processes, growth, and reproduction and makes a depraved impression on the marine environment.

In Antarctica, there are 71 research stations. Although most of the research stations have no wastewater treatment system, a few research stations have tertiary treatment systems for wastewater but are unable to remove all MP contaminants (Waller et al. 2017). MPs are distributed around every habitat through surface circulation and wind. The vertical flow of plastic debris by high wind speed traps plastic particles and distributes them along the water column. Nowadays, most of the research is focused on the MP accumulation in ocean gyres. Recent studies have estimated that the total number of MPs in sea water is about 3.3×10^9 pieces (Ikenoue et al. 2023). The concentration of MPs differs, depending on several parameters, including sea water; in the Chukchi Sea, the concentration is 0.009 pieces m^{-3}; 0.04 pieces m^{-3} in the Beaufort Sea; 0.02 pieces m^{-3} in the Canadian Arctic Archipelago; 0.11 pieces m^{-3} in the Greenland Sea; 0.46 pieces m^{-3} in the Barents Sea; and 0.15 pieces m^{-3} in the Kara Sea (Mu et al. 2019). It is documented that plastic debris gathered in the northeast Atlantic Ocean was deliberately estimated to be 2.46 units m^{-3} (Lusher et al. 2014). About 1.8×10^{10} pieces y^{-1} of MPs flow from the Pacific Ocean to the Arctic Ocean. MP is classified into primary MPs and secondary MPs on the basis of their origin. The primary MPs are originally based on plastic microbeads and powder of about < 5 mm in size. The term 'microbeads' is used in industries to define the MP pieces present in cosmetics and other personal care products. Secondary MPs are large, fragile particles, broken into small pieces through biodegradation by microbes, photodegradation by ultraviolet light, and mechanical abrasion by weave action. The physical nature—i.e., size, texture, color, shape, etc.—and chemical composition of secondary plastic can change at the time of their manufacture and after their use (Hale et al. 2020). During the production of MPs, several stabilizers are added that contain some toxic metals such as Pb, Sn, and Cd, etc. (Amelia et al. 2021). The variation of the surface charge of MPs leads to heavy metal adsorption and their transport and accumulation in the ocean. If the metal contaminant MPs are mistakenly eaten by marine animals, it can disturb their metabolic processes, growth, reproduction, and immune systems (Zhang et al. 2020). MP contamination in the marine ecosystem is represented in Figure 10.1. Those MPs particles that are too fragile then produce nanoplastic particles. These nanoplastic fragments are easily consumed by marine organisms and their accumulation makes a detrimental impression on the marine environment as well as on public health in upcoming times (Yang et al. 2021).

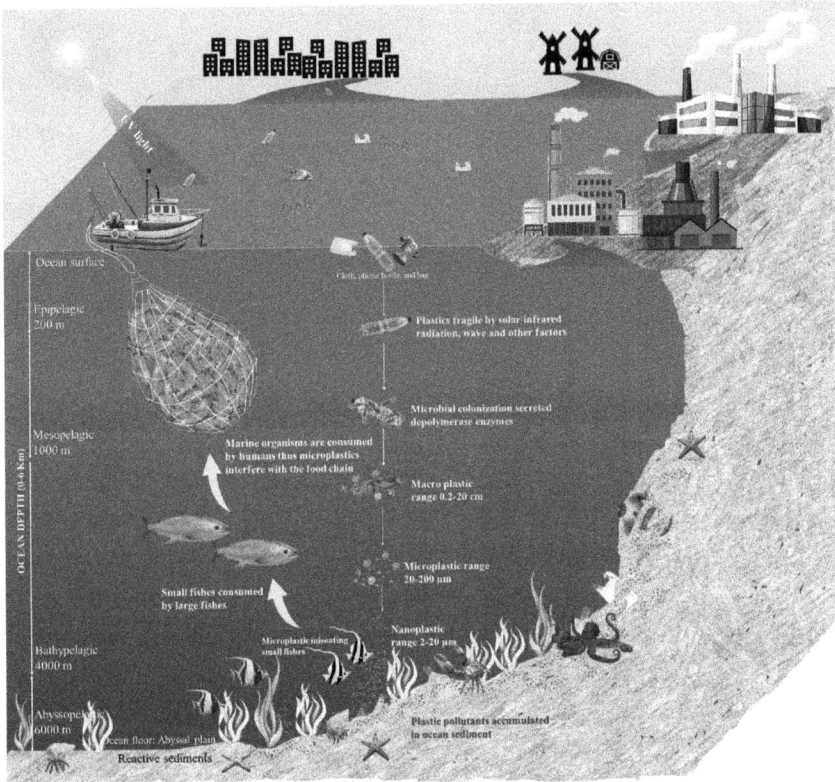

FIGURE 10.1 Microplastics in the marine ecosystem.

10.3 ARRANGEMENT OF MICROPLASTICS

MPs are present globally in aquatic environments. Smaller plastic particles are most abundant in marine environments ranging from 10 μm to 5 mm. On the basis of size, MPs is classified as mega-size plastic (20–200 cm size), macro-size plastic (0.2–20 cm size), meso-size plastics (200–2000 μm size), micro-size plastic (20–200 μm size), nano-size plastics (2–20 μm size), pico-size plastic (0.2–2 μm size), femto-size plastic (0.02–0.2 μm size) (Bermúdez and Swarzenski 2021). Browne et al. (2007) classified plastic fragments in three categories, according to their size—i.e., macroplastic size range is > 5 mm, MPs size range is between 1 μm and 1 mm, and nanoplastic size range is between 1 nm and 100 nm. Díaz-Mendoza et al. (2020) again categorize plastic particles into three classes—i.e., MP's range is > 2.5 mm, mesoplastic's size ranges between 5 and 2.5 mm, and MP's range is < 5 mm. Hartmann et al. (2015) again classified plastic into four categories—i.e., macroplastic's size range is > 1 cm, mesoplastic's size ranges between 1 mm and 1 cm, MP's size range is between 1 μm and 1 mm, and nanoplastic's size range is between 1 nm and 1 μm.

Plastic is classified on the basis of its monomer composition, uses, and size-based organization, represented in the Figure 10.2. On the basis of origin, MPs have two

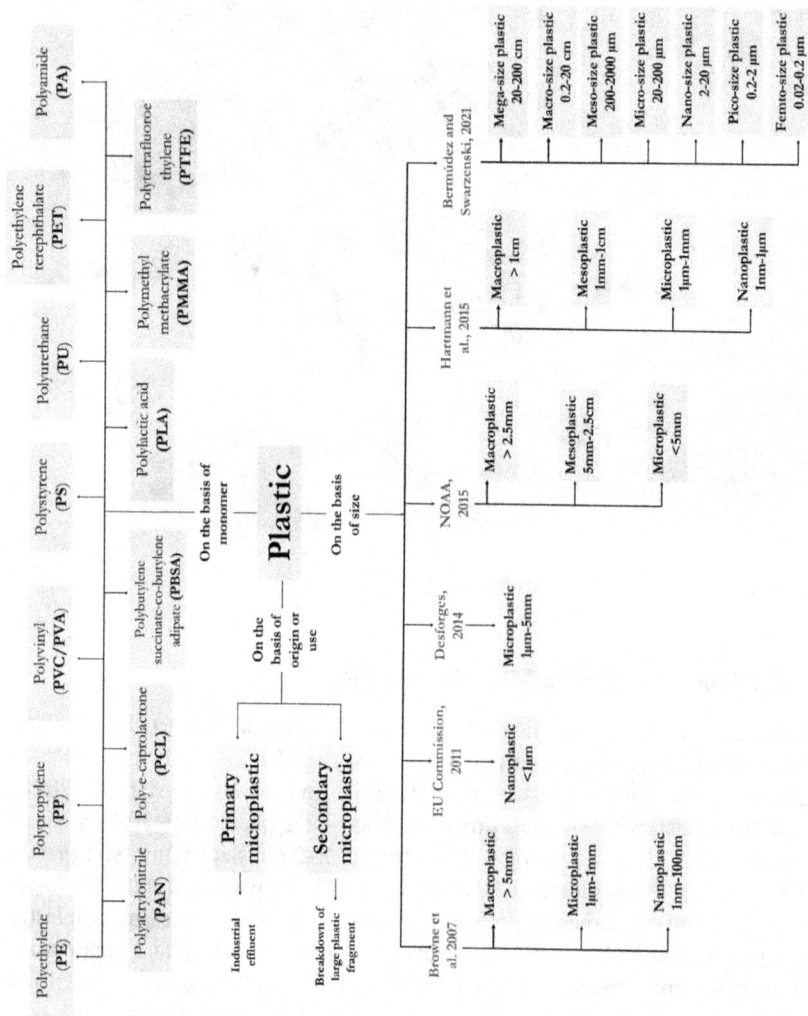

FIGURE 10.2 Microplastics classification on the basis of monomer composition, use, origin, and size (different authors have classified MPs in a diverse manner on the basis of size).

categories—i.e., primary MPs and secondary MPs. According to the monomer composition, plastic is classified in several types—i.e., polyethylene, polystyrene, polypropylene, polyvinyl, polyurethane, etc.

10.4 CHEMICAL CONSTITUENTS AND USES OF MICROPLASTICS

Numerous types of MPs such as polyvinyl, polyethylene, polypropylene, polystyrene, and polyurethane are produced in very large amounts due to their heavy commercial and industrial use. The possibility is that all the plastic particles finally end up in the ocean (Andrady 2011). The molecular structure of several types of MPs components is represented in Figure 10.3.

The C and H atom comprises polyethylene, the alcohol (-OH) and isocyanate (NCO) group is present in polyurethane, the benzene molecule is present in polystyrene, the Cl atom is present in polyvinyl chloride, and each alternative carbon molecule is involved with a methyl group in polypropylene.

10.4.1 POLYETHYLENE (PE)

Polyethylene is hydrophobic in nature and made up of a long chain of branched or unbranched ethylene monomers. It is produced as a result of polymerization of petrochemical feedstock. PE is a useful polymer product in the food, medicine, and beverage industries due to its elasticity, clearness, chemical confrontation, and durability. PE is impermeable to water vapor and moisture, and heat stable up to 123°C. Polyethylene is not an efficient barrier for oil, fat, and gases

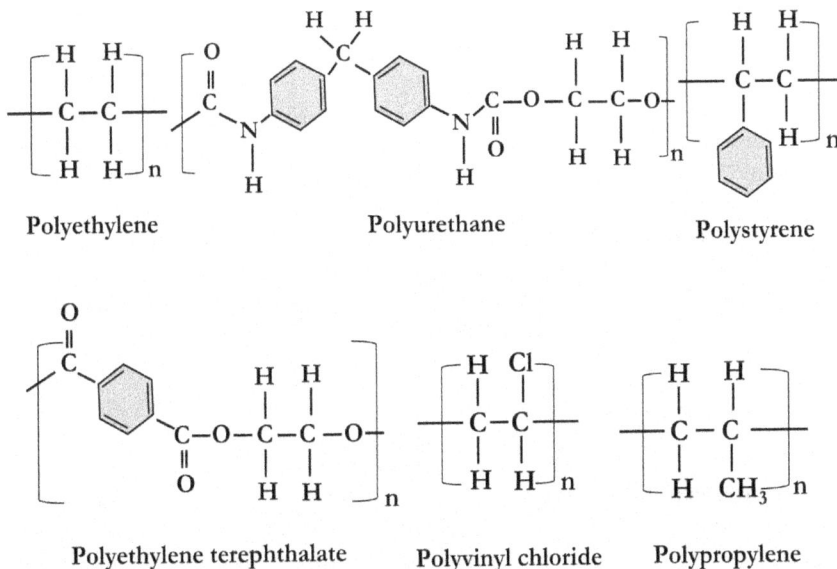

Polyethylene   Polyurethane   Polystyrene

Polyethylene terephthalate   Polyvinyl chloride   Polypropylene

FIGURE 10.3 Molecular composition of common MP monomers.

(Huerta Lwanga et al. 2018; Scalenghe 2018; Oliveira et al. 2020). On the basis of flexibility, clarity, and density, polyethylene is divided into HDPE (high-density polyethylene) and LDPE (low-density polyethylene). The HDPE is a linear chain with a large density-to-strength ratio. LDPE is highly branched and unaffected by alcohol, base, and acid but unstable in the presence of aromatic and aliphatic hydrocarbons. The conservative chain structure of PEG [HO–(CH$_2$–CH$_2$–O)$_n$–H] is non-polar, non-toxic, non-ionic, bio-compatible, and water-soluble; thus, it is used in medical, cosmetic, and pharmacy products (Parray et al. 2020).

10.4.2 POLYSTYRENE (PS)

PS is a linear, amorphous polymer. The styrene monomer is polymerized to produce polystyrene. Ethyl alcohol is united with hydrogen chloride to form ethyl chloride. The ethyl chloride mixes with the benzene molecule to produce ethyl benzene, which is superheated in a nickel tube to yield styrene. In another way, in the presence of aluminium chloride, the ethylene gas combines with benzene to produce ethylbenzene. The ethylene benzene is united with chlorine to form chloroethyl benzene, then passes to the nickel tube to form styrene monomer. In presence of sulphuric acid, styrene monomer mixes with acetic acid to produce distyrene. The distyrene is polymerized to produce matastyrene. After being super-heated, the methyl group is broken from matastyrene and undergoes polymerization processes that ultimately form polystyrene (Arfin et al. 2015). PS is thermostable, resistant to biodegradation, and can pass through water vapor and gases. PS is used for packaging purposes due to its solid, hard, and lightweight properties. Durable polystyrene is present in three forms: high crystalline, expended, and high impact. Now, dull plastic is obtained through the combination of polystyrene and rubber (McKeen 2014; Oliveira et al. 2020; Abhijith et al. 2018).

10.4.3 POLYETHYLENE TEREPHTHALATE (PET)

Polycondensation of terephthalate and ethylene glycol produces polyethylene terephthalate. PET is a clear, strong, thermotolerant (up to 260°C), crystal-like, and chemically unchanging polymer that is commonly used in beverages, microwave cooking, steam sterilization, and food vessels. PET is manufactured through the transesterification process, using dimethyl terephthalate and ethylene glycol. It is thermostable, tolerating 260°C temperatures; thus, it is used in microwave cooking and steam sterilization (Hiraga et al. 2019; Oliveira et al. 2020).

10.4.4 POLYPROPYLENE (PP)

Polymerization of propene or propylene, which is a downstream petrochemical product, produces polypropylene. In 1954, polypropylene was first formed and created greater attention than the other plastic monomer due to its lowest density. Polypropylene is a vinyl polymer where each alternative carbon is conjugated with a methyl group (Maddah 2016). Polypropylene is hard-wearing, lightweight, very clear, low-cost, and easy to manufacture, rather than polyethylene. In natural conditions, polypropylene is present in three forms—i.e., crystalline, semi-crystalline, and amorphous. It is resistant to high temperatures and immiscible to oil, fat, and gases; thus, it is used in the packaging of snacks, cookies, etc. Due to its low surface tension property, it easy for printing and coating. It is also appropriate for used in making trays, funnels, bottles, etc. (Spoerk et al. 2020; Oliveira et al. 2020).

10.4.5 POLYVINYL CHLORIDE (PVC)

Polycondensation of vinyl chloride monomer produces polyvinyl chloride. The PVC is thermostable (its melting temperature is 445°C) and is present in two varieties—i.e., in the form of film and hard structure. The PVC is impervious to fat, oil, and inorganic molecules, whereas it is permeable to gas and water vapor, etc. The permeability of PVC depends on the polycondensation processes. PVC is mostly used in construction and structure commerce (Crawford and Quinn 2017; Oliveira et al. 2020).

10.4.6 POLYURETHANE (PU)

Polyurethanes have great mechanical properties and biocompatibility. The key repetition group in polyurethane is urethane. Polyurethane has two segments: one is the crystalline, or hard, segment (A), and the other is the rubbery, or soft, segment (B). Repetitive segments of A and B is polycondensed to produce polyurethane. Polyurethane requires two steps and three precursors for synthesis—i.e., diisocyanates, diols, and chain extenders (Oliveira et al. 2020). The chain extender and diisocyanates produce a solid segment, and diols produce a soft, or rubbery, segment. The isocyanate group is essential to comprising the chain of isocyanate groups $[R-(N=C=O)_{n\geq 2}]$, and the polyol group is essential to comprising the chain of hydroxyl clusters $[R^{-}-(OH)_{n\geq 2}]$. The reaction between alcohol (-OH) and isocyanate (NCO) produces PU. Polyurethane also comprises ester, ether, urea, and some aromatic composites (Akindoyo et al. 2016). Polyurethane synthesis occurs on the basis of its application like inflexible, elastic, binder, covering, and elastomers. It has also been used in medical devices and automotive tenders (Akindoyo et al. 2016; Oliveira et al. 2020).

10.5 MICROPLASTICS' TOXICITY ON AQUATIC ORGANISMS AND THE FOOD CHAIN

Since they exist in both pelagic and benthic habitats, toxic monomers of PS, PP, and PVC function as MPs and pose a threat to aquatic animals. Aquatic organisms take various types and sizes of MPs, or MPs can enter into their circulatory systems through the intestinal tract, and they are found in specific locations within tissues.

Toxic effects of MPs are also found in the immune systems, nervous systems, reproductive systems, and endocrine systems of fish. Biomagnification is a serious threat due to transfer of those pollutants from water to higher trophic levels. Algae produce phycotoxins by MP consumption that indirectly affects human health and the economy. Algal toxins are produced by MP exposure, effectively accumulating

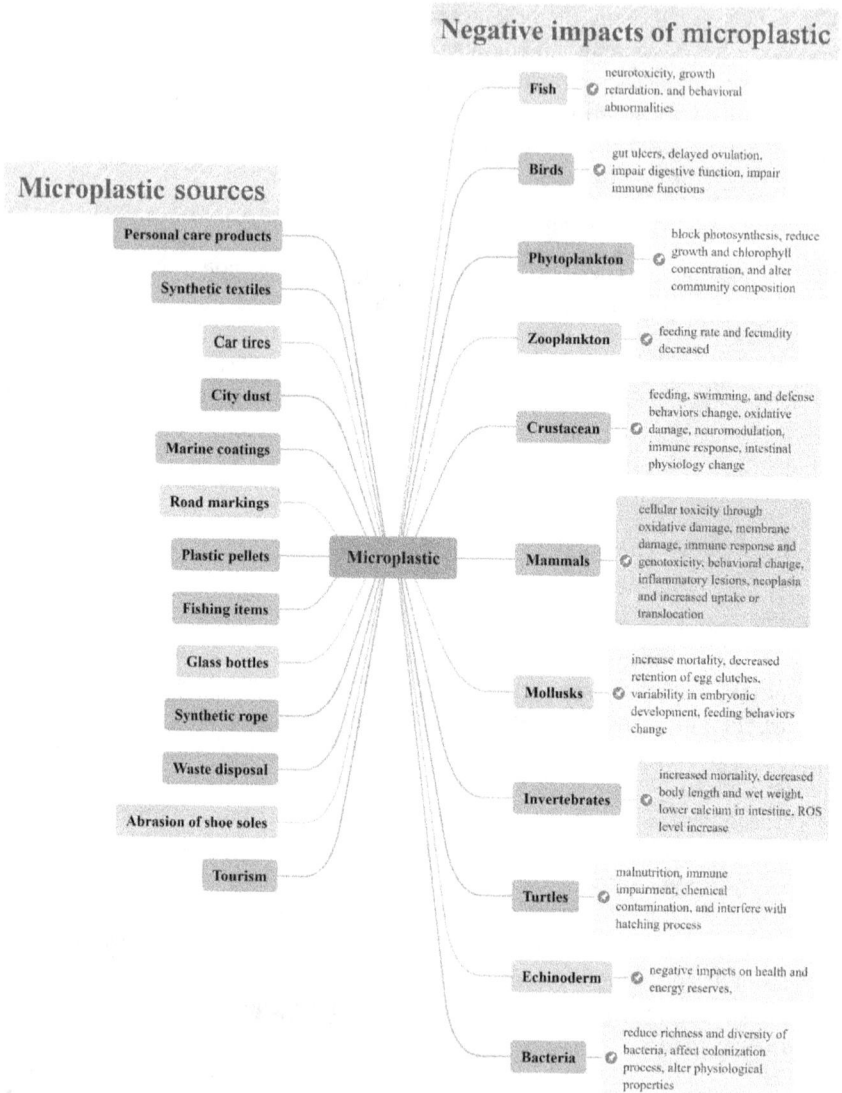

FIGURE 10.4 Negative impacts of microplastics on aquatic organisms.

in shellfish, and can spread through the food chain and indirectly cause dangerous symptoms (cause diarrhoeic and paralytic shellfish poisonings) in people.

After penetrating cell walls and membranes, MPs can reduce chlorophyll concentration in algae. Apart from phytoplankton, zooplankton (including holoplankton and meroplankton) diversity is the main contributor to marine food chain. In several studies, it was found that zooplankton potentially absorb minute plastic latex beads, or MPs (< 5 nm), by filter-feeding mode. Zooplankton (e.g., copepods, cladocerans, shrimps, worms, ciliates, and polychaete), bivalves, benthic invertebrates, fish, and large marine mammals consume MPs of different shapes intentionally as food (Botterell et al. 2020). Thirteen zooplankton species were shown to be able to consume 1.7–30.6-m polystyrene beads, with intake varying by taxa, life stage, and bead size (Cole et al. 2013). Ingestion of MPs affects the digestive system and produces obstructions, which leads to a false satiety. Among zooplankton, daphnids and copepods are most sensitive to MPs, and as a result, their feeding rate and fecundity decreased significantly (Yu et al. 2020). The study discovered that the characteristics and sinking rates of zooplankton fecal pellets can be changed by MPs, which can make it easier for plastics to reach coprophagous biota (Cole et al. 2016). MPs can also move from the gut to other intestinal organs and stay there for an extended period of time. There have been reports of MP consumption by benthic invertebrates, including oysters, barnacles, blue mussels, and lobsters. Each of them is an important contributor to the marine food chain. For example, high lipid content of the benthic worm, *Arenicola marina*, makes it an essential component of marine food chains. Unfortunately, a large percentage of MPs are also indirectly ingested by this worm during feeding. MP exposure results in reduction of feeding aptitude and reduction in weight (Besseling et al. 2013; Sharma and Chatterjee 2017). According to a study, MPs are present in about 30% of the different fish species. Because they spend more time in fishes' guts, MP beads of the 5 mm size are more hazardous to marine fish. The build-up of MPs in the stomach causes hunger, malnutrition, and finally, death in fish (Benson et al. 2022). There have also been reports of small plastic bits in sea birds. It has been noted that juvenile birds are exposed to more MPs during feeding, which then build up in their intestines. Seabirds that consume plastic suffer detrimental effects, including hunger and fitness loss (Charlton-Howard et al. 2023). MPs also have an impact on large marine mammals like whales, turtles, polar bears, etc. Due to their large fat and lipid content, whales have a significant capacity to absorb and store MPs in their stomachs and intestines. Investigations found that death of stranded whales was due to MP litter in their gut. In addition, MPs can interfere with nutrients, resulting in physiological stress and affecting stability and composition of the ecosystem. Although the long-term effect of MPs on human health is little known, they can seriously affect human health. Although MPs are present in many products used by the general public (including cosmetics, toothpaste, scrubs, and hand washes), MPs in seafood also have a number of negative effects on the health of consumers. Commercial sea creatures such as mussels, oysters, crabs, sea cucumbers, and fish have the capacity to absorb and transport MPs across the food chain. Negative impact of MPs on marine organism is represented in Figure 10.4. Bioaccumulation of MPs in crustaceans is more serious than fish, as, in crustaceans, the digestive tract is consumed. The indicator species like mussels and molluscs can shed light on the level of seafood pollution and human consumption, as they are consumed wholly and are a significant source of MPs. A study also reported the presence of MPs of size < 200 μm in sea salt. Different pathogenic microbes can attach to the surface of MPs; consuming those seafoods enhances human exposure to these bacteria. Micro- and nanoplastics of a larger size are expelled in feces, whereas MPs smaller than 150 m are absorbed by the intestinal epithelium. These MPs cause systemic exposure, whereas larger MPs can only cause local immune system effects when exposed. Nanoplastics can pass across the placenta and blood-brain barrier, and MPs of less than 1.5 m can reach deep inside organs. By creating ROS during an inflammatory response, MPs cause oxidative stress, which has lethal effects. Consuming MPs can change chromosomes, leading to the development of cancer, obesity, and infertility. Additionally, the presence of MPs in seafood poses a serious threat to food safety (Van Cauwenberghe and Janssen 2014).

10.6 MICROPLASTICS, METAL, AND OTHER CONJUGATES

The adsorption capacity of MPs depends on pore size, surface area, and lipophilicity. Interaction of MPs with toxic compounds occurs through environmental degradation and water cycles. MPs, upon degradation, form smaller plastics that enhance contaminants adsorption due to increased surface area and chemical reactivity. The kinetics of pollutant adsorption to MPs may be significantly impacted by additional environmental parameters such as weathering, sunshine, prolonged exposure, pH, and the hydrophobicity of POPs. MP molecules form sustainable bonds (hydrogen and covalent bonds) by pairing with other hydrophobic substances.

When additional organic molecules are introduced into organisms via MPs, they accumulate and have a long-term negative impact on the organism. Numerous persistent organic pollutants, including bisphenol A, polybrominated diphenyl ethers, phenanthrene, pyrene, polychlorinated biphenyls (PCBs), polycyclic aromatic hydrocarbons (PAHs), per-/polyfluroalkyl substances (PFAS), polycyclic aromatic hydrocarbons, petroleum hydrocarbons, and others, can conjugate with MPs. MP features (such as composition, structure, binding energy, and surface property), environmental parameters (such as pH, temperature, salinity, and ionic strength), and contamination factors (such as solubility, redox state, charges, and stability) all have a role in the adsorption of pollutants by MPs.

According to Du et al. (2021), the pH level of the water and the amount of time MPs spend in their environment are the determining factor regarding MPs, and metal ions adsorption before adsorption are the deciding factors. By lowering pH, contaminants are more likely to be absorbed by MPs. Surface roughness or presence of filler in MPs may increase the uptake of PFAS (Joo et al. 2021). Heavy metals are attached with MPs through electrostatic interactions. Common heavy metals like cadmium, lead, arsenic, zinc, chromium, nickel, and copper are adsorbed by MPs. Positively charged metals can bind with negatively charged MPs to neutralize them. By adhering to the surface of MPs, various inorganic minerals can change the MPs' surface characteristics and produce ligand binding sites for metal ions (Rochman et al. 2013). Studies have confirmed metal adsorption on polyethylene pellets up to several hundred micrograms per gram. Different environmental factors, including microbes, influence MPs and heavy metal interactions in water environments (Liu et al. 2021). Several types of MPs conjugate and are represented in the Figure 10.5. Persistent organic pollutants (POPs) are conjugated with MPs and, upon ingestion by marine biota, they are transferred to marine food chain. MPs, upon interaction with antibiotics, cause a combination of effects. A study reported that, when benzopyrene is adsorbed on polyvinyl chloride (PVC), the highest toxicity is found compared with bare MPs or benzopyrene alone. This indicated a significant role of MPs as a vector for organic pollutants.

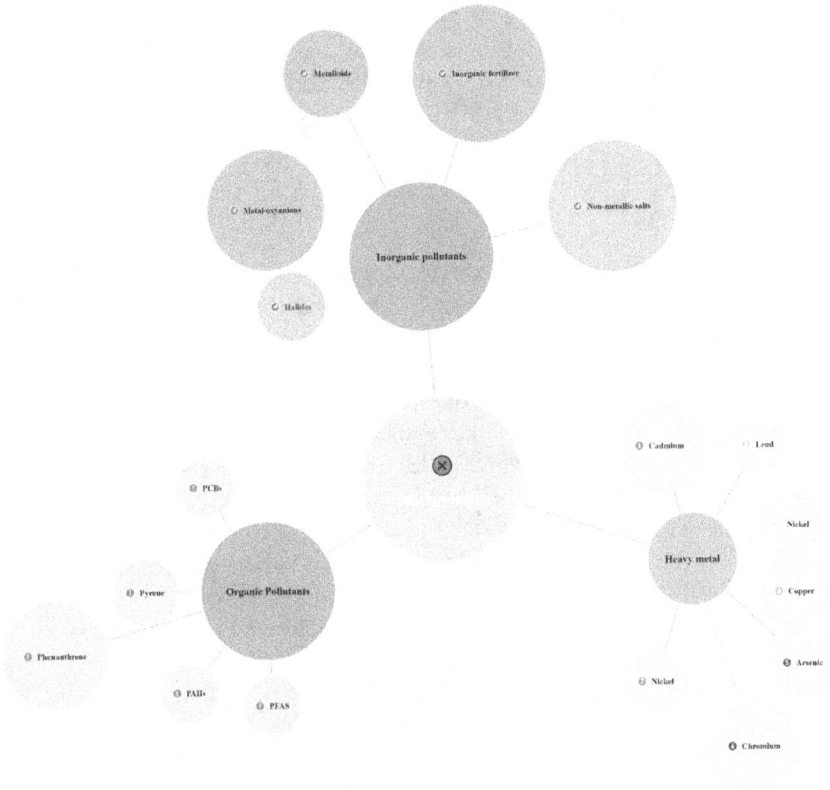

FIGURE 10.5 Microplastics conjugate with other pollutants.

10.7 MICROPLASTICS IDENTIFICATION TECHNIQUE

Despite being a nascent subject, technique development for MP analysis draws on expertise from domains including nanoscience, particle studies, and material science on synthesized or naturally occurring particles. The difficulty arising in analysing MPs is that they don't have a unique molecular structure; rather, it is made up of a range of polymer components with various shapes, sizes, and additives (Zarfl 2019). MPs are identified by two categories of methods: one is the physical method and another one is the chemical method. In case of physical identification, a microscope is used. Visual inspection is an important step for researchers to detect microplastics that have been removed from the matrix. MPs are grouped according to their type (pieces, fibers, and foam), size, and color. Besides spectroscopy, a Scanning Electron Microscope (SEM) is also used for examining the particle morphology. Due to high resolution images, SEM can identify the impurities within MPs. Additionally, SEM can detect potential alteration in the surface of MPs after the process of chemical digestion. Analytical techniques used to detect the different types of MPs include FTIR, pyrolysis, gas chromatography, Raman spectroscopy, etc. These methods provide the sample's accurate chemical analysis, allowing separation between MPs and non-MPs on the basis of available polymer and additives (BretasAlvim et al. 2020). Several methods are also present to determine and identify the chemical configuration of MPs, including NMR spectroscopy, NIR, SEM-EDS, FTIR, and Raman spectroscopy.

10.7.1   SEM-EDS

In Scanning Electron Microscopy (SEM), a high-intensity electron beam is used for sample imaging. Interaction between sample and electron generates distinctive X-ray photons that are element specific. These X-rays distinguish using an Energy Dispersive Spectroscopy (EDS) detector. SEM-EDS aid in the quick differentiation between MPs and non-MPs and have the ability to find microscopic particles that are overlooked by optical inspection (Tirkey and Upadhyay 2021). Surface characteristics and elemental makeup were used in SEM/EDS scanning to assess potential plasticity of each particle. On aluminium SEM stubs, samples are placed on carbon tabs by using double-sided adhesive. A Thermo Fisher Scientific Norn System 7 EDS System and a FEI XL 30 Environmental SEM and were used to perform SEM/EDS studies. Water vapor with a pressure of 0.6 Mbar was introduced to reduce sample charging under an electron beam for wet mode photography into the SEM chamber. Contrary to other options, like sample-covering with metal or carbon, which might have added chemical artifacts, this wet-mode approach minimized contamination. Using BSE (Backscattered electron) detector on the SEM, samples were scanned between 50 to 10,000-times superior resolution to prove surface features and elemental-traces composition. Using this data, non-plastics were ruled out, while probable MPs were screened for further use (Wang et al. 2017).

10.7.2   FOURIER-TRANSFORM INFRARED SPECTROSCOPY (FTIR)

Following visual pre-sorting, FTIR may be used to assess all separated possible plastic granules and fibers of the larger portion. ATR (Attenuated Total Reflection) is the principal method that is used in FTIR spectroscopy. It only examines the samples near the top. The sample is filtered after being enzymatically or chemically processed in order to assess the smaller MP portion. Infrared radiation may vary, depending upon chemical makeup of the sample. Mainly, a 400–4000 cm^{-1} range of wave number is measured and absorbed by the transmission process or by the reflection process, depending upon the molecular structure of MPs. There are three different FTIR imaging modes: ATR, transmission, and specular reflection. There is a possibility that, during the experiment, particles would stick to the ATR crystal, so ATR-FTIR imaging seems to be impractical. In addition, since each particle is being focused on individually, it takes a long time (Käppler et al. 2016). Some other FTIR optimizing technologies are Focal Plane Array (FPA), micro FTIR, Attenuated Total Reflection (ATR), etc. Micro FTIR can be used to cover the restriction of FTIR, which can only identify MPs up to 20 µm. Smaller samples (10 µm) may be characterized using micro FTIR (Tirkey et al. 2021).

10.7.3   Near-Infrared Spectroscopy (NIR)

The quantity of the light reflected from the surface in the 350–2500 nm wavelength range is measured using a vis-NIR Spectrometer, which provides a reflected proportion regarding each wavelength. This data may be used to forecast the chemical makeup of fresh sample sets, since it can be connected with the chemical composition of the sample, as with the direct examination of elemental composition using vis-NIR methods. This technique is helpful for quantifying the number of MPs. Thus, by employing vis-NIR spectrometric methods, plastic polymer can potentially be identified (Corradini et al. 2019).

10.7.4   Raman Spectroscopy

Raman spectroscopy is a kind of vibrational spectroscopy that uses inelastic dispersion of light to produce a vibrational spectrum that contains data on the system's molecular vibration. In comparison to FTIR spectroscopy, Raman methods exhibit improved spatial resolution (as small as 1 µm vs. 10–20 µm, in the case of FTIR), larger spectrum coverage, and stronger sensitivity to non-polar functional groups. Raman spectroscopy has a number of drawbacks, including the potential for fluorescence interference and the potential for sample heating due to employment of laser light source. These drawbacks may sometimes result in background emission and polymer deterioration (Araujo et al. 2018).

10.7.5   Nuclear Magnetic Resonance (NMR)

Usually, lengthy MP particle monitoring in the sample is reported for quantitative identification. As a result, it is clear that there is a huge demand for study in the area of size-independent quantitative evaluation of MP particles. A novel technique for the quantitative and qualitative characterization of MP in solution is the use of quantitative ^{1}H NMR spectroscopy (qNMR). Using polyethylene terephthalate (PET) fibers, calibration curve method, polystyrene (PS) and polyethylene (PE) granules with size distributions ranging from 0.5 to 1 mm were analyzed quantitatively and qualitatively using ^{1}H NMR (Peez et al. 2019).

10.8   BIOACCUMULATION OF MICROPLASTICS IN THE AQUATIC ENVIRONMENT

Saltwater and freshwater environments are both plagued by MP debris and plastic trash. Negative impacts have been linked to birds, fish, benthic invertebrates, mammals, and turtles. They are being entangled in plastic wires or ingesting it. It is generally accepted that MPs may serve as a vector for the transportation of chemicals linked to plastic monomers or oligomers (Koelmans 2015). Sorption and desorption are the main starting points for MPs' potential to affect the organic contaminant's bioaccumulation. There is still a need for proof that MPs may move across tissues and then organs or even into live cells. On the contrary, bioaccumulated pollutants should be evaluated to know the form of the sorbed and desorbed compound from ingested MPs, which is also called the 'releaser effect'. It has been shown that the surfactants, present in gut, quicken MPs' release of organic contaminants, which are hydrophobic. On the other hand, the existence of MPs may result in a reduction in the amount of dissolved organic contaminants in nature, particularly in aquatic habitats. Even though the MPs that are ingested may be loaded with the organic pollutants, if primarily collected organic pollutants by the organisms in the environment are in a dissolved phase, the presence of MPs may control their bioaccumulation process. This is called a 'diluting effect' and is caused by the concentration of free organic pollutants after a significant quantity of them were adsorbed on MPs. Additionally, the 'diluting effect' depends on the amount of MPs present in the system (Wang et al. 2020). Many authors have disputed the notion that MPs might play a significant part in the biomagnifications and bioaccumulation of harmful substance like Persistent Organic Pollutants (POPs) by marine organisms. These studies claim that the amount of plastic waste already existing in the seas is insufficient to prevent POPs from being partitioned to Dissolve Organic Matter (DOM) and water. Three different possible outcomes will be discussed to correctly determine the actual potential of MPs as transporters of dangerous substances and characterize the impact on bioavailability and bioaccumulation of these compounds via consumption of MPs.

- **Scenario 1**

  In this case, clean pellets of MPs that do not contain sorbed environmental contaminants are consumed by polluted marine biota. These pristine beads of MPs help the animal's body to reduce the chemical contaminants. In a nutshell, MPs could serve as a reservoir for the decline in bioaccumulation.

- **Scenario 2**

  In contrast to the previous scenario where uncontaminated marine creatures consume contaminated MPs, this scenario involves the exact reverse. According to Granby et al., relative to feed that included simple contaminants, seabass had a lower removal efficiency when pollutant sorbed MPs were present.

- **Scenario 3**

  This scenario describes the interaction between MPs and biota. Biota are likely to bioaccumulate or biomagnify organic contaminants like POPs via their respiration or nutrition. Before being consumed by mammals, marine biota and MPs apparently achieved the sorption optimum (Menéndez-Pedriza 2020).

10.9 PLASTIC-DEGRADING MARINE MICROBES AND THE FATE OF MICROPLASTICS

Once MPs come in contact with the marine water body, it can stay for a long time, as most of them are impervious to degradation (Duis and Coors 2016; Galloway 2015; Li 2018). Several biological and chemical methods are being used to lessen the hazardous effects of MPs (Du et al. 2021). Advanced oxidation processes such as photochemical, photocatalytic, and electrochemical oxidation, along with the involvement of microbes, enhance MP degradation. Photodegradation—i.e., prolonged exposure to UV light—generates ROS that break down polystyrene type of MPs. Photocatalytic degradation of plastics typically relies on semiconductor materials like TiO_2 and ZnO. The TiO_2 have a valance band and a conduction band; when the intensity of sunlight is higher than band gap energy, the electron is transported from the valance to the conduction band; thus, a positive hole (h^+) is formed in valance band. O_2 receives the electron in the conduction band that results in the formation of $\bullet O_{2^-}$ and, in the valance band, the H_2O or OH^- reacts with h^+ formation of hydroxycles $\bullet OH-$ that is responsible for ROS formation and, gradually, the degradation of plastic particles. Because the catalyst is floating in the water and partially degrades MPs, it cannot be recycled (Du et al. 2021). Microorganisms, including algae, fungi, and bacteria with metabolite activities, are better suited for the bioremediation of plastic pollution represented in Table 10.1 (Webb et al. 2012). For the production of biofilms, microorganisms employ MP as a substrate. The MP structure becomes softer as the biofilm grows, creating cavities and causing chemical changes. The extracellular enzymes lipase, keratinase, cutinase, and esterase produced by microorganisms can hydrolyse the polymeric surface of plastics and release monomers and small molecules (CO_2, N_2, CH_4, H_2O, and H_2S) (Othman et al. 2021; Du et al. 2021). The harmless organic by-products of MP degradation may be ingested by microbes, converting them into other valuable products like protein, biofuels, and sugar that can be applicable for people in a sustainable and ecologically acceptable manner (Du et al. 2021). For the production of biofilms, microorganisms employ MP as a substrate. The MP structure becomes softer as the biofilm grows, creating cavities and causing chemical changes. The extracellular enzymes lipase, keratinase, cutinase, and esterase produced by microorganisms can hydrolyze the polymeric surface of plastics and release monomers and small molecules (CO_2, N_2, CH_4, H_2O, and H_2S), which are then used for energy production and eventually come back to the atmosphere (Othman et al. 2021; Du et al. 2021). The harmless organic by-products of MP degradation may be ingested by microbes, converting them into other valuable products like protein, biofuels, and sugar that can be applicable for people in a sustainable and ecologically acceptable manner (Du et al. 2021). Chemical nature, shape, size, and physical property of the MP determined their impact on the marine environment. Deposition and dispersal pattern of MPs are mediated by different parameters like shape, size, thickness, and chemical properties of the MP (Haque and Fan 2022). MPs having lighter density than sea water remain floating on sea surface, while high-density MPs are immersed (Li 2018). Depending on wind energy and geostrophic equilibrium, buoyant MPs are transported on the oceanic surface. Neutral MPs float on the sub-surface layer of the ocean. The gathering of MPs is high in the surface region and gradually declines with the depth of the water (Song et al. 2018). The accumulation of MPs in marine snow makes them more prone to sink and deposition in the benthic zone. Benthic organism *Mytilus edulis* intakes more snow-associated MPs than free (Porter et al. 2018). At the time of ice formation, MPs are trapped within it. During the defrosting of the ice, these plastics become accessible for deep oceanic communities and they are more prone to face the toxic effect of MPs (Caruso et al. 2022). Vertical as well as horizontal transmission of MPs takes place through temporal

TABLE 10.1

Microplastics Degrading Bacteria Reported from the Aquatic Environment

Microorganism	Isolated Source	Types of MPs	Amount of Degradation	References
Exiguobacterium sp., Halomonas sp., Ochrobactrum sp.	Upper tidal locations in the Huiquan Bay	Polyethylene terephthalate (PET), polyethylene (PE)	2.7% and 19.6%	Gao and Sun 2021
Glaciecolalipolytica, Aestuaribacter halophilus	Sea water	Poly(3-hydroxybutyrate-*co*-3-hydroxyhexanoate)	–	Morohoshi et al. 2018
Zalerion maritimum	–	Polyethylene (PE)	–	Paço et al. 2017
Pseudomonas pachastrellae	Okinoshima Park, Chiba, Japan	Polyester poly (ε-caprolactone)	–	Suzuki et al. 2018
Proteobacteria, Betaproteobacteria, Deltaproteobacteria, Gammaproteobacteria	Marine environment	Polyethylene terephthalate (PET)	–	Danso et al. 2018
Alcanivorax sp., Hyphomonas sp., Cycloclasticus sp.	Torre Faro	Polyethylene terephthalate (PET)	–	Denaro et al. 2020
Alcanivorax sp., Marinobacter sp., Arenibacter sp.	Northern Corsica (Calvi Bay, Mediterranean Sea	Low-density polyethylene (LDPE), polyethylene terephthalate (PET)	–	Delacuvellerie et al. 2019
Vibrio parahaemolyticus, Bacillus licheniformis, Paenibacillus woosongensis, Vibrio fluvialis	Coastal environments of Andaman Island	Low-density polyethylene (LDPE)	47.07 ± 6.67%	Joshi et al. 2022

Microorganism(s)	Location	Plastic type	Degradation	Reference
Synedropsis sp., *Chaetoceros* sp., *Melosira* sp., *Thalassiosira cf. antarctica* Comber, *Thalassiosira* sp., *Navicula* sp.	Antarctic Peninsula	Polyurethane (PU), polyamide (PA), polyethylene (PE) polypropylene (PP), polystyrene (PS)	–	Lacerda et al. 2019
Shewanella sp., *Moritella* sp., *Psychrobacter* sp., *Pseudomonas* sp.	Kurile and Japan Trenches	Poly ε-caprolactone (PCL)	–	Sekiguchi et al. 2011a
Pseudomonas sp., *Arthrobacter* sp.	Gulf of Mannar, India	Polyethylene (PE)	12 and 15%	Balasubramanian et al. 2010
Cobetia sp., *Halomonas* sp., *Exigobacterium* sp., *Alcanivorax* sp	Marine environment	Polypropylene (PP)	1.40%, 1.72%,1.26%, 0.97%	Khandare et al. 2021
Pseudomonas sp., *Rhodococcus* sp.	Victoria Land (Ross Sea sector Antarctica) soil	Polypropylene (PP)	17.3% and 7.30%	Habib et al. 2020
Kocuria palustris, Bacillus pumilus, Bacillus subtilis.	Arabian Sea, India	Low-density polyethylene (LDPE)	1%, 1.5% and 1.75%	Harshvardhan and Jha 2013
Bacillus spp., *Pseudomonas* spp.	Coastal regions of Tamil Nadu, India	Polyethylene (PE)	–	Devi et al. 2019
Pseudomonas sp., *Alcanivorax* sp., *Tenacibaculum* sp.	Deep sea water of Rausu, Kume and Toyama	Polyesters poly(ε-caprolactone)	–	Sekiguchi et al. 2011b
Bacillus sp.	Coastal area of Arambhada, Gujarat	Low-density polyethylene (LDPE), high-density polyethylene (HDPE), polyvinyl chloride (PVC)	0.26±0.02, 0.96±0.02, and 1.0±0.01%	Kumari et al. 2019

(Continued)

TABLE 10.1 *(Continued)*

Microplastics Degrading Bacteria Reported from the Aquatic Environment

Microorganism	Isolated Source	Types of MPs	Amount of Degradation	References
Alcaligenes faecalis LNDR-1	Bay of Bengal near Puri, India	Polyethylene (PE)	15.25 ± 1%; 21.72 ± 2.1%	Nag et al. 2021
Amycolatopsis, Collimonas, Kribbella, Psychrobacter, Streptomyces sp., *Lachnellula* sp., *Neodevriesia* sp., *Thelebolus* sp.	Alpine and Arctic habitats	Polyurethane (PU)	–	Rüthi et al. 2023
Pseudogymnoascus pannorum, Lachnellula sp.	Alpine and Arctic habitats	Polybutylene adipate-co-terephthalate (PBAT), polylactic acid (PLA)	34.9% and 25.8%	Rüthi et al. 2023
Lachnellula sp., *Neodevriesia* sp.	Alpine and Arctic habitats	Biodegradable plastics (BI-OPL)	18% and 10%	Rüthi et al. 2023
Sclerotinia sp., *Fusarium* sp.	King George Island, South Shetland Islands, Antarctica	Poly-e-caprolactone (PCL)	33.7% and 49.65%	Urbanek et al. 2021
Sclerotinia sp. B11IV and *Fusarium* sp. B30 M	Arctowski Polish Antarctic Station	Polybutylene succinate-co-butylene adipate (PBSA)	49.68% and 45.99%	Urbanek et al. 2021
Bacillus cereus, Bacillus sphericus, Bacillus furnisii, Vibrio furnisii, Brevundimonas vesicularis.	Indian Ocean	Polyamide (PA)	42%, 31%, 7% and 2%	Sudhakar et al. 2007
Aestuariibacter halophilus	Osaka, Yokohama, Okinawa-1 (Naha port), Okinawa-2 (Kume Island)	Polyamide (PA)	30%	Yamano et al. 2019

succession of different microorganisms (bacteria, fungi, algae, virus, archaea, protozoa) on the MP surface. Pioneer microorganisms start to aggregate around the MP's surface, depending on the environmental factor. Gradually, they form biofilm, known as 'plastisphere', where other microorganisms accumulate through their secreted, adhesive protein and pili. In this way, hetero-aggregating microorganisms increase MP density with the help of an extracellular polymeric substance and then slowly sink down (Zhai et al. 2023). However, the reverse phenomena is also taking place. The density of biofouling plastic may be diminished by benthic animals' mediated discharge, elimination, and uptake, through which it may recover buoyancy, which results in sediment-to-surface movement (Rummel et al. 2017). Harmful microbes attached on MPs can reach to a new area and negatively influence several species in the marine ecosystem (Zhai et al. 2023). Some organisms ingest MP that can digest and absorb through their tissue, or they can release it into sea water through feces. MPs digested by organisms eaten by other sea animals enter into the food chain, and ultimately, accumulate in other, higher animals.

10.10 CONCLUSION

Microplastic is considered as a type of marine litter—and a hazardous one—that has now become a big issue for the ocean environment. It accumulates and interferes in the marine ecosystem due to inappropriate disposal. Identification and quantification methodology for MP detection from biological samples creates great interest among many researchers for MP characterization around the globe. Those techniques help to determine the particle composition and size for classification. Injection of MP residue by aquatic species poses a serious threat to the marine food web and also to terrestrial flora and fauna. Most of the microbial community is able to degrade PE, PP, PS, and PA types of plastic monomers. Advanced processes, along with microbes, are now being used for partial degradation and management to reduce the harmful effects of this monomer micro-toxin.

ACKNOWLEDGMENT

The authors are thankful to the UGC Centre for the Advanced Study, Department of Botany, The University of Burdwan. RS is thankful to CSIR-JRF for financial support. DK, MG, and MN are thankful to the UGC-JRF.

REFERENCES

Abhijith, R., Ashok, A., and Rejeesh, C.R., 2018. Sustainable packaging applications from mycelium to substitute polystyrene: A review. *Materials Today: Proceedings*, 5 (1), 2139–2145.

Akindoyo, J.O., Beg, M.D.H., Ghazali, S., Islam, M.R., Jeyaratnam, N., and Yuvaraj, A.R., 2016. Polyurethane types, synthesis and applications-a review. *RSC Advances*, 6 (115), 114453–114482.

Amelia, S.T.M., Mohd, W., Wan, A., Khalik, M., Ong, M.C., and Shao, Y.T., 2021. Marine microplastics as vectors of major ocean pollutants and its hazards to the marine ecosystem and humans. *Progress in Earth and Planetary Science*, 4.

Andrady, A.L., 2011. Microplastics in the marine environment. *Marine Pollution Bulletin*, 62 (8), 1596–1605.

Araujo, C.F., Nolasco, M.M., Ribeiro, A.M.P., and Ribeiro-Claro, P.J.A., 2018. Identification of microplastics using Raman spectroscopy: Latest developments and future prospects. *Water Research*, 142, 426–440.

Arfin, T., Mohammad, F., and Yusof, N.A., 2015. Applications of polystyrene and its role as a base in industrial chemistry. In *Polystyrene: Synthesis, Characteristics and Applications*. Nova Science Publishers, 269–280.

Balasubramanian, V., Natarajan, K., Hemambika, B., Ramesh, N., Sumathi, C.S., Kottaimuthu, R., and Rajesh Kannan, V., 2010. High-density polyethylene (HDPE)-degrading potential bacteria from marine ecosystem of Gulf of Mannar, India. *Letters in Applied Microbiology*, 51 (2), 205–211.

Barboza, L.G.A., and Gimenez, B.C.G., 2015. Microplastics in the marine environment: Current trends and future perspectives. *Marine Pollution Bulletin*, 97 (1–2), 5–12.

Benson, N.U., Agboola, O.D., Fred-Ahmadu, O.H., De-la-Torre, G.E., Oluwalana, A., and Williams, A.B., 2022. Micro(nano)plastics prevalence, food web interactions, and toxicity assessment in aquatic organisms: A review. *Frontiers in Marine Science*, 9.

Bermúdez, J.R., and Swarzenski, P.W., 2021. A microplastic size classification scheme aligned with universal plankton survey methods. *MethodsX*, 8, 10–15.

Besseling, E., Wegner, A., Foekema, E.M., van den Heuvel-Greve, M.J., and Koelmans, A.A., 2013. Effects of microplastic on fitness and PCB bioaccumulation by the lugworm Arenicola marina (L.). *Environmental Science & Technology*, 47 (1), 593–600.

Botterell, Z.L.R., Beaumont, N., Cole, M., Hopkins, F.E., Steinke, M., Thompson, R.C., and Lindeque, P.K., 2020. Bioavailability of microplastics to marine zooplankton: Effect of shape and infochemicals. *Environmental Science & Technology*, 54 (19), 12024–12033.

BretasAlvim, C., Mendoza-Roca, J.A., and Bes-Piá, A., 2020. Wastewater treatment plant as microplastics release source—Quantification and identification techniques. *Journal of Environmental Management*, 255, 109739.

Browne, M.A., Galloway, T., and Thompson, R., 2007. Microplastic-an emerging contaminant of potential concern? *Integrated Environmental Assessment and Management*, 3(4), 559–561.

Caruso, G., Bergami, E., Singh, N., and Corsi, I., 2022. Plastic occurrence, sources, and impacts in Antarctic environment and biota. *Water Biology and Security*, 100034.

Cesa, F.S., Turra, A., and Baruque-Ramos, J., 2017. Synthetic fibers as microplastics in the marine environment: A review from textile perspective with a focus on domestic washings. *Science of the Total Environment*, 598, 1116–1129.

Charlton-Howard, H.S., Bond, A.L., Rivers-Auty, J., and Lavers, J.L., 2023. 'Plasticosis': Characterising macro- and microplastic-associated fibrosis in seabird tissues'. *Journal of Hazardous Materials*, 450, 131090.

Cincinelli, A., Scopetani, C., Chelazzi, D., Lombardini, E., Martellini, T., Katsoyiannis, A., Fossi, M.C., and Corsolini, S., 2017. Microplastic in the surface waters of the Ross Sea (Antarctica): Occurrence, distribution and characterization by FTIR. *Chemosphere*, 175, 391–400.

Cole, M., Lindeque, P., Fileman, E., Halsband, C., Goodhead, R., Moger, J., and Galloway, T.S., 2013. Microplastic ingestion by zooplankton. *Environmental Science & Technology*, 47(12), 6646–6655.

Cole, M., Lindeque, P.K., Fileman, E., Clark, J., Lewis, C., Halsband, C., and Galloway, T.S., 2016. Microplastics alter the properties and sinking rates of zooplankton faecal pellets. *Environmental Science & Technology*, 50 (6), 3239–3246.

Corradini, F., Bartholomeus, H., Huerta Lwanga, E., Gertsen, H., and Geissen, V., 2019. Predicting soil microplastic concentration using vis-NIR spectroscopy. *Science of the Total Environment*, 650, 922–932.

Crawford, C.B., and Quinn, B., 2017. *Physiochemical Properties and Degradation.* Microplastic Pollutants.

da Costa, J.P., Santos, P.S.M., Duarte, A.C., and Rocha-Santos, T., 2016. (Nano)plastics in the environment—Sources, fates and effects. *Science of the Total Environment*, 566–567, 15–26.

Danso, D., Schmeisser, C., Chow, J., Zimmermann, W., Wei, R., Leggewie, C., Li, X., Hazen, T., and Streit, W.R., 2018. New insights into the function and global distribution of polyethylene terephthalate (PET)-degrading bacteria and enzymes in marine and terrestrial metagenomes. *Applied and Environmental Microbiology*, 84 (8), e02773.

Delacuvellerie, A., Cyriaque, V., Gobert, S., Benali, S., and Wattiez, R., 2019. The plastisphere in marine ecosystem hosts potential specific microbial degraders including Alcanivorax borkumensis as a key player for the low-density polyethylene degradation. *Journal of Hazardous Materials*, 380, 120899.

Denaro, R., Aulenta, F., Crisafi, F., Di Pippo, F., Viggi, C.C., Matturro, B., Tomei, P., Smedile, F., Martinelli, A., Di Lisio, V., and Venezia, C., 2020. Marine hydrocarbon-degrading bacteria breakdown poly (ethylene terephthalate) (PET). *Science of the Total Environment*, 749, 141608.

Devi, R.S., Ramya, R., Kannan, K., Antony, A.R., and Kannan, V.R., 2019. Investigation of biodegradation potentials of high density polyethylene degrading marine bacteria isolated from the coastal regions of Tamil Nadu, India. *Marine Pollution Bulletin*, 138, 549–560.

Díaz-Mendoza, C., Mouthon-Bello, J., Pérez-Herrera, N.L., and Escobar-Díaz, S.M., 2020. Plastics and microplastics, effects on marine coastal areas: A review. *Environmental Science and Pollution Research*, 27 (32), 39913–39922.

Du, H., Xie, Y., and Wang, J., 2021. Microplastic degradation methods and corresponding degradation mechanism: Research status and future perspectives. *Journal of Hazardous Materials*, 418, 126377.

Duis, K., and Coors, A., 2016. Microplastics in the aquatic and terrestrial environment: Sources (with a specific focus on personal care products), fate and effects. *Environmental Sciences Europe*, 28 (1), 2.

Eriksen, M., Liboiron, M., Kiessling, T., Charron, L., Alling, A., Lebreton, L., Richards, H., Roth, B., Ory, N.C., Hidalgo-Ruz, V., Meerhoff, E., Box, C., Cummins, A., and Thiel, M., 2018. Microplastic sampling with the AVANI trawl compared to two neuston trawls in the Bay of Bengal and South Pacific. *Environmental Pollution*, 232, 430–439.

Fendall, L.S., and Sewell, M.A., 2009. Contributing to marine pollution by washing your face: Microplastics in facial cleansers. *Marine Pollution Bulletin*, 58 (8), 1225–1228.

Galloway, T.S., 2015. Micro-and nano-plastics and human health. *Marine Anthropogenic Litter*, 343–366.

Gao, R., and Sun, C., 2021. A marine bacterial community capable of degrading poly (ethylene terephthalate) and polyethylene. *Journal of Hazardous Materials*, 416, 125928.

Habib, S., Iruthayam, A., Abd Shukor, M.Y., Alias, S.A., Smykla, J., and Yasid, N.A., 2020. Biodeterioration of untreated polypropylene microplastic particles by Antarctic bacteria. *Polymers*, 12 (11), 2616.

Hale, R.C., Seeley, M.E., La Guardia, M.J., Mai, L., and Zeng, E.Y., 2020. A Global Perspective on Microplastics. *Journal of Geophysical Research: Oceans*, 125 (1), 1–40.

Haque, F., and Fan, C., 2022. *Microplastics in the Marine Environment: A Review of Their Sources, Formation, Fate, and Ecotoxicological Impact.* intechopen.

Harshvardhan, K., and Jha, B., 2013. Biodegradation of low-density polyethylene by marine bacteria from pelagic waters, Arabian Sea, India. *Marine Pollution Bulletin*, 77 (1–2), 100–106.

Hartmann, N.B., Nolte, T., Sørensen, M.A., Jensen, P.R., and Baun, A., 2015. Aquatic ecotoxicity testing of nanoplastics: Lessons learned from nanoecotoxicology. In *ASLO Aquatic Sciences Meeting*, Vol. 2015.

Hiraga, K., Taniguchi, I., Yoshida, S., Kimura, Y., and Oda, K., 2019. Biodegradation of waste PET. *EMBO Reports*, 20 (11), 1–5.

Huerta Lwanga, E., Thapa, B., Yang, X., Gertsen, H., Salánki, T., Geissen, V., and Garbeva, P., 2018. Decay of low-density polyethylene by bacteria extracted from earthworm's guts: A potential for soil restoration. *Science of the Total Environment*, 624, 753–757.

Ikenoue, T., Nakajima, R., Fujiwara, A., Onodera, J., Itoh, M., Toyoshima, J., Watanabe, E., Murata, A., Nishino, S., and Kikuchi, T., 2023. Horizontal distribution of surface microplastic concentrations and water-column microplastic inventories in the Chukchi Sea, western Arctic Ocean. *Science of the Total Environment*, 855 (November 2022), 159564.

Joo, S.H., Liang, Y., Kim, M., Byun, J., and Choi, H., 2021. Microplastics with adsorbed contaminants: Mechanisms and treatment. *Environmental Challenges*, 3 (December 2020), 100042.

Joshi, G., Goswami, P., Verma, P., Prakash, G., Simon, P., Vinithkumar, N.V., and Dharani, G., 2022. Unraveling the plastic degradation potentials of the plastisphere-associated marine bacterial consortium as a key player for the low-density polyethylene degradation. *Journal of Hazardous Materials*, 425, 128005.

Käppler, A., Fischer, D., Oberbeckmann, S., Schernewski, G., Labrenz, M., Eichhorn, K.-J., and Voit, B., 2016. Analysis of environmental microplastics by vibrational microspectroscopy: FTIR, Raman or both? *Analytical and Bioanalytical Chemistry*, 408 (29), 8377–8391.

Khandare, S.D., Chaudhary, D.R., and Jha, B., 2021. Marine bacterial biodegradation of low-density polyethylene (LDPE) plastic. *Biodegradation*, 32 (2), 127–143.

Koelmans, A.A., 2015. Modeling the role of microplastics in bioaccumulation of organic chemicals to marine aquatic organisms. A critical review. *Marine Anthropogenic Litter*, 309–324.

Kumari, A., Chaudhary, D.R., and Jha, B., 2019. Destabilization of polyethylene and polyvinylchloride structure by marine bacterial strain. *Environmental Science and Pollution Research*, 26, 1507–1516.

Kunz, A., Walther, B.A., Löwemark, L., and Lee, Y.C., 2016. Distribution and quantity of microplastic on sandy beaches along the northern coast of Taiwan. *Marine Pollution Bulletin*, 111 (1–2), 126–135.

Lacerda, A.L.D.F., Rodrigues, L.D.S., Van Sebille, E., Rodrigues, F.L., Ribeiro, L., Secchi, E.R., Kessler, F., and Proietti, M.C., 2019. Plastics in sea surface waters around the Antarctic Peninsula. *Scientific Reports*, 9 (1), 3977.

Li, W.C., 2018. The occurrence, fate, and effects of microplastics in the marine environment. In *Microplastic Contamination in Aquatic Environments*. Elsevier, 133–173.

Liu, G., Dave, P.H., Kwong, R.W.M., Wu, M., and Zhong, H., 2021. Influence of microplastics on the mobility, bioavailability, and toxicity of heavy metals: A review. *Bulletin of Environmental Contamination and Toxicology*, 107 (4), 710–721.

Lusher, A.L., Burke, A., O'Connor, I., and Officer, R., 2014. Microplastic pollution in the Northeast Atlantic Ocean: Validated and opportunistic sampling. *Marine Pollution Bulletin*, 88 (1–2), 325–333.

Maddah, H.A., 2016. Polypropylene as a promising plastic: A review. *American Journal of Polymer Science*, 6 (1), 1–11.

McKeen, L.W., 2014. *Plastics Used in Medical Devices. Handbook of Polymer Applications in Medicine and Medical Devices*. Elsevier Inc.

Menéndez-Pedriza, A., and Jaumot, J., 2020. Interaction of environmental pollutants with microplastics: A critical review of sorption factors, bioaccumulation and ecotoxicological effects. *Toxics*, 8 (2), 40.

Morohoshi, T., Ogata, K., Okura, T., and Sato, S., 2018. Molecular characterization of the bacterial community in biofilms for degradation of poly (3-hydroxybutyrate-co-3-hydroxyhexanoate) films in seawater. *Microbes and Environments*, 33 (1), 19–25.

Mu, J., Zhang, S., Qu, L., Jin, F., Fang, C., Ma, X., Zhang, W., and Wang, J., 2019. Microplastics abundance and characteristics in surface waters from the Northwest Pacific, the Bering Sea, and the Chukchi Sea. *Marine Pollution Bulletin*, 143, 58–65.

Nag, M., Lahiri, D., Dutta, B., Jadav, G., and Ray, R.R., 2021. Biodegradation of used polyethylene bags by a new marine strain of Alcaligenes faecalis LNDR-1. *Environmental Science and Pollution Research*, 28, 41365–41379.

Naji, A., Esmaili, Z., and Khan, F.R., 2017. Plastic debris and microplastics along the beaches of the strait of Hormuz, Persian gulf. *Marine Pollution Bulletin*, 114 (2), 1057–1062.

Obbard, R.W., Sadri, S., Wong, Y.Q., Khitun, A.A., Baker, I., and Richard, C., 2014. Who Where Why—wordpress blog—Community mapping examples. *Earth's Future*, 2, 315–320.

Oliveira, J., Belchior, A., da Silva, V.D., Rotter, A., Petrovski, Ž., Almeida, P.L., Lourenço, N.D., and Gaudêncio, S.P., 2020. Marine environmental plastic pollution: Mitigation by microorganism degradation and recycling valorization. *Frontiers in Marine Science*, 7 (December).

Othman, A.R., Hasan, H.A., Muhamad, M.H., Ismail, N.I., and Abdullah, S.R.S., 2021. Microbial degradation of microplastics by enzymatic processes: A review. *Environmental Chemistry Letters*, 19, 3057–3073.

Paço, A., Duarte, K., da Costa, J.P., Santos, P.S., Pereira, R., Pereira, M.E., Freitas, A.C., Duarte, A.C., and Rocha-Santos, T.A., 2017. Biodegradation of polyethylene microplastics by the marine fungus Zalerion maritimum. *Science of the Total Environment*, 586, 10–15.

Parray, Z.A., Hassan, M.I., Ahmad, F., and Islam, A., 2020. *Amphiphilic Nature of Polyethylene Glycols and Their Role in Medical Research*. Polymer Testing. Elsevier Ltd.

Peez, N., Janiska, M.-C., and Imhof, W., 2019. The first application of quantitative 1H NMR spectroscopy as a simple and fast method of identification and quantification of microplastic particles (PE, PET, and PS). *Analytical and Bioanalytical Chemistry*, 411 (4), 823–833.

Pereao, O., Opeolu, B., and Fatoki, O., 2020. Microplastics in aquatic environment: Characterization, ecotoxicological effect, implications for ecosystems and developments in South Africa. *Environmental Science and Pollution Research*, 27 (18), 22271–22291.

Porter, A., Lyons, B.P., Galloway, T.S., and Lewis, C., 2018. Role of marine snows in microplastic fate and bioavailability. *Environmental Science & Technology*, 52 (12), 7111–7119.

Rochman, C.M., Hoh, E., Kurobe, T., and Teh, S.J., 2013. Ingested plastic transfers hazardous chemicals to fish and induces hepatic stress. *Scientific Reports*, 3 (1), 3263.

Rummel, C.D., Jahnke, A., Gorokhova, E., Kühnel, D., and Schmitt-Jansen, M., 2017. Impacts of biofilm formation on the fate and potential effects of microplastic in the aquatic environment. *Environmental Science & Technology Letters*, 4 (7), 258–267.

Rüthi, J., Cerri, M., Brunner, I., Stierli, B., Sander, M., and Frey, B., 2023. Discovery of plastic-degrading microbial strains isolated from the alpine and Arctic terrestrial plastisphere. *Frontiers in Microbiology*, 14, 1306.

Scalenghe, R., 2018. Resource or waste? A perspective of plastics degradation in soil with a focus on end-of-life options. *Heliyon*, 4 (12), e00941.

Sekiguchi, T., Saika, A., Nomura, K., Watanabe, T., Watanabe, T., Fujimoto, Y., Enoki, M., Sato, T., Kato, C., and Kanehiro, H., 2011b. Biodegradation of aliphatic polyesters soaked in deep seawaters and isolation of poly (ε-caprolactone)-degrading bacteria. *Polymer Degradation and Stability*, 96 (7), 1397–1403.

Sekiguchi, T., Sato, T., Enoki, M., Kanehiro, H., Uematsu, K., and Kato, C., 2011a. Isolation and characterization of biodegradable plastic degrading bacteria from deep-sea environments. *JAMSTEC Report of Research and Development*, 11, 33–41.

Shahul Hamid, F., Bhatti, M.S., Anuar, N., Anuar, N., Mohan, P., and Periathamby, A., 2018. Worldwide distribution and abundance of microplastic: How dire is the situation? *Waste Management and Research*, 36 (10), 873–897.

Sharma, S., and Chatterjee, S., 2017. Microplastic pollution, a threat to marine ecosystem and human health: A short review. *Environmental Science and Pollution Research*, 24 (27), 21530–21547.

Shaw, D.G., and Day, R.H., 1994. Colour- and form-dependent loss of plastic micro-debris from the North pacific ocean. *Marine Pollution Bulletin*, 28 (1), 39–43.

Shaw, R., Dutta, B., Ghosh, D., and Bandopadhyay, R., 2023. Hepatitis C—route of asymptomatic to symptomatic switch in raising hepatocarcinogenesis: Revisiting nobel prize 2020 in physiology and medicine. *National Academy Science Letters*, 46 (1), 51–54.

Song, Y.K., Hong, S.H., Eo, S., Jang, M., Han, G.M., Isobe, A., and Shim, W.J., 2018. Horizontal and vertical distribution of microplastics in Korean coastal waters. *Environmental Science & Technology*, 52 (21), 12188–12197.

Spoerk, M., Holzer, C., and Gonzalez-Gutierrez, J., 2020. Material extrusion-based additive manufacturing of polypropylene: A review on how to improve dimensional inaccuracy and warpage. *Journal of Applied Polymer Science*, 137 (12), 1–16.

Sruthy, S., and Ramasamy, E.V., 2017. Microplastic pollution in Vembanad Lake, Kerala, India: The first report of microplastics in lake and estuarine sediments in India. *Environmental Pollution*, 222, 315–322.

Sudhakar, M., Priyadarshini, C., Doble, M., Murthy, P.S., and Venkatesan, R., 2007. Marine bacteria mediated degradation of nylon 66 and 6. *International Biodeterioration & Biodegradation*, 60(3), 144–151.

Suzuki, M., Tachibana, Y., Oba, K., Takizawa, R., and Kasuya, K.I., 2018. Microbial degradation of poly (ε-caprolactone) in a coastal environment. *Polymer Degradation and Stability*, 149, 1–8.

Tirkey, A., and Upadhyay, L.S.B., 2021. Microplastics: An overview on separation, identification and characterization of microplastics. *Marine Pollution Bulletin*, 170, 112604.

Urbanek, A.K., Strzelecki, M.C., and Mirończuk, A.M., 2021. The potential of cold-adapted microorganisms for biodegradation of bioplastics. *Waste Management*, 119, 72–81.

Van Cauwenberghe, L., and Janssen, C.R., 2014. Microplastics in bivalves cultured for human consumption. *Environmental Pollution*, 193, 65–70.

Waller, C.L., Griffiths, H.J., Waluda, C.M., Thorpe, S.E., Loaiza, I., Moreno, B., Pacherres, C.O., and Hughes, K.A., 2017. Microplastics in the Antarctic marine system: An emerging area of research. *Science of the Total Environment*, 598, 220–227.

Wang, T., Wang, L., Chen, Q., Kalogerakis, N., Ji, R., and Ma, Y., 2020. Interactions between microplastics and organic pollutants: Effects on toxicity, bioaccumulation, degradation, and transport. *Science of the Total Environment*, 748, 142427.

Wang, Z.-M., Wagner, J., Ghosal, S., Bedi, G., and Wall, S., 2017. SEM/EDS and optical microscopy analyses of microplastics in ocean trawl and fish guts. *Science of the Total Environment*, 603–604, 616–626.

Webb, H.K., Arnott, J., Crawford, R.J., and Ivanova, E.P., 2012. Plastic degradation and its environmental implications with special reference to poly (ethylene terephthalate). *Polymers*, 5 (1), 1–18.

Yamano, N., Kawasaki, N., Ida, S., and Nakayama, A., 2019. Biodegradation of polyamide 4 in seawater. *Polymer Degradation and Stability*, 166, 230–236.

Yang, H., Chen, G., and Wang, J., 2021. Microplastics in the marine environment: Sources, fates, impacts and microbial degradation. *Toxics*, 9 (2), 1–19.

Yu, Q., Hu, X., Yang, B., Zhang, G., Wang, J., and Ling, W., 2020. Distribution, abundance and risks of microplastics in the environment. *Chemosphere*, 249, 126059.

Yu, X., Ladewig, S., Bao, S., Toline, C.A., Whitmire, S., and Chow, A.T., 2018. Occurrence and distribution of microplastics at selected coastal sites along the southeastern United States. *Science of the Total Environment*, 613–614, 298–305.

Zarfl, C., 2019. Promising techniques and open challenges for microplastic identification and quantification in environmental matrices. *Analytical and Bioanalytical Chemistry*, 411 (17), 3743–3756.

Zhai, X., Zhang, X.H., and Yu, M., 2023. Microbial colonization and degradation of marine microplastics in the plastisphere: A review. *Frontiers in Microbiology*, 14.

Zhang, B., Chen, L., Chao, J., Yang, X., and Wang, Q., 2020. Research progress of microplastics in freshwater sediments in China. *Environmental Science and Pollution Research*, 27 (25), 31046–31060.

Zhao, S., Zhu, L., and Li, D., 2015. Microplastic in three urban estuaries, China. *Environmental Pollution*, 206, 597–604.

Zhao, S., Zhu, L., Wang, T., and Li, D., 2014. Suspended microplastics in the surface water of the Yangtze estuary system, China: First observations on occurrence, distribution. *Marine Pollution Bulletin*, 86 (1–2), 562–568.

11 Plastic Degradation Using Chemical/Solar/ Biodegradation

Poushali Chakraborty, Sampad Sarkar,
Arkaprava Roy, Kesang Tamang and Papita Das

11.1 INTRODUCTION

Plastic is widely used all over the world in various industries and agricultural activities due to its low weight, low price, transparency, resistance to harsh weather, user-friendly design, electrical insulation, and chemical resistance (Jenkins *et al.*, 2019; Amobonye *et al.*, 2021). The majority of plastics are not degradable and even take centuries to degrade completely. As a consequence, plastics frequently accumulate in disposal areas or the environment, despite breaking down (Matjašič *et al.*, 2021). However, for the synthesis of plastic polymers (e.g., propylene and ethylene), the most commonly used monomers are fossil-derived hydrocarbons; due to their polymeric structure, they resist microbial degradation. This accumulation of plastics leads to serious problems in the long term. Degradation of the accumulated plastic in several biotic and abiotic ways leads to solving this problem. Some of the synthetic polymers can be degraded by thermo-oxidative and photo-oxidative reactions by absorbing ultraviolet (UV) radiation. Another efficient route to degrade plastic is biodegradation which involves the biochemical transformation of the polymers by microorganisms, depending on the abundance of oxygen.

Polyethylene (PE), polyethylene terephthalate (PET), polystyrene (PS), polypropylene (PP), and polyvinylchloride (PVC) are currently the most extensively used polymers. They are further classified into heteroatomic polymers and the backbone of C-C polymers. Among them, heteroatomic polymers (PET and PU) can be transformed by biological degradation, photo-oxidation, and hydrolysis, whereas the backbone of C-C polymers (PP, PS, PVP, and PE) are resistant to biodegradation and hydrolysis but have susceptibility to thermal oxidation (Gewert *et al.*, 2015).

Single use of plastic items over the years leads to the global rise in plastic pollution, and for this reason, it's predicted that the global market will reach USD 722.6 billion by the year 2027 (Dar *et al.*, 2022). Numerous hazardous gaseous chemicals, including hydrogen cyanide (HCN), carbon dioxide (CO_2), carbon monoxide (CO), and nitrogen oxides which are extremely dangerous to the biotic system are released into the air throughout the plastic production process. The outbreak of COVID-19 increases the demand for polymer materials like gloves, face shields, and single-use, disposable face masks, food containers, and plastic bags. Due to the pandemic, around 52 billion single-use, disposable face masks were made in 2020 alone, and

DOI: 10.1201/9781003449133-11

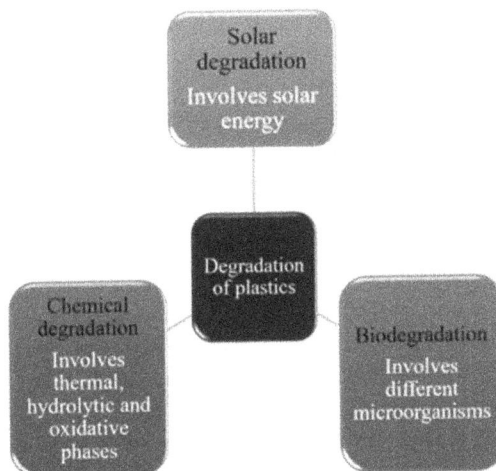

FIGURE 11.1 Different techniques used to degrade plastics.

among them, about 1.6 billion ended up in the oceans, where they will take 40–50 centuries to degrade completely (Lee and Li, 2021). Because of improper disposal processes and management, huge amounts of plastic wastes are introduced into the living world, where they degrade into small pieces through physical, chemical, and biological processes. It leads to the formation of microplastics, which become more toxic for aquatic and land animals. During the natural breakdown of these masks, microplastic will be discharged into water bodies and eventually enter into our food chain. Among these plastic wastes, only 10% are recycled, and in nature, most of the waste is accumulated (Liang *et al.*, 2021).

This chapter has covered various plastics sources, their potentially harmful impacts on the living world, and the mechanisms underlying the different kinds of polymer degradation (photo-oxidative degradation, thermal degradation, chemical degradation, and biological degradation). (Refer to Figure 11.1.)

11.2 SOURCES OF PLASTICS

Plastics are a type of organic polymer which are synthetic and semi-synthetic in nature and have multiple properties like long durability, low cost, high tensile-strength, corrosion resistance, and being lightweight (Thompson *et al.*, 2009). Some examples of commonly used polymers that cover up to 90% of the plastic production worldwide are polystyrene (PS), polyethylene terephthalate (PET), high-density and low-density polyethylene (HDPE and LDPE), polyvinyl chloride (PVC), and polypropylene (PP) (Andrady and Neal, 2009). This huge production rate of plastics has led the way in contaminating all aspects of nature, especially aquatic lives. From different sources, plastic wastes are added to aquatic ecosystems (Engler, 2012). In land sites, sources of plastics are household wastes, various industries, markets, etc., whereas parts of fishing vehicles, ships, aquaculture facilities, oil spillage, and boats are responsible for the pollution caused by plastics in the marine environment (Andrady, 2011). According to size variations, plastics can be categorized as meso-plastic (size < 2.5 cm), macroplastic (size < 1 m), microplastic (size < 5 mm), and megaplastic (size > 1 m) (Wang *et al.*, 2020). In recent findings, it has been seen that plastics degrades in nature and forms nano-sized plastic particles, sized < 100 nm (Mattsson *et al.*, 2018). According to Teuten *et al.* (2009), effects of micro- and nano-plastics are a subject of concern in present times for marine lives. As they can easily penetrate living cells of organisms due to their small size, their adverse effects on human health and every living organism's body are immense (Sharma *et al.*, 2022). Of the two types, primary microplastics can be defined as microscopic-sized plas-tics. Their sources include air-blasting media, facial cleansers and cosmetics, vir-gin plastic production pellets and vectors for drugs, etc. (Patel *et al.*, 2009; Cole *et al.*, 2011; Auta *et al.*, 2017). These particles flow down into urban drainage systems and meet water bodies. Due to physical, chemical, and biological weathering, large plastic particles get embrittled into finer particles and form secondary microplastics (Auta *et al.*, 2017; Sundt *et al.*, 2014). These fragments may include polymer fibres, which are released from synthetic textiles during fabric washing, fibre-manufacture discharge, household items, consumer products, and wear and tear of plastic items (Mason *et al.*, 2016). Fragmentation increases the surface area and number of par-ticles per unit mass of the plastics. In marine waters, primary causes of fragmenta-tion are both wave action and exposure to sunlight. Fragmentation of plastics occurs readily on land, especially at the soil surface, as the particles are subjected to direct UV radiation exposure. Temperature fluctuation also plays a significant role in this process (Andrady, 2011). Transportation of marine microplastics from land to the oceans can be traced back to anthropological problems of littering and inadequate waste management practices (Pruter, 1987).

11.3 TOXICITY CAUSED BY PLASTICS

The most harmful environmental impact of plastics involves global warming as they release CO_2 while burning and cause problems like land filling due to poor waste-disposal systems (Ali *et al.*, 2009). Emissions while burning the plastics includes toxic gases that cause respiratory disorders, including cancer. Improper dis-posal often leads to the leaching of toxins, thereby contaminating both the soil and the ground water. These leachate often contaminate drinking water sources, either directly or through run-off contamination, thereby entering into the food web in dif-ferent ecosystems (Kale *et al.*, 2015). Entering the food chain, microplastics continue to accumulate in the biomass at each trophic level, increasing the toxicity through the phenomenon of bioaccumulation from zooplankton to the largest organisms in the marine environment (Matsuguma *et al.*, 2017; Hipfner *et al.*, 2018). Consumption of these organisms such as fish, oysters, etc., containing microplastics embedded in their tissues causes microplastics to enter the human body, thereby effecting us severely. Effects on human brain cells, the cardiovascular system, and the digestive system are quite common (Schirinzi *et al.*, 2017). Toxicity from microplastics can also be considered to be carcinogenic nature. From the recent findings of Osman *et al.* (2023), the presence of microplastics has been observed in different human body fluids. Similar studies on different lower-level organisms have also shown same

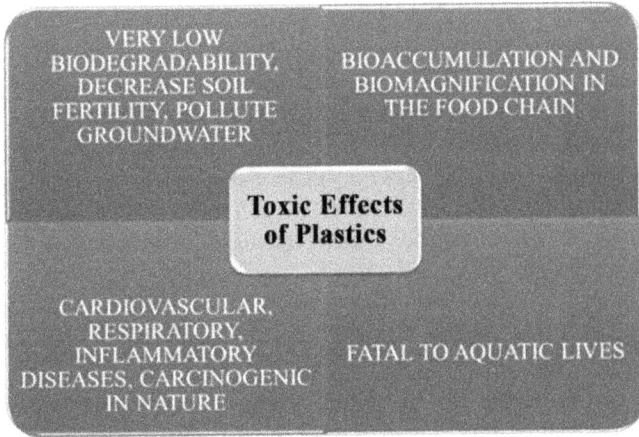

FIGURE 11.2 Toxic effects caused by plastics in nature.

effects. To mitigate these issues, proper waste management at a grassroots level, coupled with environmentally friendly policies, along with proper and cost-effective technologies—viz. photodegradation, chemical degradation, and biodegradation—should be implemented. (Refer to Figure 11.2.)

11.4 PHOTODEGRADATION

Photodegradation is one of the most important processes that initiates the abiotic degradation pathway which involves reactions of free radicals mediated by the absorption of photons. The process of photodegradation changes the physical as well as chemical structure of the plastic. There is a wide range of solar irradiation, from the ultraviolet spectrum to the infrared spectrum, and the sun emits energy with decreasing energy. Among the ultraviolet (UV) irradiation UV-C (100–280 nm) is absorbed by the ozone in the stratosphere (Lyon, 2014). In the presence of oxygen, UV-B (280–320 nm) and UV-A (320–400 nm) is mainly responsible for the degradation, which includes chain scission, deterioration of polymeric properties, and change in molecular weight (Liu $et\ al.$, 2019).

Several factors are responsible for the process of photodegradation, which includes internal impurities (hydroperoxide, carbonyl, unsaturated bonds, charge transfer complexes with O_2, catalyst residue) and external impurities (a compound from pollutants, traces solvents, metal and metal oxides, additives-pigments, dyes) (Yousif and Haddad, 2013). The plastic degradation mechanism consists of three steps, which include photoinitiation, propagation, and termination.

11.4.1 PHOTOINITIATION

The chromophoric group is important for the initiation of photochemical reactions. In this step, the chromophoric group of the plastic absorbs light and breaks the chemical bonds of the polymeric chains, which leads to the production of polymeric radicals and hydrogen radicals. Some of the plastic, like PP and PE, are considered as non-absorbing polymers, as they only have bonds of C-H and C-C, which make them stable above the wavelength of 290 nm (Fairbrother *et al.*, 2019). Though they are devoid of any unsaturated chromophores to absorb light, the presence of impurities (external) and/or some structural defect can lead them to initiate photodegradation.

Different initiation steps are present in different environmental conditions to produce free radicals. Hydroperoxide (POOH) and ketone are the potential initiators which can generate radicals that can remove hydrogen atoms from polymers and lead to photodegradation (Amin *et al.*, 1975).

11.4.2 PROPAGATION

In the propagation reaction, all the polymers which have carbon backbone display the same auto-oxidation cycle. As a result of this process, hydroperoxide is created, which is the key intermediate in subsequent reactions but does not immediately lead to backbone cleavage. The generation of hydroperoxide, produced during the propagating step, leads the backbone to degrade by cleaving the hydroperoxide O-O bond and initiating β-scission (Tyler, 2004). Cleavage of the polymer backbone is governed by Norrish Type I and Norrish Type II reactions of the chromophores, which are a common mechanism for photo-oxidative degradation (Mark *et al.*, 1986). This scission takes place in the amorphous domains of semi-crystalline polymers. As oxidative degradation progresses, the scission process produces two chain ends, which are free to reconstruct and increase crystallinity frequently. In the end, chain scission is caused by propagation to create an oxygen-containing functional group, such as ketones and olefin (Yousif and Haddad, 2013).

11.4.3 TERMINATION

The free radicals generated in the propagation step then interact freely with one another and undergo the crosslinking reaction, resulting in the production of inert chemicals. Polymer peroxyl radicals interact with one another under conditions of high oxygen pressure to produce oxygen and inert products. But, in the absence of adequate oxygen, a reaction between the polymer's peroxyl radicals and macro radicals occurs. As the fate of the termination reaction, ketones and olefins are expected (Ali *et al.*, 2021).

Polyethylene (PE) lacks chromophores, which leads to resistance to photodegradation, but structural defects or impurities during manufacture can lead to degradation. The presence of the carbonyl group of the PE backbone can function as chromophores (Fairbrother *et al.*, 2019). Down to several reactions, end-vinyl, radicals, and ketone groups are formed, which results in a scission of the main chain (Karlsson and Albertsson, 2002). Peroxy radicals are produced from the free radicals and oxygen reactions. By hydrogen abstraction, peroxide moiety is formed by the conversion of peroxy radicals. This peroxide moiety breaks into hydroxyl radicals and macro alkoxy. Aldehydes, ketones, esters, carboxylic acid, and alcohols may be produced throughout the reaction sequence, and crosslinking and chain scission take place (Torikai *et al.*, 1986).

In the case of photodegradation, PP has a similar mechanism as polyethylene (PE), in which the impurities in PP allow it to form radicals in the presence of UV radiation. This leads to chain scission and cross-linking, and finally, lower molecular weight of the degraded product generated (Su *et al.*, 2019; He *et al.*, 2019).

In the case of PVC, UV irradiation short-sequence unsaturated polymers are formed by the mechanism of rapid dehydrochlorination. Next, the unsaturated double bonds of carbon are photodegraded. The presence of impurities in the PVC absorbs UV radiation, which generates free radicals and leads to the formation of hydroperoxides. These hydroperoxides can break more double bonds, and finally, produce the degraded product of PVC (Yang *et al.*, 2018).

The presence of a phenyl ring in polystyrene (PS) gets excited under UV radiation and forms a triple state. The phenyl group's dissociation, or transfer, to the closest C-H or C-C bonds are two possible outcomes of the excited benzene's triplet energy. A polystyryl radical is produced in the absence of oxygen. The conversion of a polystyryl radical into a peroxy radical is initiated in the presence of oxygen, which then reacts with the nearby polystyrene molecule. Olefins styrene monomer and carbonyl compounds can be formed as the end product of the process of chain scission and cross-linking (Dris *et al.*, 2017; Kumar *et al.*, 2020)

11.5 CHEMICAL DEGRADATION

Degradation of plastic wastes by means of chemical processes is one of the most eminent methods used by either a complete or partial depolymerisation process and forms monomers (due to complete polymerization), oligomers (due to partial polymerization), and other products along with it. Depolymerization, induced due to chemical reactions in different mechanisms, is depicted here in detail, which includes a solvo-thermal mechanism, a hydrolytic mechanism, and an oxidative mechanism instigated thermally. In order to provide a better and clearer picture of the thermal degradation of polymers, suitable reaction kinetics are also discussed.

11.5.1 SOLVO-THERMAL PROCESS

The solvo-thermal process for plastic degradation is considered as one of the emerging and most novel chemical techniques of degradation. The plastic-laden pollutant feedstock is reacted with sub-critical (i.e., thermodynamic states such as pressure and temperature are below the critical point) solvent and super-critical solvent (Saha *et al.*, 2022). The whole process consists of four major steps such as segregation, dissolving, reprecipitation (devolatilization), and polymer separation (Zhao *et al.*, 2018). Initially, the plastic impurities are removed or segregated mechanically for decreasing the heterogeneity of the pollutant-laden solution. Thereafter, the dissolution is done using solvents at different sub-critical and super-critical conditions. For this purpose, organic solvents such as toluene are preferred (Saha *et al.*, 2022). Also, solvents such as methanol, propanol, and xylene are used as well. The main criteria for choosing the solvents are RED (i.e., relative energy difference) and boiling point (Zhao *et al.*, 2017).

$$RED = \frac{R_a}{R_0}$$

$$\left(R_a\right)^2 = 4\,(\delta_{Db} - \delta_{Da})^2 + (\delta_{Pb} - \delta_{Pa})^2 + (\delta_{Hb} - \delta_{Ha})^2$$

Where

R_0 : *Experimentally measured radius of the polymer solubilty sphe..*
δ_D : *Dispersion solubility parameter.*
δ_P : *Polar solubility parameter.*
δ_H : *Hydrogen bonding solubility parameter.*
a: Solvent chosen for solvo-thermal process
b: Plastic laden solution

Solvents were chosen on the basis of less toxicity, easy operability, and less environmental impact (Hansen, 2007). The recovered polymers are collected by reducing the solubility of plastics, done by adding non-solvents or by increasing the temperature or decreasing pressure by adding supercritical solvents.

11.5.2 HYDROLYTIC DEGRADATION PROCESS

Hydrolytic degradation method is considered one of the most recognized chemical processes for plastic waste treatment and degradation. Plastic contains unstable and weak polymeric bonds which react with water molecules and degrade to form shorter chains (i.e., monomers, oligomers) to produce new polymeric chain ends. The active site vulnerable to water molecules contains oxygen, nitrogen, sulphur, phosphorous, and other non-carbon atoms (Lyu and Untereker, 2009).

PEG (polyethylene glycol) and PET (polyethylene terephthalate) are the O-containing polymers with huge industrial applicability which cause adverse effects on environment. N-containing polymers such as polyamides, polyurethanes, polyureas, and nylon (6, 6) also have huge significance in daily life but are harmful for the environment. S-containing polymers are also incorporated with N molecules. The main configuration looks like (-S-N-S-N-), known as polythiazyls. Similarly, phosphorous forms polymers forming a similar structure like polythiazyls (-P-N-P-N-P-), known as polyphosphazenes. The polymers form cleaved product due to the attack of water molecules on electron-deficient carbon atoms adjacent to high-electronegative oxygen, sulphur, nitrogen, and phosphorous atoms (Lyu and Untereker, 2009). There are several reaction mechanisms which depict the process. A popular model depicts few steps such as diffusivity and dissolution of water molecules in the polymeric structure, generation of pore inside the polymeric wall structure, release of degraded oligomers from the polymer, and finally, the remaining degraded polymer porous skeleton (Batycky, R.P. *et al.*, 1997). Autocatalytic degradation of polylactic acids such as PLGA follows the heterogeneous model, which follows steps such as penetrative diffusion of water into the polymeric core; the released carboxylic chains, in turn, favours the degradation process, making it more of an autocatalytic reaction

(Grizzi I. *et al.*, 1995). The hydrolytic process can also be differentiated in bulk erosion and surface erosion; for the bulk erosion process, the size of the particle changes w.r.t time, whereas the degraded plastic remains in the main polymeric structure, showing no effects in the change in radius and particle size for the surface erosion process (Siepmann, 2001).

11.5.3 Thermo-Oxidative Process

Permanent degradative changes in polymers by chemical means in the presence of oxygen at higher elevated temperatures are another popular process, which is known as a thermo-oxidative process. The whole process mechanism is studied in TGA. It is seen that the influence of oxygen reduces the processing temperature for plastic degradation. The process follows steps such as the initiation process, the chain propagation process, and monomolecular decomposition (limiting step) (Peterson, J.D. *et al.*, 2001).

$$\text{Initiation: } RH \rightarrow R^* + H^*$$

The thermal degradation process starts by producing free radicals. The addition of chemical initiators such as peroxides and hydroperoxides enhances the initiation process for this mechanism.

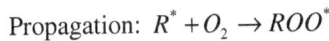

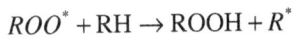

$$\text{Propagation: } R^* + O_2 \rightarrow ROO^*$$

$$ROO^* + RH \rightarrow ROOH + R^*$$

In this step, the alkyl free radical released from the previous step reacts with oxygen present in the reaction environment and forms per-oxy radical intermediate. The highly reactive per-oxy intermediate attacks the polymeric carbon to eliminate hydrogen and forms carboxylic acid with polymeric free radical.

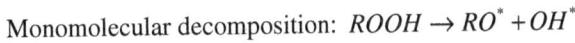

$$\text{Monomolecular decomposition: } ROOH \rightarrow RO^* + OH^*$$

This step is a highly energy-demanding process which is provided by the previous exothermic step (i.e., production of carboxylic acid).

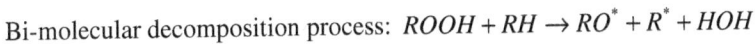

$$\text{Bi-molecular decomposition process: } ROOH + RH \rightarrow RO^* + R^* + HOH$$

However, sometimes the reaction of carboxylic acid is considered to be a bi-molecular process which requires less energy than the mono-molecular process.

Non-biodegradable polymers such as polyethylene, polypropylene, and polystyrene could be degraded effectively using the previously mentioned mechanism under different thermodynamic conditions. Even sulphur-based plastics such as polyphenylene sulphide (PPS), polysulfone (PSU), and polyether sulphone (PES) can also be effectively degraded using the thermo-oxidative method (Kumagai S. *et al.*, 2022).

The kinetics of the thermal and thermo-oxidative process can be best described using the iso-conversional methods (Das and Tiwari, 2017). The conversion due to degradation (α) is represented as →

$$\alpha = \frac{W_o - W}{(W_o - W_\infty)}$$

Where W_o: Weight of initially measured polymer sample
  W: Weight of polymer sample at a particular temperature.
  W_∞: Weight of the polymer sample after complete degradation.

The kinetic equation suggests that the reaction rate is directly proportional to a function of conversion. Which looks like

$$\frac{d\alpha}{dT} = k(T)f(\infty)$$

Taking the time-dependent term of temperature into concern (i.e., $T = T_0 + \beta t$)

where $(\beta = \frac{dT}{dt})$ we can write the equation as

$$\beta \frac{d\alpha}{dT} = A \exp\left(\frac{-E}{RT}\right) f(\alpha)$$

The solvo-thermal process for degradation requires an additional, suitable solvent, whereas other processes have no such requirements, so the solvo-thermal process may be an efficient and novel one, but it is, indeed, hard to execute for its complex experimentation. The hydrolytic degradation only needs the presence of water molecules, and the thermo-oxidative process requires an aerobic reaction environment for its completion. Analyzing the degraded polymers for each process shows that thermo-oxidative process can degrade a wide range of polymers such as LDPE and also a polymeric substance containing electro-negative atoms such as sulphur in its structure. Whereas the hydrolytic process requires the presence of oxygen, sulphur, and phosphorous molecules, polymeric substances like polyethylene, which is one of the most common plastics existing in nature, can't be degraded using this method. On the basis of the requirement of energy, the thermo-oxidative process requires a high amount of energy, whereas the other two processes are very energy-efficient.

11.6 BIODEGRADATION

Biodegradation of plastics depends upon the development of biofilms taking the surface layer of the plastics as substrates, then the breakdown these plastics into molecules of lower molecular weights through the help of extracellular enzymes present in the microorganisms, followed by ingestion of these smaller molecules within the microorganism's body and its metabolization of them. Plastics are a mixture of different compositions of polymer and associated additives. Microbial enzyme systems have a significant role in the degeneration of plastics (Krueger *et al.*, 2015). For oxidizing polymers like PP, PS and PE, several enzymes like mono and di- oxygenases, peroxidases, laccases, and cutinases are used (Jacquin *et al.*, 2019). These enzymes break the plastic polymers into smaller compounds with low molecular weight and more hydrophilicity. These oligomers of plastics are then subjected to microbial breakdown by various enzymes that lead to the formation of end metabolite products such as succinyl coA and acetyl coA (Jacquin *et al.*, 2019). Then these molecules are metabolized in the cells of microbes entering in the TCA cycle and convert to CO_2 and H_2O.

The process of biodegradation of plastics indicates the deterioration of various plastic polymers by different microbes which can use plastics as their main source of carbon and, as a result, organic compounds covert into residual biomass and biogas (Ali *et al.*, 2019). As the physiochemical properties of different plastic polymers act as weak substratum for microbial growth, widely used synthetic plastics available in markets like polyvinyl, polyamide (PA), PES, etc., poorly biodegrade in nature (Biffinger *et al.*, 2014).

The detailed procedure of microbial degradation of plastics can be explained in following five steps. (Refer to Figure 11.3.)

Colonization: At first, microorganisms colonize on the surface of the plastics, which leads to the formation of consortia and biofilm, as a result several physiochemical properties of the polymer changes like the pores' size alternates (Cappitelli *et al.*, 2006).

Biodeterioration: Various filamentous fungi and bacteria then invade the plastic material and result in more cracks and increase the pores. Chemolithotrophic microbes can chemically deteriorate the polymers by causing oxido-reductive reactions for up taking minerals like iron (Fe^{3+}) and manganese (Pelmont, 2005). Peroxidases are the group of extracellular enzymes that can be effective in biodeterioration. But for the resistant polymers like PU and PVC, several other enzymes like ureases, lipases, and proteases are secreted by the microbes to break down the crystal structure as much possible (Ameen *et al.*, 2015; Shah *et al.*, 2008).

Biofragmentation: Biofragmentation is the process in which lysis of polymers into monomers, dimers, and oligomers occurs by microbial activity. Enzymes like oxidoreductases and hydrolases are used to cleave the plastic polymers. Microbes that have the potential to degrade lignocellulosic compounds could also be responsible for cleaving bonds between plastic monomers, as they are a similar type in nature (Chen *et al.*, 2020). However, the

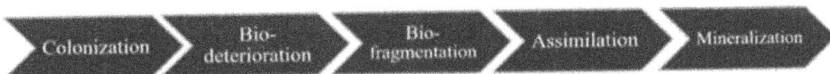

FIGURE 11.3 Schematic diagram of the biodegradation process.

microbial breakdown of plastic polymers take a longer period, as enzymes' activity increases with time. Soil isolated *Streptomyces sp.* can take 180 days to degrade polyethylene (PE) plastics up to 46.7% (Deepika and Jaya, 2015); *Pantoea sp.* isolated from soil can decrease the initial weight of LDPE strips by 81% (Skariyachan *et al.*, 2016). Enzymes like mono- and dioxygenases play an important role, as they tend to change the structure polarity of plastic polymers by forming –OH and peroxyl groups. Peroxidases then help to catalyze the reaction between peroxyl molecules. Several microbes can produce hydrogen peroxide that can increase oxidation in microbial cells and subsequently increase microbial attack.

Assimilation: After producing such monomers, microbes tend to assimilate them in their bodies by obtaining energy from the monomers and continue to grow. Monomers can enter the cells directly due to the permeability of microbial cell membrane or in association with a carrier. Different metabolites are then produced for preparing the final stage of mineralization.

Mineralization: As the monomers enter in the cells, they are oxidized, enter in three respiration processes—aerobic, anaerobic, and fermentation—and finally, produce H_2O and CO_2 after completing TCA cycle (Ali *et al.*, 2020; Hong and Gu, 2009).

Biodegradation of plastics can be assessed by weight loss, change in dimensions, tensile strength, and physical and chemical properties of plastic polymers. Production of carbon dioxide and bacterial activity in soil are also observed to characterize the degradation of plastics (Kathiresan K, 2003). In the following Table 11.1, some methods for biodegradation are described.

TABLE 11.1
Methods and Conditions for Biodegradation

Method	Conditions
1. Pure culture method	In laboratory conditions, plastic films are disinfected first, then added to specific growth medium, inoculated with previously isolated microorganisms. After being incubated for 3–4 weeks in a shaker, plastic films are washed and dried and final weight is compared with initial weight (Lee *et al.*, 1991).
2. Soil burial method	Plastic wastes are buried under soil for a specific interval of time, after which they are collected, washed, dried, and the final weight is measured, which is then compared with initial weight for observing degradation (Lopez-Llorca and Valiente, 1993).
3. Composting method	Dry plastic wastes are mixed up with manure compost, incubated at 58° C, and 65% moisture content. The amount of carbon molecules changed into CO_2 gas indicates the extent of biodegradation (Corti *et al.*, 1992).
4. Sewage sludge method	In the presence of sewage sludge enriched with different minerals, aerobic degradation of plastics occurs by diverse microbes (Ishigaki *et al.*, 1999).

11.6.1 Fungal Biodegradation of Plastics

In nature, some fungi are present that have the potential to degrade plastic debris. *Aspergillus*, *Penicillium*, and *Fusarium* are the most potential plastic degraders till date (Raghavendra *et al.*, 2016). Ameen *et al.* (2015) described the efficiency of *Alternaria alternata*, isolated from the mangrove forest, in degrading the LDPE in 28 days by observing the emission of CO_2, increasing biomass in the culture with LDPE. According to Deepika and Jaya (2015), *Aspergillus flavus*, isolated from the soil sample collected from waste disposal site, showed the potential of degrading PE by 16.2% in 180 days. Sangeetha Devi *et al.* (2015) also depicted the role of *Aspergillus flavus* VRKPT2, isolated from the coastal area of Gulf of Mannar, in weight reduction of HDPE by 8.5% in 30 days. Several strains of *Penicillium* were shown as potential PE plastic-degraders. The potential of *Leptosphaeria sp.* in degrading PE and PU by giving positive results in the Halo test in several days was described by Brunner *et al.* (2018). However, low crystalline structure of PET polymer showed 97% weight loss and degradation into TPA in only 6 days by *Thermomyces insolens* (Ronkvist *et al.*, 2009).

Several enzymes are associated with the process of degradation of plastics by fungi. Lignin and manganese peroxidases are commonly involved in the degradation of PE and nylon that are secreted by several strains like *Alternaria alternate*, *Trichoderma harzianum*, *Aspergillus terreus*, etc. (Xu *et al.*, 2013; Ameen *et al.*, 2015). Laccase enzyme activity was seen to degrade PE and PP polymers. In the marine conditions, these enzymes were shown as effective degraders of different plastics. Polyesterase and cutinase enzymes are mainly associated in the process of biodegradation of PET polymers.

11.6.2 Bacterial Degradation of Plastics

The potential of a wide range of bacteria in using various plastics as their carbon source and eventually degrading them has been observed since 1961 by Fuhs (1961). Nanda *et al.* (2010) showed the biodegradation capacity of *Pseudomonas sp.* isolated from sewage sludge by 29.1% for synthetic polythene bags, whereas Thakur (2012) observed 30% weight loss of polythene bags in 30 days by *Bacillus amylolyticus*. Rajandas *et al.* (2012) reported the capability of degrading polythene by *Microbacterium paraoxydans* and *Pseudomonas aeruginosa* up to 61% and 50% respectively. The role of different enzymatic systems has been discussed already. Hence, different strains of *Pseudomonas*, *Bacillus*, and *Lysinibacillus* are potential plastic degraders in the nature.

Lee *et al.* (1991) showed the efficiency of different strains of *Streptomyces* in degrading PE polymers in 20 days. *Vibrio sp.* was an efficient degrader of PET polymer by showing weight reduction of 35% in 42 days (Sarkhel *et al.*, 2020).

11.7 CONCLUSION

In recent times, plastic materials have become an important part of our daily needs. Properties like long durability, high tensile strength, and cheap rate have increased their usage in every sector. But limitless production and usage have posed a threat for living organisms and the environment, as many of the synthetic polymers are comprised of toxic chemicals and are non-biodegradable in nature. In this book chapter, sources and harmful effects of toxic plastics and microplastics have been evaluated, followed by the different kinds of degradation of these synthetic polymers. Chemical and solar degradation techniques, along with biodegradation, have proved an efficient tool for solving the plastic pollution problems. In this process, microbes play a significant part, as the process of biodegradation is an eco-friendly method. Toxic compounds produced after partial degradation of plastics and the presence of microplastics everywhere in nature are an emerging threat in recent times. Hence, more advanced technology in accordance with advanced biodegradation techniques should be evaluated by different research teams.

REFERENCES

Ali, M.I., Perveen, Q., Ahmad, B., Javed, I., Razi-Ul-Hussnain, R., Andleeb, S., Atique, N., Ghumro, P.B., Ahmed, S., and Hameed, A., 2009. Studies on biodegradation of cellulose blended polyvinyl chloride films. *International Journal of Agriculture and Biology*, *11*(5), pp. 577–580.

Ali, S.S., Al-Tohamy, R., Manni, A., Luz, F.C., Elsamahy, T., and Sun, J., 2019. Enhanced digestion of bio-pretreated sawdust using a novel bacterial consortium: Microbial community structure and methane-producing pathways. *Fuel, 254*, p. 115604.

Ali, S.S., Elsamahy, T., Koutra, E., Kornaros, M., El-Sheekh, M., Abdelkarim, E.A., Zhu, D., and Sun, J., 2021. Degradation of conventional plastic wastes in the environment: A review on current status of knowledge and future perspectives of disposal. *Science of the Total Environment, 771*, p. 144719.

Ali, S.S., Kornaros, M., Manni, A., Sun, J., El-Shanshoury, A.E.R.R., Kenawy, E.R., and Khalil, M.A., 2020. Enhanced anaerobic digestion performance by two artificially constructed microbial consortia capable of woody biomass degradation and chlorophenols detoxification. *Journal of Hazardous Materials, 389*, p. 122076.

Ameen, F., Moslem, M., Hadi, S., and Al-Sabri, A.E., 2015. Biodegradation of low density polyethylene (LDPE) by Mangrove fungi from the red sea coast. *Progress in Rubber Plastics and Recycling Technology, 31*(2), pp. 125–143.

Amin, M.U., Scott, G., and Tillekeratne, L.M.K., 1975. Mechanism of the photo-initiation process in polyethylene. *European Polymer Journal, 11*(1), pp. 85–89.

Amobonye, A., Bhagwat, P., Singh, S., and Pillai, S., 2021. Plastic biodegradation: Frontline microbes and their enzymes. *Science of the Total Environment, 759*, p. 143536.

Andrady, A.L., 2011. Microplastics in the marine environment. *Marine Pollution Bulletin, 62*(8), pp. 1596–1605.

Andrady, A.L., and Neal, M.A., 2009. Applications and societal benefits of plastics. *Philosophical Transactions of the Royal Society B: Biological Sciences, 364*(1526), pp. 1977–1984.

Auta, H.S., Emenike, C.U., and Fauziah, S.H., 2017. Distribution and importance of microplastics in the marine environment: A review of the sources, fate, effects, and potential solutions. *Environment International, 102*, pp. 165–176.

Batycky, R.P. *et al.*, 1997. A theoretical model of erosion and macromolecular drug release from biodegrading microspheres. *Journal of Pharmaceutical Sciences, 86*(12), pp. 1464–1477. Available at: https://doi.org/10.1021/js9604117.

Biffinger, J.C., Barlow, D.E., Pirlo, R.K., Babson, D.M., Fitzgerald, L.A., Zingarelli, S., Nadeau, L.J., Crookes-Goodson, W.J., and Russell Jr, J.N., 2014. A direct quantitative agar-plate based assay for analysis of Pseudomonas protegens Pf-5 degradation of polyurethane films. *International Biodeterioration & Biodegradation, 95*, pp. 311–319.

Brunner, I., Fischer, M., Rüthi, J., Stierli, B., and Frey, B., 2018. Ability of fungi isolated from plastic debris floating in the shoreline of a lake to degrade plastics. *PLoS One, 13*(8), p. e0202047.

Cappitelli, F., Principi, P., and Sorlini, C., 2006. Biodeterioration of modern materials in contemporary collections: Can biotechnology help? *Trends in Biotechnology, 24*(8), pp. 350–354.

Chen, C.C., Dai, L., Ma, L., and Guo, R.T., 2020. Enzymatic degradation of plant biomass and synthetic polymers. *Nature Reviews Chemistry, 4*(3), pp. 114–126.

Cole, M., Lindeque, P., Halsband, C., and Galloway, T.S., 2011. Microplastics as contaminants in the marine environment: A review. *Marine Pollution Bulletin, 62*(12), pp. 2588–2597.

Corti, A., Vallini, G., Pera, A., Cioni, F., Solaro, R., and Chiellini, E., 1992. Composting microbial ecosystem for testing the biodegradability of starch-filled polyethylene films. *Special Publication-Royal Society of Chemistry, 109*, pp. 245–245.

Dar, M.A., Dhole, N.P., Pawar, K.D., Xie, R., Shahnawaz, M., Pandit, R.S., and Sun, J., 2022. Ecotoxic Effects of the plastic waste on marine fauna: An overview. *Impact of Plastic Waste on the Marine Biota*, pp. 287–300.

Das, P., and Tiwari, P., 2017. Thermal degradation kinetics of plastics and model selection. *Thermochimica Acta, 654*, pp. 191–202. Available at: https://doi.org/10.1016/j.tca.2017.06.001.

Deepika, S., and Jaya, M.R., 2015. Biodegradation of low density polyethylene by microorganisms from garbage soil. *Journal of Experimental Biology and Agricultural Sciences, 3*, pp. 1–5.

Devi, R.S., Kannan, V.R., Nivas, D., Kannan, K., Chandru, S., and Antony, A.R., 2015. Biodegradation of HDPE by Aspergillus spp. from marine ecosystem of Gulf of Mannar, India. *Marine Pollution Bulletin, 96*(1–2), pp. 32–40.

Dris, R., Gasperi, J., Mirande, C., Mandin, C., Guerrouache, M., Langlois, V., and Tassin, B., 2017. A first overview of textile fibers, including microplastics, in indoor and outdoor environments. *Environmental Pollution, 221*, pp. 453–458.

Engler, R.E., 2012. The complex interaction between marine debris and toxic chemicals in the ocean. *Environmental Science & Technology, 46*(22), pp. 12302–12315.

Fairbrother, A., Hsueh, H.C., Kim, J.H., Jacobs, D., Perry, L., Goodwin, D., White, C., Watson, S., and Sung, L.P., 2019. Temperature and light intensity effects on photodegradation of high-density polyethylene. *Polymer Degradation and Stability, 165*, pp. 153–160.

Fuhs, G.W., 1961. Der mikrobielle Abbau von Kohlenwasserstoffen. *Archiv für Mikrobiologie, 39*, pp. 374–422.

Gewert, B., Plassmann, M.M., and MacLeod, M., 2015. Pathways for degradation of plastic polymers floating in the marine environment. *Environmental Science: Processes & Impacts, 17*(9), pp. 1513–1521.

Grizzi, I. *et al.*, 1995. Hydrolytic degradation of devices based on poly(Dl-lactic acid) size-dependence. *Biomaterials, 16*(4), pp. 305–311. Available at: https://doi.org/10.1016/0142-9612(95)93258-F.

Hansen, C.M., 2007. *Hansen Solubility Parameters: A User's Handbook.* 2nd ed. Boca Raton: CRC Press.

He, P., Chen, L., Shao, L., Zhang, H., and Lü, F., 2019. Municipal solid waste (MSW) landfill: A source of microplastics?-Evidence of microplastics in landfill leachate. *Water Research, 159*, pp. 38–45.

Hipfner, J.M., Galbraith, M., Tucker, S., Studholme, K.R., Domalik, A.D., Pearson, S.F., Good, T.P., Ross, P.S., and Hodum, P., 2018. Two forage fishes as potential conduits for the vertical transfer of microfibres in Northeastern Pacific Ocean food webs. *Environmental Pollution, 239*, pp. 215–222.

Hong, Y., and Gu, J.D., 2009. Bacterial anaerobic respiration and electron transfer relevant to the biotransformation of pollutants. *International Biodeterioration & Biodegradation, 63*(8), pp. 973–980.

Ishigaki, T., Kawagoshi, Y., Ike, M., and Fujita, M., 1999. Biodegradation of a polyvinyl alcohol-starch blend plastic film. *World Journal of Microbiology and Biotechnology*, *15*, pp. 321–327.

Jacquin, J., Cheng, J., Odobel, C., Pandin, C., Conan, P., Pujo-Pay, M., Barbe, V., Meistertz-heim, A.L., and Ghiglione, J.F., 2019. Microbial ecotoxicology of marine plastic debris: A review on colonization and biodegradation by the "Plastisphere". *Frontiers in Microbiology*, *10*, p. 865.

Jenkins, S., Quer, A.M.I., Fonseca, C., and Varrone, C., 2019. Microbial degradation of plastics: New plastic degraders, mixed cultures and engineering strategies. *Soil Microenvironment for Bioremediation and Polymer Production*, pp. 213–238.

Kale, S.K., Deshmukh, A.G., Dudhare, M.S., and Patil, V.B., 2015. Microbial degradation of plastic: A review. *Journal of Biochemical Technology*, *6*(2), pp. 952–961.

Karlsson, S., and Albertsson, A.C., 2002. Techniques and mechanisms of polymer degradation. *Degradable Polymers: Principles and Applications*, pp. 51–69.

Kathiresan, K., 2003. Polythene and plastics-degrading microbes from the mangrove soil. *Revista de Biologia Tropical*, *51*(3–4), pp. 629–633.

Krueger, M.C., Harms, H., and Schlosser, D., 2015. Prospects for microbiological solutions to environmental pollution with plastics. *Applied Microbiology and Biotechnology*, *99*, pp. 8857–8874.

Kumagai, S. *et al.*, 2022. A comprehensive study into the thermo-oxidative degradation of sulfur-based engineering plastics. *Journal of Analytical and Applied Pyrolysis*, *168*, p. 105754. Available at: https://doi.org/10.1016/j.jaap.2022.105754.

Kumar, M., Xiong, X., He, M., Tsang, D.C., Gupta, J., Khan, E., Harrad, S., Hou, D., Ok, Y.S., and Bolan, N.S., 2020. Microplastics as pollutants in agricultural soils. *Environmental Pollution*, *265*, p. 114980.

Lee, B., Pometto III, A.L., Fratzke, A., and Bailey Jr, T.B., 1991. Biodegradation of degradable plastic polyethylene by Phanerochaete and Streptomyces species. *Applied and Environmental Microbiology*, *57*(3), pp. 678–685.

Lee, Q.Y., and Li, H., 2021. Photocatalytic degradation of plastic waste: A mini review. *Micromachines*, *12*(8), p. 907.

Liang, Y., Tan, Q., Song, Q., and Li, J., 2021. An analysis of the plastic waste trade and management in Asia. *Waste Management*, *119*, pp. 242–253.

Liu, K., Wang, X., Wei, N., Song, Z., and Li, D., 2019. Accurate quantification and transport estimation of suspended atmospheric microplastics in megacities: Implications for human health. *Environment International*, *132*, p. 105127.

Lopez-Llorca, L.V., and Valiente, M.F.C., 1993. Study of biodegradation of starch-plastic films in soil using scanning electron microscopy. *Micron*, *24*(5), pp. 457–463.

Lyon, F., 2014. IARC monographs on the evaluation of carcinogenic risks to humans. *World Health Organization, International Agency for Research on Cancer*. Available at: publication@iarc.fr.

Lyu, S., and Untereker, D., 2009. Degradability of polymers for implantable biomedical devices. *International Journal of Molecular Sciences*, *10*(9), pp. 4033–4065. Available at: https://doi.org/10.3390/ijms10094033.

Mark, H.F., Bikales, N.M., Overberger, C.G., and Menges, G., 1986. *Encyclopedia of Polymer Science and Engineering*. Vol. 4. New York: Wiley-Interscience, 605 Third Avenue, 10158.

Mason, S.A., Garneau, D., Sutton, R., Chu, Y., Ehmann, K., Barnes, J., Fink, P., Papazissimos, D., and Rogers, D.L., 2016. Microplastic pollution is widely detected in US municipal wastewater treatment plant effluent. *Environmental Pollution*, *218*, pp. 1045–1054.

Matjašič, T., Simčič, T., Medvešček, N., Bajt, O., Dreo, T., and Mori, N., 2021. Critical evaluation of biodegradation studies on synthetic plastics through a systematic literature review. *Science of the Total Environment*, *752*, p. 141959.

Matsuguma, Y., Takada, H., Kumata, H., Kanke, H., Sakurai, S., Suzuki, T., Itoh, M., Okazaki, Y., Boonyatumanond, R., Zakaria, M.P., and Weerts, S., 2017. Microplastics in sediment cores from Asia and Africa as indicators of temporal trends in plastic pollution. *Archives of Environmental Contamination and Toxicology, 73*, pp. 230–239.

Mattsson, K., Jocic, S., Doverbratt, I., and Hansson, L.A., 2018. Nanoplastics in the aquatic environment. *Microplastic Contamination in Aquatic Environments*, pp. 379–399.

Nanda, S., Sahu, S., and Abraham, J., 2010. Studies on the biodegradation of natural and synthetic polyethylene by Pseudomonas spp. *Journal of Applied Sciences and Environmental Management, 14*(2).

Osman, A.I., Hosny, M., Eltaweil, A.S., Omar, S., Elgarahy, A.M., Farghali, M., Yap, P.S., Wu, Y.S., Nagandran, S., Batumalaie, K., and Gopinath, S.C., 2023. Microplastic sources, formation, toxicity and remediation: A review. *Environmental Chemistry Letters*, pp. 1–41.

Patel, M.M., Goyal, B.R., Bhadada, S.V., Bhatt, J.S., and Amin, A.F., 2009. Getting into the brain: Approaches to enhance brain drug delivery. *CNS Drugs, 23*, pp. 35–58.

Pelmont, J., 2005. *Biodégradations et métabolismes: les bactéries pour les technologies de l'environnement*. Les Ulis, France: EDP Sciences.

Peterson, J.D., Vyazovkin, S., and Wight, C.A., 2001. Kinetics of the thermal and thermo-oxidative degradation of polystyrene, polyethylene and poly(Propylene). *Macromolecular Chemistry and Physics, 202*(6), pp. 775–784. Available at: https://doi.org/10.1002/1521-3935(20010301)202:6<775::AID-MACP775>3.0.CO;2-G.

Pruter, A.T., 1987. Sources, quantities and distribution of persistent plastics in the marine environment. *Marine Pollution Bulletin, 18*(6), pp. 305–310.

Raghavendra, V.B., Uzma, M., Govindappa, M., Vasantha, R.A., and Lokesh, S., 2016. Screening and identification of polyurethane (PU) and low density polyethylene (LDPE) degrading soil fungi isolated from municipal solid waste. *International Journal of Current Research, 8*(7), pp. 34753–34761.

Rajandas, H., Parimannan, S., Sathasivam, K., Ravichandran, M., and Yin, L.S., 2012. A novel FTIR-ATR spectroscopy based technique for the estimation of low-density polyethylene biodegradation. *Polymer Testing, 31*(8), pp. 1094–1099.

Ronkvist, Å.M., Xie, W., Lu, W., and Gross, R.A., 2009. Cutinase-catalyzed hydrolysis of poly (ethylene terephthalate). *Macromolecules, 42*(14), pp. 5128–5138.

Saha, N., Banivaheb, S., and Toufiq Reza, M., 2022. Towards solvothermal upcycling of mixed plastic wastes: Depolymerization pathways of waste plastics in sub- and supercritical toluene. *Energy Conversion and Management: X, 13*, p. 100158. Available at: https://doi.org/10.1016/j.ecmx.2021.100158.

Sarkhel, R., Sengupta, S., Das, P., and Bhowal, A., 2020. Comparative biodegradation study of polymer from plastic bottle waste using novel isolated bacteria and fungi from marine source. *Journal of Polymer Research, 27*, pp. 1–8.

Schirinzi, G.F., Pérez-Pomeda, I., Sanchís, J., Rossini, C., Farré, M., and Barceló, D., 2017. Cytotoxic effects of commonly used nanomaterials and microplastics on cerebral and epithelial human cells. *Environmental Research, 159*, pp. 579–587.

Shah, A.A., Hasan, F., Hameed, A., and Ahmed, S., 2008. Biological degradation of plastics: A comprehensive review. *Biotechnology Advances, 26*(3), pp. 246–265.

Sharma, V.K., Ma, X., Lichtfouse, E., and Robert, D., 2022. Nanoplastics are potentially more dangerous than microplastics. *Environmental Chemistry Letters*, pp. 1–4.

Siepmann, J., 2001. Mathematical modeling of bioerodible, polymeric drug delivery systems. *Advanced Drug Delivery Reviews, 48*(2–3), pp. 229–247. Available at: https://doi.org/10.1016/S0169-409X(01)00116-8.

Skariyachan, S., Manjunatha, V., Sultana, S., Jois, C., Bai, V., and Vasist, K.S., 2016. Novel bacterial consortia isolated from plastic garbage processing areas demonstrated enhanced degradation for low density polyethylene. *Environmental Science and Pollution Research, 23*, pp. 18307–18319.

Su, Y., Zhang, Z., Wu, D., Zhan, L., Shi, H., and Xie, B., 2019. Occurrence of microplastics in landfill systems and their fate with landfill age. *Water Research, 164*, p. 114968.

Sundt, P., Schulze, P.E., and Syversen, F., 2014. Sources of microplastic-pollution to the marine environment. *Mepex for the Norwegian Environment Agency, 86*, p. 20.

Teuten, E.L., Saquing, J.M., Knappe, D.R., Barlaz, M.A., Jonsson, S., Björn, A., Rowland, S.J., Thompson, R.C., Galloway, T.S., Yamashita, R., and Ochi, D., 2009. Transport and release of chemicals from plastics to the environment and to wildlife. *Philosophical Transactions of the Royal Society B: Biological Sciences, 364*(1526), pp. 2027–2045.

Thakur, P., 2012. *Screening of Plastic Degrading Bacteria from Dumped Soil Area* (Doctoral dissertation).

Thompson, R.C., Swan, S.H., Moore, C.J., and Vom Saal, F.S., 2009. Our plastic age. *Philosophical Transactions of the Royal Society B: Biological Sciences, 364*(1526), pp. 1973–1976.

Torikai, A., Takeuchi, A., Nagaya, S., and Fueki, K., 1986. Photodegradation of polyethylene: Effect of crosslinking on the oxygenated products and mechanical properties. *Polymer Photochemistry, 7*(3), pp. 199–211.

Tyler, D.R., 2004. Mechanistic aspects of the effects of stress on the rates of photochemical degradation reactions in polymers. *Journal of Macromolecular Science, Part C: Polymer Reviews, 44*(4), pp. 351–388.

Wang, C., Zhao, L., Lim, M.K., Chen, W.Q., and Sutherland, J.W., 2020. Structure of the global plastic waste trade network and the impact of China's import Ban. *Resources, Conservation and Recycling, 153*, p. 104591.

Xu, J.Z., Zhang, J.L., Hu, K.H., and Zhang, W.G., 2013. The relationship between lignin peroxidase and manganese peroxidase production capacities and cultivation periods of mushrooms. *Microbial Biotechnology, 6*(3), pp. 241–247.

Yang, H., Ma, M., Thompson, J.R., and Flower, R.J., 2018. Waste management, informal recycling, environmental pollution and public health. *Journal of Epidemiology and Community Health, 72*(3), pp. 237–243.

Yousif, E., and Haddad, R., 2013. Photodegradation and photostabilization of polymers, especially polystyrene. *SpringerPlus, 2*(1), pp. 1–32.

Zhao, Y.-B., Lv, X.-D., and Ni, H.-G., 2018. Solvent-based separation and recycling of waste plastics: A review. *Chemosphere, 209*, pp. 707–720. Available at: https://doi.org/10.1016/j.chemosphere.2018.06.095.

Zhao, Y.-B. *et al.*, 2017. Laboratory simulations of the mixed solvent extraction recovery of dominate polymers in electronic waste. *Waste Management, 69*, pp. 393–399. Available at: https://doi.org/10.1016/j.wasman.2017.08.018.

12 Fate of Micro/Nano Plastic Pollutants in the Marine Ecosystem

Jesse Joel Thathapudi, Levin Anbu Gomez, Vishruth Vijay, Vani Chandrapragasam, Ritu Shepherd, Subhankar Paul, Meng-Jen Lee and Prathap Somu

12.1 INTRODUCTION

In the past century, the rapid growth of plastic production and consumption has brought about significant environmental challenges. While plastic materials have provided immense benefits to society, the improper disposal and inadequate management of plastic waste have led to a concerning global issue. One particularly worrisome consequence is the proliferation of micro- and nano-sized plastic particles, which have become ubiquitous pollutants within marine ecosystems. Understanding the fate of micro-/nano-particle plastic pollutants in the marine environment is a critical scientific endeavor that requires meticulous investigation. These particles—microparticles measuring 100 nm to 5 nm in diameter—and nanoplastics measuring > 100 nm pose unique challenges due to their small size and widespread distribution. Their presence in marine ecosystems has been linked to detrimental effects on marine organisms as well as potential risks to human health through the food chain. (Souza et al., 2022)

The ecological impacts of micro- and nanoparticle plastics on marine organisms are of great concern. These particles can be ingested by a wide array of marine organisms, ranging from microorganisms to apex predators. The ingestion of plastic particles can cause physical harm, internal injuries, and potential toxicity due to the adsorption and release of chemical additives or contaminants associated with the plastics. Digestive blockages and reduced feeding efficiency are common consequences of plastic ingestion. Additionally, the accumulation of plastics in the tissues of marine organisms can disrupt physiological functions and impair reproductive success.

Micro- and nanoparticle plastics not only affect individual organisms but also exert indirect effects on marine habitats. These particles can alter habitat structures and physical processes by accumulating in sensitive habitats such as coral reefs, seagrass meadows, and estuaries. Microplastics can smother benthic habitats and disrupt sediment stability, affecting bottom-dwelling organisms and altering the

DOI: 10.1201/9781003449133-12

composition of benthic communities. Moreover, plastic particles can act as vectors for the transport of other pollutants, facilitating the transfer of harmful substances throughout the marine food web.

The presence of micro- and nanoparticle plastics in the marine environment also disrupts biogeochemical cycles. Plastics can serve as surfaces for microbial colonization and biofilm formation, potentially influencing nutrient cycling and microbial community dynamics. The degradation of plastic particles can release associated chemicals and alter nutrient availability in marine ecosystems, impacting primary productivity and ecosystem functioning.

12.2 MICRO/NANO PLASTIC POLLUTANTS

Plastics are affordably priced and have versatile application, which results in the increasing demand for plastics. The global plastic industry witnesses a consequent 4% rise annually (Li et al., 2021). Plastics, having the properties of being non-biodegradable, effect the globe immensely and the results is their accumulation. Plastic fails to exit the cycle and poses as an immense threat to the environment. Research suggests that 85% of plastics end up in the marine niche. Plastics act as a pollutant and clutter environmental ecosystems all around the globe. The complex nature of plastic becomes an alarming problem for the aquatic biota and humans (Sarma et al., 2022). Researchers found plastics and microplastics in aquatic ecosystems like polypropylene, polystyrene, polyvinylchloride, polyethylene terephthalate, and polyethylene. Additionally, the production and manufacturing process of plastics is a complex one and includes the use of many chemicals, making it life threatening when consumed (Bhuyan et al., 2022). Plastics have pores present in them that provide spaces for chemicals to pile up, which showcase toxic traits (Sarma et al., 2022). Plastics harbor environmental pollutants like polycyclic aromatic hydrocarbons and heavy metals, as they have the characteristics of having enormous surface area to volume ratio (Bhuyan et al., 2022). Plastics do not entirely decompose, but rather, break down into smaller particles like micro- and nanoplastics. Micro- and nanoplastics are plastic debris that are produced in an attempt at disposal and breakdown of consumer products and industrial-waste-containing plastics. Apart from this, plastics degradation, giving micro-/nanoplastics, can also be induced by the production of various biomedical instruments, coatings, drug delivery, medical diagnostics instruments, electronics machinery, magnetic, and optoelectronics machinery. The decrease in size shows an inverse relation with its adsorption properties toward potent contaminants and becomes, furthermore, reactive. This leads to changes in the physical and chemical properties, which subsequently showcases a change in the biological effects on marine organisms (Ferreira et al., 2019).

These unbroken forms of plastics act as an agent to draw organic pollutants to itself, like dioxins and 1,1-dichloro-2,2-bisethylene, also known as DDE (Bhuyan et al., 2022). This suggests that the characteristics of nano- and microparticles show variations due to the change in the adsorption properties. The nano and micro size of plastics provides the capability to journey over cellular boundaries effortlessly and aggregate in them (Ferreira et al., 2019). Their extremely small size makes them a potential threat as they enter the food chain after their intake by small aquatic species.

Research also suggests that microplastic entering the aquatic organism's lifecycle can cause serious complications, like growth inhibition, reduced immune system responses, and oxidative stresses (Bai et al., 2021) Microplastics found in marine ecosystem are further divided into two types: primary microplastics and secondary microplastics (Bhuyan et al., 2022). Primary microplastics are the minute plastic particles that join the aquatic ecosystem with industrial effluents, sewage wastes, and spills. To list a few examples of primary microplastics, we have, fragments, fibers, films, wind turbines, resin pellets, and catalysts (Bhuyan et al., 2022). Additionally, mechanical exfoliants like microfiber clothing, adhesive scrub sheets, and skincare products and cosmetics—namely, facewash, handwash, perfumes, scents, eyeliner, hair care products, sunscreen, nail polish, and toothpaste—are also considered to constitute primary microplastics. Furthermore, the presence of microplastics is also observed in clinical usage, as in acting as carriers for various medicines and also its presence in products used to clean teeth by dental surgeons. Studies show that all these sources of primary microplastics are used without caution and discarded freely, and they then end up in the marine ecosystem (Sarma et al., 2022). Fragmentation of plastic particles that are present in the ecosystem from a long time with the assistance of natural processes like inorganic deposition, structural decay, microbial treatment, or photodegradation results in the formation of secondary microplastics. Secondary microplastics are introduced to aquatic systems via hard plastics, synthetic fibers, plastic sheets, clothing pipes, nets, and bottles. Synthetic polymers of plastic—namely, polyethylene, polypropylene, polystyrene, polyvinyl chloride, polyester, polyamide, polyether, polyurethane, polyethylene, terephthalate, acrylic polymer, and cellophane—constitute the secondary microplastics in marine ecosystems. With increasing deposition of secondary microplastics in marine ecosystems due to processes like mechanical abrasion, oxidation, and photothermal degradation, the formation of microplastic fiber is observed in the marine ecosystems. Ultraviolet radiation and physiological tide erosion facilitate plastic size reduction in coastal regions (Sarma et al., 2022)

12.2.1 Mechanism of Generation

The exponentially increasing trend of usage of plastic is the outcome of an increase in the human population and its activities. Estimations suggests that 60 to 90 million tons of mishandled waste consisting of plastic was generated in the year 2015, and it is also suggested that, by the year 2060, the levels could elevate to 155 to 260 million tons. (Soares et al., 2020). Microplastics enter the ecosystems through pathways listed here:

1. Drainage system
2. Wastes from ships and recreational events
3. Washing and cleaning of cloths
4. Agricultural polythene materials dissolution
5. Car tire abrasions
6. Fertilizer runoff
7. Wastewater treatment plant

The breakdown of plastics to give microplastics can take place at any level of the cycle. It could take place at the time of its production or at the time of decomposition too. As formerly mentioned, car tire abrasion is a major area of microplastic generation. The presence of microplastics produced due to wear and tear of automobile tires is witnessed in every compartment of the ecosystem and has negative impacts. Another noteworthy microplastic source is fibers from clothes and textiles. During the process of cloth washing, cloth fibers are washed off, which aids in the exploitation of the environment. This is observed because the chemicals used in the manufacture of the textiles contribute to being the main agent in the formation of an enormous amount microplastics.

Furthermore, low-grade plastics made for single-time use undergo a considerable amount of abrasion during use, giving rise to microplastics in the environment (Syberg et al., 2022). It is evidently noted that an enormous amount of plastic debris and various marine plants and animals come in contact as the former is ingested by or is entangled with the latter. Bigger pieces of plastics are fragmented by the natural processes like UV degradation, wave action, and physical abrasions, which finally give rise to microplastics. Microplastics are present in cosmetics in the form of microbeads. Often, some microplastics are stuck in sewage sludge while undergoing treatment. Such microplastics are later used as fertilizer on agricultural land. Another major source of microplastic worthy to be noted is plastic pellets which is also called nurdle. Plastic pellets are the predecessors of larger plastics, which are often mishandled and spilt while transporting. Abrasive air blasting and coating for boats is carried out with the assistance of microbeads, which, in turn, contributes to the formation of microplastics. Through the methods of wind and water erosion, the microplastics present in terrestrial ecosystems enter aquatic ecosystems. Rainfall acts as an active agent in washing off microplastics present after the car-tire abrasion process and bringing them to drainage systems. Coastal regions, having industrial set-ups nearby them, see a high percentage of microplastics. Given the changing climatic circumstances on the Earth, the microplastics captive in glaciers and ice form melt and eventually end up in the oceans. Variation in the climate could also result in changes of oceanic currents, which, in turn, would alter the distribution and accumulation of microplastics in a particular place (Botterell et al., 2019).

12.2.2 Spread to the Environment

Out of the total plastic waste produced globally, 10% of waste eventually get deposited in marine systems through various pathways. Plastics constitute 60% of marine waste, and in some areas, they amount to 90 to 95%. Research shows that 8 million tonnes of plastic end up in the ocean bodies annually. Microplastics are dispersed near shore of beaches, water surface, and in the bed of the water body (Li et al., 2021). The presence of microplastics is evidently found near water bodies due to intensive human activities around the coasts. Research suggests that the amount of microplastics present in coastal regions is in elevated amounts, as compared to that in deep sea sediments. The physical attributes of microplastic such as shape, size, and surface shows irregularities, as they are subjected to environmental forces of various kinds. Microplastics have cracks and often are seen associated with crude oil, organic pollutants, iron oxides, bacteria, and viruses. Due to environmental factors like ocean currents and the variation in seasons, microplastics accumulate in different regions of the sea. Properties of plastic such as size and type determine the density of the material, which, in turn, effect the buoyancy of the plastics. The dispersion and distribution of plastics are attributed by the property of the buoyancy of plastic. Studies show that low-density plastic polymers are found in sea surfaces, whereas the denser plastic polymers are found in deeper levels of the water body. Research also suggests that, among buoyant plastics, the negatively buoyant ones show inertial effects, and subsequently, drift in horizontal distance to a few wavelengths, by the effect of sea waves. To the contrary of this, positively buoyant plastics are governed by Stokes drift phenomena and, in turn, remain in a superficial layer where intense sea waves or drifts are observed. Processes like biofouling prove helpful in changing the density and also the settling velocity for the previously mentioned cases (Soares et al., 2020).

12.3   MARINE ECOSYSTEM AND POLLUTANTS

The marine ecosystem is a complex and diverse environment that plays a vital role in the health of our planet. It encompasses a wide range of habitats, including coral reefs, seagrass beds, mangrove forests, and open ocean waters. Unfortunately, this delicate ecosystem is under constant threat from various pollutants, which can have devastating effects on marine life and the overall balance of the ecosystem (Smith et al., 2021). One of the most significant sources of pollutants in the marine environment is human activities. Industrial discharge, oil spills, sewage, agricultural runoff, and plastic waste are some of the major contributors to marine pollution (Jones & Johnson, 2020). These pollutants introduce harmful substances, such as heavy metals, pesticides, fertilizers, petroleum byproducts, and plastics, into the marine ecosystem. The impacts of these pollutants on marine life are far-reaching. They can contaminate water, sediments, and organisms, causing widespread damage to marine species and habitats. For example, the accumulation of plastic debris in the ocean poses a significant threat to marine animals, as they can become entangled in it or mistake it for food, leading to injury, suffocation, or starvation. Additionally, the release of chemical pollutants into the water can disrupt the reproductive, immune, and endocrine systems of marine organisms, affecting their survival and reproductive success (Li et al., 2018). Moreover, pollutants can have cascading effects on the entire marine food web. Toxic substances can bioaccumulate in the tissues of organisms, leading to biomagnification as predators consume contaminated prey. This can result in higher concentrations of pollutants in top predators, such as sharks, whales, and seabirds, which can ultimately threaten their populations and disrupt the ecological balance (Chen et al., 2022). In order to mitigate the impact of pollutants on the marine ecosystem, it is crucial to implement effective pollution-prevention measures. This includes improving waste management practices, reducing the use of harmful chemicals, implementing stricter regulations on industrial discharges, and promoting sustainable fishing and aquaculture practices (Laffoley et al., 2020). Additionally, public awareness and education play a vital role in fostering a sense of responsibility and promoting individual actions to reduce pollution and protect the marine environment.

12.3.1 Pollutants and Their Course to the Marine System

Marine ecosystems are vital components of the Earth's biosphere, providing a habitat for diverse organisms and supporting various ecological processes. Plastics accumulate and break down via various natural factors such as sunlight, thermal aging, oxidation, chemical or biological degradation, etc. The small size, shape, surface area, crystallinity, and density of microplastics allows them to be easily transported by various mechanisms. Ocean currents play a crucial role in their dispersal, carrying these particles over long distances. Surface waves and wind-driven processes can also transport microplastics, causing them to be suspended in the water column or washed up onto coastal areas. Furthermore, the buoyancy and density of microplastics influence their vertical movement within the water column, with some particles floating at the surface while others sink to deeper depths or settle in sediments. In addition to their transport mechanisms, microplastics exhibit diverse physical and chemical properties that contribute to their persistence in the marine environment. The durability of plastic polymers enables them to withstand degradation processes, leading to their long-term presence in marine habitats. Microplastics can be composed of various polymer types, such as polyethylene, polypropylene, and polystyrene, each with its own characteristics that influence their fate and interactions within the environment. Surface runoffs, direct breakdown, and deposition via travelling cargos, etc., carried by wind, animals, industrial disposing, etc., are the common pathways for microplastic accumulation in the marine biome. Surface properties of microplastics, including roughness and hydrophobicity, can influence their interactions with other substances in the marine environment. These properties can enhance the adsorption of organic pollutants and other toxic substances onto microplastic surfaces, potentially leading to the transfer of contaminants to marine organisms upon ingestion.

The different types of plastics having impact on the marine environment are as shown in Figure 12.1 (Ashrafy et al., 2023).

However, these ecosystems are increasingly threatened by pollution from various sources. Understanding the types of pollutants and their pathways to marine systems is crucial for effective management and conservation efforts. In this chapter, we will explore different pollutants and their courses to marine systems.

12.3.1.1 Oil and Petroleum Products

Oil spills and leakage from shipping accidents or offshore drilling activities are significant sources of oil pollution in marine systems (Smith et al., 2018). These spills can have devastating effects on marine life, leading to habitat degradation, reduced biodiversity, and long-term ecological impacts.

12.3.1.2 Heavy Metals

Industrial activities, including mining, smelting, and manufacturing, release heavy metals such as mercury, lead, and cadmium into the environment (Santos et al., 2019). These metals can accumulate in marine ecosystems through various routes, including atmospheric deposition and direct discharge, posing risks to marine organisms and human health.

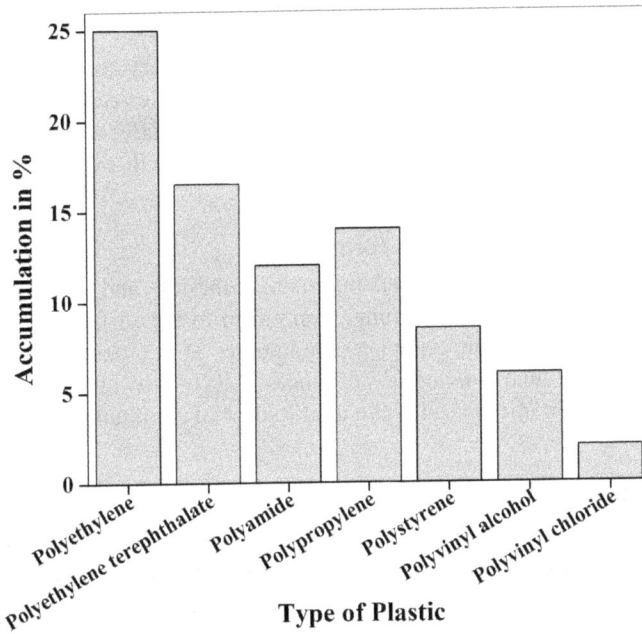

FIGURE 12.1 Graphical representation of types of plastics and their percentage accumulation in the marine environment (based on the data in Ashrafy et al., 2023).

12.3.1.3 Nutrient Runoff

Excessive nutrient runoff from agricultural activities and urban areas can enter marine systems through rivers and streams (Diaz & Rosenberg, 2008). The high nutrient inputs, particularly nitrogen and phosphorus, can lead to eutrophication, harmful algal blooms, and oxygen depletion, negatively impacting marine organisms and ecosystems.

12.3.1.4 Plastics and Microplastics

Plastics, including single-use items and microplastics (tiny plastic particles), are pervasive pollutants in marine systems (Rochman et al., 2013). They enter the oceans through improper waste management, stormwater runoff, and direct littering. Marine organisms can ingest or become entangled in plastics, resulting in injury, suffocation, and disruption of marine food webs.

12.3.1.5 Pesticides and Chemicals

Pesticides used in agriculture and chemicals from industrial activities can contaminate marine systems through runoff and direct discharge (Vighi et al., 2018). These pollutants can have adverse effects on marine organisms, including developmental abnormalities, reproductive impairments, and population declines. To mitigate the impacts of these pollutants on marine systems, it is essential to adopt sustainable practices, implement strict regulations, and promote environmental awareness and education.

12.3.2 MICROBIAL INTERACTIONS

Microplastics, small plastic particles less than 5 millimeters in size, have become a pervasive pollutant in marine environments. These particles can interact with various microorganisms, leading to both direct and indirect effects on microbial communities and ecosystem dynamics. In this chapter, we will explore the microbial interactions associated with microplastics.

12.3.2.1 Adhesion and Biofilm Formation

Microplastics provide surfaces for microbial adhesion and biofilm formation. Microbes, including bacteria and fungi, can attach to the surface of microplastics and form biofilms, which are complex communities of microorganisms encased in a matrix of extracellular polymeric substances (EPS) (Keswani et al., 2016). This biofilm formation on microplastics can alter microbial community composition and function.

12.3.2.2 Microbial Degradation

Certain microorganisms possess the ability to degrade plastics through enzymatic activity. Studies have identified bacterial strains capable of breaking down specific types of plastics, such as polyethylene terephthalate (PET) and polyurethane (Karan et al., 2018). This microbial degradation of plastics represents a potential pathway for plastic waste management.

12.3.2.3 Transfer of Pathogens and Harmful Microorganisms

Microplastics can act as vectors for the transport of pathogens and harmful microorganisms in marine environments. Bacteria and other microorganisms can attach to the surface of microplastics and be transported over long distances (Zettler et al., 2013). This transport mechanism may contribute to the spread of microbial pathogens in marine ecosystems.

12.3.2.4 Disruption of Microbial Communities

The presence of microplastics can disrupt the structure and function of microbial communities in marine environments. Studies have shown that microplastics can induce changes in microbial diversity, community composition, and metabolic activities (Kesy et al., 2019). These disruptions can have cascading effects on ecosystem processes and functions.

12.3.2.5 Influence on Nutrient Cycling

Microplastics can impact nutrient cycling processes in marine ecosystems. Microbes associated with microplastics may influence the breakdown and release of organic compounds from plastic particles, potentially affecting the availability of nutrients for other organisms (Guzzetti et al., 2018). These alterations in nutrient cycling can have implications for the overall functioning of marine ecosystems.

Understanding the microbial interactions with microplastics is crucial for comprehending the ecological consequences of microplastic pollution. Further research is needed to elucidate the complex relationships between microplastics and microbial communities and their implications for marine ecosystems.

The degradation products of microplastics can participate in biotic or abiotic reactions within the aquatic environment, influencing biogeochemical processes such as nutrient cycling and elemental transformations. Understanding the dynamics of microorganisms and microplastics in aquatic environments, their interactions, and their role in biogeochemical cycles is crucial for comprehending the ecological impacts of microplastic pollution and developing effective strategies for mitigating its effects. Further scientific investigation is needed to unravel the complex relationship between microplastics and microorganisms and to assess the potential consequences of microplastic pollution in marine ecosystems (Rogers et al., 2020). The microorganisms predominantly associated with the degradation of microplastics are primarily found within terrestrial strata. These microorganisms include *Achromobacter, Acinetobacter, Arthrobacter, Aspergillus* (fungus), *Bacillus, Comamonas, Delftia, Micrococcus, Nesiotobacter, Paenibacillus, Pseudomonas, Rahnella, Staphylococcus,* and *Stenotrophomonas.* The specific microbial communities involved in microplastic degradation may vary depending on the environmental context and the type of microplastic present (Zhai et al., 2023).

12.3.3 Effects on the Marine Environment and Health (Chronic and Acute Effects)

Pollution in marine ecosystems has significant impacts on both the environment and human health. Understanding the chronic and acute effects of pollution is essential for addressing these issues and implementing effective mitigation strategies. In this chapter, we will explore the effects of pollution on the marine environment and health.

12.3.3.1 Chronic Effects on the Marine Environment

Chronic pollution, resulting from continuous or long-term exposure to pollutants, can have several detrimental effects on marine ecosystems. Some key effects include:

Habitat Degradation: Pollution can cause the degradation of habitats such as coral reefs, seagrass beds, and mangrove forests. This degradation disrupts the balance of the ecosystem, affecting various organisms that rely on these habitats for food, shelter, and reproduction (Jackson et al., 2001).

Biodiversity Loss: Pollution can lead to a decline in species diversity and abundance, as certain organisms are more sensitive to pollutants than others. The loss of key species can disrupt ecological interactions, affecting the stability and functioning of marine ecosystems (Lotze et al., 2006).

Bioaccumulation and Biomagnification: Some pollutants, such as heavy metals and persistent organic pollutants, can bioaccumulate and biomagnify in the food web. This means that they accumulate in the tissues of organisms and increase in concentration as they move up the food chain, posing risks to higher trophic levels, including marine mammals and humans (Oehlmann et al., 2009).

12.3.3.2 Acute Effects on Marine Organisms and Health

Acute pollution events, such as oil spills or chemical leaks, can have immediate and severe impacts on marine organisms and human health. Some notable effects include:

Toxicity and Mortality: Pollutants can directly poison marine organisms, leading to acute toxicity and mortality. Oil spills, for example, can coat the feathers or fur of marine birds and mammals, impairing their ability to regulate body temperature and causing suffocation (Peterson et al., 2003).

Reproductive Impairments: Certain pollutants, including endocrine-disrupting chemicals, can interfere with the reproductive systems of marine organisms. This can result in reduced fertility, developmental abnormalities, and population declines (Hinck et al., 2009).

Human Health Risks: Marine pollution can pose risks to human health through the consumption of contaminated seafood or exposure to polluted coastal areas. Contaminants such as mercury and PCBs can accumulate in fish and shellfish, leading to potential health problems in humans, including neurological disorders and cardiovascular diseases (Mahaffey et al., 2009). To protect the marine environment and human health, it is crucial to reduce pollution through strict regulations, sustainable practices, and the promotion of environmental awareness.

REFERENCES

Ashrafy, A., Liza, A. A., Islam, M. N., Billah, M. M., Arafat, S. T., Rahman, M. M., & Rahman, S. M. (2023). Microplastics pollution: A brief review of its source and abundance in different aquatic ecosystems. *Journal of Hazardous Materials Advances*, *9*, 100215. https://doi.org/10.1016/j.hazadv.2022.100215.

Bai, Z., Wang, N., & Wang, M. (2021). Effects of microplastics on marine copepods. In *Ecotoxicology and Environmental Safety* (Vol. 217). Academic Press. https://doi.org/10.1016/j.ecoenv.2021.112243.

Bhuyan, M. S., Rashed-Un-Nabi, M., Alam, M. W., Islam, M. N., Cáceres-Farias, L., Bat, L., Saiyad Musthafa, M., Senapathi, V., Chung, S. Y., & Alfaro Núñez, A. (2022). *Environmental and Morphological Detrimental Effects of Microplastics on Marine Organisms to Human Health*. https://doi.org/10.21203/rs.3.rs-1290795/v1.

Botterell, Z. L. R., Beaumont, N., Dorrington, T., Steinke, M., Thompson, R. C., & Lindeque, P. K. (2019). Bioavailability and effects of microplastics on marine zooplankton: A review. In *Environmental Pollution* (Vol. 245, pp. 98–110). Elsevier Ltd. https://doi.org/10.1016/j.envpol.2018.10.065.

Chen, L., Huang, Q., Yu, Y., Li, Y., Zhang, X., & Gao, A. (2022). Impacts of polycyclic aromatic hydrocarbons on marine organisms and ecosystems: A review. *Marine Pollution Bulletin*, *174*, 113091.

Diaz, R. J., & Rosenberg, R. (2008). Spreading dead zones and consequences for marine ecosystems. *Science*, *321*(5891), 926–929.

Ferreira, I., Venâncio, C., Lopes, I., & Oliveira, M. (2019). Nanoplastics and marine organisms: What has been studied? In *Environmental Toxicology and Pharmacology* (Vol. 67, pp. 1–7). Elsevier B.V. https://doi.org/10.1016/j.etap.2019.01.006.

Guzzetti, E., Sureda, A., Tejada, S., & Faggio, C. (2018). Microplastic in marine organism: Environmental and toxicological effects. *Environmental Toxicology and Pharmacology*, *64*, 164–171.

Hinck, J. E., Schmitt, C. J., Blazer, V. S., Denslow, N. D., Bartish, T. M., Anderson, P. J., . . . Coyle, J. J. (2009). Environmental contaminants in fish and wildlife from the Chesapeake Bay Watershed: Complex mixtures, complex responses, and the need for an integrated perspective. *Environmental Monitoring and Assessment, 157*(1–4), 17–24.

Jackson, J. B., Kirby, M. X., Berger, W. H., Bjorndal, K. A., Botsford, L. W., Bourque, B. J., . . . Hughes, T. P. (2001). Historical overfishing and the recent collapse of coastal ecosystems. *Science, 293*(5530), 629–637.

Jones, H. P., & Johnson, M. (2020). Anthropogenic marine debris: Trends and future challenges. *Journal of Environmental Management, 259*, 109565.

Karan, R., Capriles-González, E., Liew, K. J., Naidu, G., & Megharaj, M. (2018). Microbial degradation of polyethylene: A review. *Critical Reviews in Environmental Science and Technology, 48*(5), 383–414.

Keswani, A., Oliver, D. M., Gutierrez, T., & Quilliam, R. S. (2016). Microbial hitchhikers on marine plastic debris: Human exposure risks at bathing waters and beach environments. *Marine Environmental Research, 118*, 10–19.

Kesy, K., Oberbeckmann, S., Müller, F. D., Labrenz, M., & Grossart, H. P. (2019). Differential effects of microplastics on microbial community composition and functioning along the water column of a tropical estuary. *Limnology and Oceanography, 64*(3), 1079–1092.

Laffoley, D., Baxter, J. M., Blaxter, J., Clarke, J., Crevatin, L., Cucknell, A.-C., . . . Wright, K. (2020). Towards a representative network of marine protected areas in the seas of the British Isles. *Aquatic Conservation: Marine and Freshwater Ecosystems, 30*(2), 214–238.

Li, Y., Sun, Y., Li, J., Tang, R., Miu, Y., & Ma, X. (2021). Research on the influence of microplastics on marine life. *IOP Conference Series: Earth and Environmental Science, 631*(1). https://doi.org/10.1088/1755-1315/631/1/012006.

Li, Z., Zheng, W., Wang, Z., Qiu, L., Wang, Q., & Chen, C. (2018). Effects of microplastics on the toxicity of pesticides to aquatic organisms. *Environmental Pollution, 242*, 1478–1485.

Lotze, H. K., Lenihan, H. S., Bourque, B. J., Bradbury, R. H., Cooke, R. G., Kay, M. C., . . . Petrie, B. (2006). Depletion, degradation, and recovery potential of estuaries and coastal seas. *Science, 312*(5781), 1806–1809.

Mahaffey, K. R., Clickner, R. P., & Jeffries, R. A. (2009). Methylmercury and omega-3 fatty acids: Co-occurrence of dietary sources with emphasis on fish and shellfish. *Environmental Research, 107*(1), 20–29.

Oehlmann, J., Schulte-Oehlmann, U., Tillmann, M., Markert, B., & Oetken, M. (2009). Effects of endocrine disruptors on prosobranch snails (Mollusca: Gastropoda) in the laboratory. Part I: Bisphenol A and octylphenol as xeno-estrogens. *Ecotoxicology and Environmental Safety, 72*(3), 691–696.

Peterson, C. H., Rice, S. D., Short, J. W., Esler, D., Bodkin, J. L., Ballachey, B. E., & Irons, D. B. (2003). Long-term ecosystem response to the Exxon Valdez oil spill. *Science, 302*(5653), 2082–2086.

Rochman, C. M., Browne, M. A., Halpern, B. S., Hentschel, B. T., Hoh, E., Karapanagioti, H. K., . . . Thompson, R. C. (2013). Classify plastic waste as hazardous. *Nature, 494*(7436), 169–171.

Rogers, K. L., Carreres-Calabuig, J. A., Gorokhova, E., & Posth, N. R. (2020). Micro-by-micro interactions: How microorganisms influence the fate of marine microplastics. In *Limnology and Oceanography Letters* (Vol. 5, Issue 1, pp. 18–36). John Wiley and Sons Inc. https://doi.org/10.1002/lol2.10136.

Santos, E. C., Di Beneditto, A. P. M., Romão, S., & Santos, D. O. (2019). Heavy metal contamination in marine mammals: A review. *Environmental Pollution, 252*, 1348–1359.

Sarma, H., Hazarika, R. P., Kumar, V., Roy, A., Pandit, S., & Prasad, R. (2022). Microplastics in marine and aquatic habitats: Sources, impact, and sustainable remediation approaches. *Environmental Sustainability, 5*(1), 39–49. https://doi.org/10.1007/s42398-022-00219-8.

Smith, B. D., Donaldson, S. G., & Bonnell, M. (2018). Impacts of offshore oil and gas development activities on marine mammals and sea turtles. *Environmental Reviews, 26*(2), 199–214.

Smith, S. D., Fulton, E. A., Hobday, A. J., Smith, A. D., & Shoulder, P. (2021). Predicting environmental conditions and their ecological consequences in a coastal marine ecosystem. *Ecological Modelling, 447*, 109423.

Soares, J., Miguel, I., Venâncio, C., Lopes, I., & Oliveira, M. (2020). Perspectives on micro(Nano)plastics in the marine environment: Biological and societal considerations. In *Water (Switzerland)* (Vol. 12, Issue 11, pp. 1–16). MDPI AG. https://doi.org/10.3390/w12113208.

Souza, A. M. de, Santos, A. L., Araújo, D. S., Magalhães, R. R. de B., & Rocha, T. L. (2022). Micro(nano)plastics as a vector of pharmaceuticals in aquatic ecosystem: Historical review and future trends. *Journal of Hazardous Materials Advances, 6*, 100068. https://doi.org/10.1016/j.hazadv.2022.100068.

Syberg, K., Nielsen, M. B., Oturai, N. B., Clausen, L. P. W., Ramos, T. M., & Hansen, S. F. (2022). Circular economy and reduction of micro(nano)plastics contamination. *Journal of Hazardous Materials Advances, 5*, 100044. https://doi.org/10.1016/j.hazadv.2022.100044.

Vighi, M., Manenti, D., & Finizio, A. (2018). Pesticides in Italian rivers: Occurrence and environmental risk assessment. *Environmental Science and Pollution Research, 25*(34), 33663–33672.

Zettler, E. R., Mincer, T. J., & Amaral-Zettler, L. A. (2013). Life in the "plastisphere": Microbial communities on plastic marine debris. *Environmental Science & Technology, 47*(13), 7137–7146.

Zhai, X., Zhang, X. H., & Yu, M. (2023). Microbial colonization and degradation of marine microplastics in the plastisphere: A review. In *Frontiers in Microbiology* (Vol. 14). Frontiers Media S.A. https://doi.org/10.3389/fmicb.2023.1127308.

13 The Fate of Micro/Nano Plastic Pollutants in the Natural Environment

Monu Dinesh Ojha and Sinosh Skariyachan

13.1 INTRODUCTION

Plastics comprising various synthetic and semi-synthetic organic polymers are malleable substances that can be shaped into various solid objects. Plastics are utilized for various items because of their limited cost of production, stability, and adaptability. As a result, plastics manufacturing is on the rise, and in 2018, it hit a record high of 359 million tonnes (PlasticsEurope, 2019). The inappropriate disposal of plastics is a global issue that contributes to the uncontrolled environmental release of plastics (PlasticsEurope, 2019), despite advances in recycling and management solutions, thanks to their hydrophobicity and resilience to physical and chemical degradation. Plastic has shown its widespread presence at various levels, be it the air, water, or soil. It has been detected in the seafloor sediment as well as fresh water and ground water. It can also be widely found in the soil. Some researchers have also shown its presence in the atmosphere (Eerkes-Medrano et al., 2015; Carr et al., 2016; Alimi et al., 2018; Ng et al., 2018; Hurley & Nizzetto, 2018; Prata, 2018). Plastic waste of about 0.3 million tonnes has been dumped into the ocean in the last decade (Eriksen et al., 2014). The primary source of microplastics in various marine habitats (mangroves and beaches) is due to the fragmentation of the massive macro-waste (Heo et al., 2013; Fok & Cheung, 2015; Ho & Not, 2019; Martin et al., 2020; Not et al., 2020; Luo et al., 2021; Deng et al., 2021) After the fragmentation, these plastic wastes are available in all sizes and undergo primary deterioration by various means— i.e., mechanical transformation (wave and wind action), photooxidation (ultra-violet radiation), and bacterial degradation. Degradation of microplastics into nanoplastics (1–100 nm), upon discharge into the environment, occurs at varying rates, depending on the mentioned environmental conditions (Koelmans et al., 2015). It has been suggested that the ingestion of plastics by aquatic and terrestrial animals has frequent adverse effects, causing worldwide concern (Guzzetti et al., 2018; de Souza Machado et al., 2018).

When terrestrial or aquatic animals ingest plastics, their survival is highly impacted, as their reproductive capacity and energy balance are compromised (Sussarellu et al., 2016; Cole et al., 2013; Wright et al., 2013). Other impacts include oestrogenic effects (Yang et al., 2011) due to the leaching of plastic polymer additives such as bisphenol A and phthalates. The cellular membranes and organelles of these organisms are highly impacted by the bio-persistent nanoplastics (Forte et al., 2016).

DOI: 10.1201/9781003449133-13

The systematic reviews and research have been mainly focused on the direct inter-actions between plastics and organisms, their ecological impact, and their accumu-lation in the tissues (Ribeiro et al., 2019). The macrofauna and associated microbial population, which helps in the degradation of plastics and their ingestion and diges-tion, have yet to be explored. In recent studies, bio-fragmentation has been shown to increase the bioavailability of plastics, mainly due to the size reduction due to mechanical disintegration (Wright et al., 2013). This type of mechanical fragmen-tation allows smaller organisms to utilize the plastic fragments. This enhances the ratio of surface area to volume of the lesser fragments and helps in biofouling, lead-ing to an increased sinking dynamic of the microplastics in the aquatic environment (Mattsson et al., 2018). Moreover, chemical mechanisms are responsible for ingested plastic fragmentation (Song et al., 2020; Bombelli et al., 2017).

Macrofauna have been shown to alter the fate of plastics released into the environ-ment. A thorough comprehension of the driving mechanisms allow for a more accurate evaluation of the microplastic abundance in the environment and, by extension, their impact on ecosystems. It will provide new insights into how to reduce plastic contami-nation. This chapter seeks to summarize the current state of knowledge, identify areas that have not been adequately explored, and suggest future research directions.

13.2 SOURCES OF PLASTICS IN THE ECOSYSTEMS

There are various sources of plastic pollutants which enter the environmental strata; disposable items such as plastic bags, utensils, bottles, and food packaging contribute significantly to plastic pollution. Plastic bags and packaging materials are frequently discarded improperly or in water bodies and landfills due to littering (Jambeck et al., 2015; Andrady, 2017).

Microplastics are minuscule plastic particles, with a size less than 5 millimetres. They enter the environment through the disintegration of larger plastic objects, the abrasion of synthetic textiles, and the direct discharge of microplastics from personal care products (Geyer et al., 2017).

Synthetic textile fibres, like polyester and nylon, are shed during washing and infiltrate waterways via wastewater systems. These fibres are a major source of microplastic pollution in aquatic ecosystems (Hartline et al., 2016).

Plastic packaging and improper waste management: Plastic pollution is exacer-bated by improper waste management practices, such as an inadequate recycling infrastructure and a lack of appropriate disposal systems. Plastic waste can wind up in the environment, including rivers and oceans, if it is improperly managed (Lebreton et al., 2017;).

Various industries and applications, including medical applications, cosmetics, adhesives, waterborne paints, and coatings are contributors to primary microplastics. When larger plastics are broken down or degraded, they are called secondary micro-plastics. In addition to the mechanical and chemical breakdown of microplastics, the degradation may be due to various factors like radiation (UV) exposure, biological activities, wind, and tillage (Waldman & Rillig, 2020; Guo et al., 2020; Karbalaei et al., 2018).

Plastic pollution can originate from industrial activities such as the production, refining, and handling of plastics. Hermabessiere et al. (2017) report that accidental spills, waste mismanagement, and inadequate containment measures in these industries can lead to the discharge of plastic particles and chemicals into the environment.

13.3 WEATHERING OF PLASTICS

Weathering plastic polymers in the environment refers to the degradation and disintegration of plastic materials under environmental factors. Various biotic (living) and abiotic (non-living) factors contribute to it. Abiotic factors include sunlight (UV radiation), temperature, humidity, and mechanical stress. The degradation process can cause plastics to undergo physical and chemical changes, resulting in their fragmentation and the formation of microplastics, whereas microorganisms are biotic factors in plastic degradation. The process of plastic degradation is explained in Figure 13.1.

13.3.1 ABIOTIC FACTORS

i. Photodegradation: Sunlight, specifically ultraviolet (UV) radiation, plays a vital role in the degradation of plastics. UV radiation degrades polymer chains, resulting in the fragmentation of plastics. The term for this process is photodegradation (Barnes et al., 2009)

ii. Oxidative degradation: Plastics exposed to oxygen can undergo oxidative degradation. Oxygen reacts with polymer chains, resulting the scission of chains and the formation of free radicals, which accelerates the degradation process even further (Rychter et al., 2012)

iii. Thermal degradation: The plastic degradation can be accelerated by high temperatures. Thermal degradation entails the disintegration of polymer chains because of elevated temperatures, resulting in modifications to their physio-chemical features (Andrady, 2011)

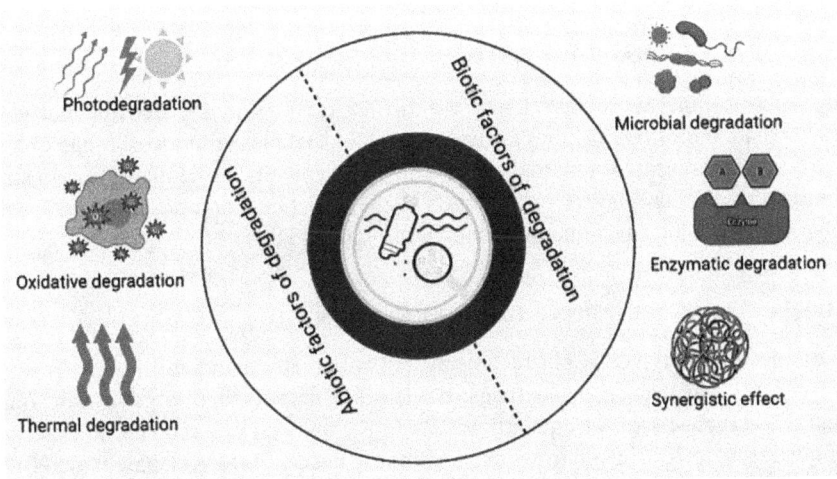

FIGURE 13.1 Process of degradation of plastics, showing the abiotic and biotic factors responsible.

13.3.2 Biotic Factors

Various microbes, such as fungi and bacteria, serve an essential role in the biodegradation of plastics. Some microorganisms produce enzymes, such as lipases and esterases, that can degrade plastics by breaking their polymer chains (Yang et al., 2014, 2015; Wei & Zimmermann, 2017).

Microbial biodegradation:

Bacteria: Several bacterial species have been identified as microorganisms capable of degrading plastic. Various bacteria, such as *Bacillus, Ideonella,* and *Pseudomonas*, contain enzymes capable of degrading polyethylene terephthalate (PET) and polyethylene (PE) (Yoshida et al., 2016).

Fungi: Certain fungi, such as *Aspergillus* and *Penicillium* species, have demonstrated the ability to degrade polymers. These fungi exude enzymes, such as esterases and lipases, that degrade plastic polymer chains (Wei et al., 2014).

Enzymatic action: Microorganisms produce specific enzymes, such as lipases, cutinases, and esterases, which can cleave the polymer chains of plastics, allowing for their degradation. These enzymes target the ester or amide bonds present in numerous types of plastics, thereby facilitating the degradation process (Shah et al., 2008).

Synergistic interactions: Plastic degradation can involve intricate microbial communities in which various microorganisms collaborate synergistically. A consortium of bacteria and fungi, for instance, may degrade plastics more effectively than individual species, as their enzymatic activities are complementary (Sivan, 2011).

Adaptation and evolution: Microorganisms can adapt and evolve in response to environmental changes, such as the presence of plastics. Over time, they can develop mechanisms to degrade and utilize plastics as a carbon source, contributing to the process of plastic biodegradation (Wei & Zimmermann, 2017).

13.4 MACRO/NANO PLASTICS

Macroplastics and microplastics are two distinct types of plastic detritus that vary in size and composition. Microplastics are minute particles with a size range of less than 5 mm, in contrast to macroplastics, which are typically larger than 5 mm in size. The sorting of microplastics on the basis of their chemical composition and breakdown pattern in the environment is shown in Figure 13.2.

13.4.1 Characterization and Differentiation

Macroplastics are typically plastic objects that are greater than 5 millimetres in size. These items can include plastic bags, bottles, fishing nets, and larger plastic fragments. Macroplastics are frequently visible to the unaided eye and can be observed in aquatic and terrestrial environments (Thompson et al., 2004; Kershaw et al., 2019). Microplastics are tiny plastic particulates between 1 nanometre and 5 millimetres

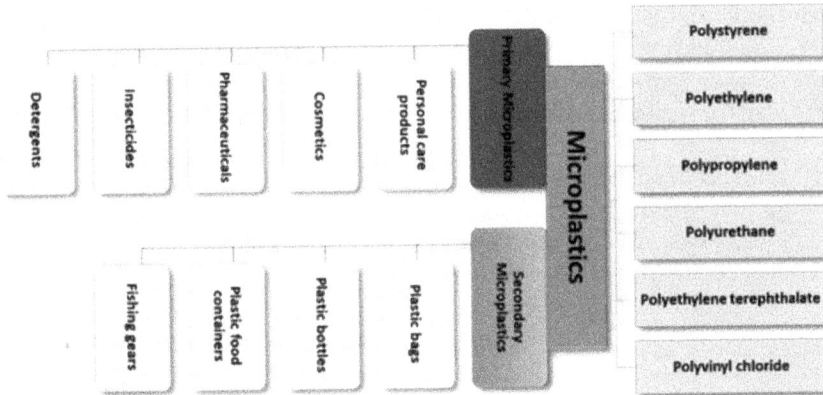

FIGURE 13.2 Classification of microplastics on the basis of their chemical constituents and degradation pattern in the environment.

in size. They can be further classified as primary microplastics—small plastic particles that are manufactured intentionally—and secondary microplastics—the plastics that result from the decomposition of larger plastic objects, such as synthetic fibres. Micro-beads used in personal care products, plastic granules (nurdles) used in plastic manufacturing, and microfibres from synthetic textiles are examples of primary microplastics (Koelmans et al., 2015). The fragmentation and deterioration of larger plastic objects produce secondary microplastics. They can originate from plastic detritus, degraded fishing nets, and river-borne plastic debris (Andrady, 2017; Wright & Kelly, 2017). Microplastics are typically classified according to their size range (Eriksen et al., 2014).

- Microplastics: Plastics smaller than 5 millimetres.
- Mesoplastics: Sized between 5 millimetres and 1 centimetre.
- Macroplastics: Plastics larger than one centimetre.

Nanoplastics are plastic particulates with nanometre-scale dimensions (less than 1 micron). They were produced by the further breakdown of microplastics or by the direct discharge of nanoplastics from products like nanocomposites and nanoparticle-based coatings (Nowack & Bucheli, 2007; Gigault et al., 2018). They are a subcategory of microplastics and are of increasing concern because of their detrimental impacts on the environment and living organisms.

13.4.2 Techniques Used for the Analysis of Macro/Microplastics

Various techniques are used to analyze macroplastics and microplastics in various samples. These techniques involve visual inspection, spectroscopic analysis, and microscopy-based methods. The techniques are frequently employed for analyzing macroplastics and microplastics, allowing researchers to evaluate the presence, quantity, composition, and characteristics of plastic detritus in various environmental samples.

i. Visual inspection and sorting: Visual inspection is a straightforward method that entails the manual classification and identification of macroplastic items based on their size, shape, and composition. This is a standard method for removing larger plastic debris (> 5 mm) from coastlines and shorelines (Arthur et al., 2009; Lusher et al., 2015; Thompson et al., 2009).
ii. Spectroscopic analysis: Various spectroscopic techniques have been employed to determine the type of plastics in the environment. The techniques such as Raman spectroscopy and Fourier-transform infrared (FTIR) spectroscopy, are used to categorize and characterize the chemical composition of microplastics and microplastics. The characterization of various peaks has been shown in Table 13.1 by Raman spectroscopy. In the case of FTIR, the infrared absorption spectra are specific to the type of polymers, thus facilitating its identification and characterization. It identifies specific functional groups, like C=O, C-H, and C-O, and provides information on the chemical nature of microplastics. Raman spectroscopy employs laser-induced light scattering to determine the chemical constituents of microplastics at the molecular level. It can identify specific chemical groups, including C-C, C-H, and C=O bonds. These techniques analyze the molecular vibrations and interactions of plastic polymers, allowing for their classification (Löder & Gerdts, 2015; Talvitie et al., 2017; Frias & Nash, 2019).

TABLE 13.1

The Peak Characteristics of Various Chemical-Bond Modifications Observed Using Raman Spectroscopy for Different Types of Plastics

Plastic Type	Chemical Bond	Peak Characteristics	Reference
Polyethene (PE)	C-C stretching.	1100 cm^{-1} and 1460 cm^{-1}	Araujo et al., 2018; Käppler et al., 2015
Polypropylene (PP)	CH$_2$ bending mode and the CH$_2$ twisting mode.	852 cm^{-1} and 997 cm^{-1}	Lim et al., 2008
Polyethylene terephthalate (PET)	The vibrations of C-C stretching and C=O stretching in the ester groups.	1600 cm^{-1} and 1720 cm^{-1}.	Li et al., 2016; Cabernard et al., 2018
Polystyrene (PS)	The aromatic ring vibrations.	1000 cm^{-1} and 1600 cm^{-1}	Araujo et al., 2018; Huang et al., 2023
Polyvinyl chloride (PVC)	C-Cl stretching vibrations.	1000 cm^{-1}, 1316 cm^{-1}, and 1596 cm^{-1}.	Anger et al., 2018

iii. Microscopy techniques: Microscopy techniques allow for the identification and characterization of microplastics via visual inspection. Optical microscopy permits the visual characterization of microplastics because of their morphology and optical properties. It helps estimate the size distribution and abundance of microplastics (Löder & Gerdts, 2015). The transmission electron microscopy (TEM) and scanning electron microscope (SEM) provide high-resolution images of microplastics (< 100 μm)., allowing for detailed morphological and surface characterization. It facilitates identifying the type of polymer and evaluating physical and chemical changes. Atomic Force Microscopy (AFM) can measure particle size, shape, and surface roughness by scanning a pointed probe across the surface of microplastics to obtain topographic information (Dehaut et al., 2016; Eerkes-Medrano et al., 2015).

iv. Analytical pyrolysis gas chromatography-mass spectrometry (Py-GC-MS): Py-GC-MS is an effective method for analyzing the chemical composition of macro- and microplastics. Thermal degradation of plastics is followed by gas chromatography-mass spectrometry analysis of the pyrolysis products. This method provides comprehensive information regarding the polymer type and additives prevalent in plastics (Peng et al., 2022; Mariano et al., 2021).

13.4.3 SORPTION OF CHEMICALS

Macro/microplastics can act as sorbents of various chemicals present in the environment, and this process is called sorption. These materials act as carriers and transporters of contaminants to various compartments of the environment. Thus, macro/microplastics play a pivotal role in the environment for the fate and transport of these contaminants. The sorption process occurs mainly by two primary mechanisms, like adsorption and absorption. Adsorption refers to accumulating chemicals on the surface of polymers due to weaker interactions, such as electrostatic interactions, van der Waals interactions, and hydrogen bonding. This process is typically reversible and is affected by the physicochemical properties of both the chemical and plastic surfaces. During absorption, chemicals penetrate the substance of the plastic material. When the plastic material is porous or contains hydrophilic regions, this process occurs. Absorption is typically irreversible and may be dependent on the chemical basis and structural composition of the plastic. There are various factors which influence the sorption process, as explained later and shown in Figure 13.3:

a. Plastic characteristics: The chemical composition, surface area, morphology, porosity, and surface charge of macro- and microplastics can influence sorption. The sorption capacities of plastics with larger surface areas and more porous structures are typically greater.

b. Chemical properties: The physico-chemical properties of substances like hydrophobicity, molecular weight, polarity, and solubility play a fundamental role in their sorption onto polymers. Due to favourable interactions with the hydrophobic plastic surface, hydrophobic substances exhibit enhanced sorption.

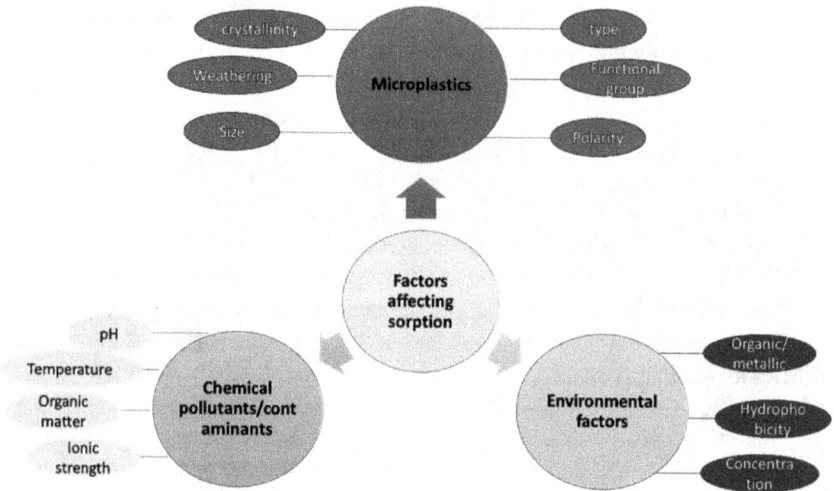

FIGURE 13.3 Factors affecting the sorption of various additives on a plastic surface.

c. Environmental conditions: Sorption can be affected by environmental factors like temperature, pH, salinity, and the presence of other substances. These conditions can influence the surface charge and chemical interactions of the plastic, thereby affecting its sorption behaviour (Tang, 2021; Prajapati et al., 2022; Costigan et al., 2022).

The sorption of polychlorinated biphenyls (PCBs) onto marine macroplastics has been studied, and it was seen that sorption capacity varied based on the type of plastic, with polyethene exhibiting a higher sorption capacity than polypropylene (Vo & Pham, 2021). Also, the sorption of hydrophobic organic compounds on macroplastics showed that additives and weathering can influence sorption behaviour. Similarly, studies conducted on the sorption of microplastics showed that the sorption of pharmaceuticals onto microplastics can function as carriers for these substances, thereby potentially facilitating their transport and exposure to organisms (Wang et al., 2020). The sorption of metals onto microplastics highlighted the role of surface chemistry and particle size in determining the capacity of sorption (Ho et al., 2022).

13.4.4 TOXIC IMPACT ON ORGANISMS

The environmental toxicity of macro- and microplastics is an increasing concern. Plastics can contain various additives, including colourants, stabilizers, and flame retardants. Plastics are derived from various raw materials, including petroleum, natural gas, and lignite. These additives can leach from plastics into the ecosystem, where they can have various adverse effects on vegetation, animals, and humans. Notably, the specific toxic effects can vary, depending on the size, shape, type and composition of the plastics, and the characteristics of the ecosystem and the organisms involved. The toxic effects associated with macro/microplastic can be physical harm to the organism/ecosystem, transfer or exposure of chemical contaminants/constituents, various ecological disruptions and their effects, bioaccumulation, and biomagnification. These pollutants have adverse effects on various ecosystems and effects on humans, as discussed in the following paragraphs.

Terrestrial ecosystem: As discussed by most researchers, the fate of the macro/microplastic is the oceans, but before they enter there, they have been produced, extensively utilized, and disposed of on the land and can enter the terrestrial ecosystem through multitude pathways (Jambeck et al., 2015; Nizzetto et al., 2016). Microplastics may first interact with flora and fauna in terrestrial systems, eliciting ecologically potential effects. According to recent studies and pieces of research evidence, it is said that microplastics, when interacting with the terrestrial ecosystem and organisms, alter various important functionalities and services like the plant pollinators, soil-dwelling invertebrates, and terrestrial fungi. The primary (household items, cosmetics and personal care products, landfills, sludge) and secondary (municipal solid waste, mulches, irrigation water, landfill disposals, soil amendments) microplastics contribute to pollute the terrestrial ecosystem (Karbalaei et al., 2018; Galafassi et al., 2019; Bradney et al., 2019).

Various activities like tillage, water infiltration due to digging, and various animal activities, like ingestion and egestion by earthworms, help the plastic particles penetrate from the soil into layers of the deep soil (Guo et al., 2020; Rillig et al., 2017). Other factors which are capable of contributing to the degrading microplastics are thermal oxidation, physical abrasions, UV radiations, microbial degradation, and various interactions with the soil colloids (Benítez et al., 2013; Krueger et al., 2015; Sen & Raut, 2015; Zhu et al., 2019; Ren et al., 2021).

A combinatorial effect of several factors leads to the degradation of microplastics; however, all the factors operate individually and differently under various environmental conditions (seasonal, regional, and time-based factors). The light/radiation during the day plays an important role, but its effect may be negligible during the night (Liu et al., 2021). The temperature during summers in an arid region may have a greater impact than that of the colder regions during winters (Liu et al., 2021). The physical and chemical features of microplastics, such as surface chemistry chemical composition, colour, and crystallinity, change because of ageing (Ren et al., 2021).

During the ageing process of microplastics, numerous chemicals are released, like retardants, phthalates, stabilizers, oligomers, pigments, and oxygenated products, such as carboxylated products, phenols, and acetophenones (Liu et al., 2021). It has been seen that the ageing of microplastics is relatively very slow. Previous studies suggested that up to 15 years after sludge application, synthetic fibres were detected at depths of up to 100 cm in soil (Zubris & Richards, 2005). The presence of microplastics have been seen in the various strata—i.e., the plant system, groundwater, as well as the terrestrial ecosystems and food web due to biogenic transportation (Huerta Lwanga et al., 2017). The discharge of microplastics is probably the direct route for freshwater ecosystem contamination via land-based microplastics (Rocha-Santos & Duarte, 2015).

Microplastics affect the physio-chemical features and microbial population of the soil, causing detrimental effects on soil health. Microplastics can hamper the soil structure and texture, and these changes can be visually observed in highly polluted areas. Studies have shown that microplastics have an impact on water-holding capacity, bulk density, soil porosity, and water-stable soil aggregates (de Souza MacHado et al., 2018; Wang et al., 2021, 2022; Lehmann et al., 2021). The presence of microplastics alters the pH of the soil, thus impacting the nutrient availability to the plant and impacting productivity (Boots et al., 2019; Qi et al., 2020; Wang et al., 2021).

Soil carbon availability may be affected by the inert carbon of the microplastics. This influences the carbon availability of the soil microbiota and the plant, impacting the biogeochemical carbon cycling (Rillig et al., 2019).

Microbial activity is driven primarily by the changes in the soil habitat (Zhou et al., 2020). Microplastics in the soil alter the soil porosity and moisture uptake capacity, which impact the concentration of oxygen in the soil. This, in turn, highly affects the abundance of anaerobic and aerobic microorganisms in the soil. Decreased soil respiration rates caused by microplastics can affect soil microbial population and the community structure of various microorganisms (Judy et al., 2019).

Various soil enzymes like catalases, urease, diacetate hydrolase, fluorescein, and phenol oxidase undergo alterations (Huang et al., 2019). It has been observed in various types of research that microplastics have an extreme potential to affect microbial growth and reproduction. This has emerged as a threat to soil microbial biodiversity. However, contradictory results have shown that microplastics showed a negligible effect on the soil microbes (Wang et al., 2019; Rodriguez-Seijo et al., 2017).

13.4.4.1 Aquatic Ecosystem

The migration of microplastics from terrestrial to the aquatic environment is generally due to industrial discharges, sewage sludge, rainwater discharge, littering of plastic waste, farmland wastes, plastic films from the farm used for mulching, irrigation runoff, etc. (Kay et al., 2018; Hurley et al., 2018; Wagner et al., 2019; Yu et al., 2017; Zhaorong et al., 2020). Microplastics are ubiquitous in natural and anthropogenic water cycles (Carr et al., 2016).

Microplastics can result in physical damage due to entanglement and ingestion of aquatic organisms. They are swallowed and ingested, thus creating a pseudo-sense of fullness. This affects the appetite and may also cause blockage or digestive damage. These microplastics may even form aggregates and enter the bloodstream (Wang et al., 2019; Du et al., 2021). This series of events finally affects the development and reproduction of living organisms. The size of microplastics highly influences its assimilation and aggregation in the organism. The smaller microplastics tend to be more readily ingested and accumulated, thereby reducing the growth rate, fertility, and longevity (Chang-Bum et al., 2016). Microplastics are even capable of penetrating the intestinal epithelium and entering biological tissue, causing increased harm to organisms (Bouwmeester et al., 2015). The residual fibres of microplastics are reported to cause the organism's death as it entangles in the intestine. These fibres cannot be completely excreted by the body, leading to long-term retention in the body (Gray & Weinstein, 2017). Polychlorinated biphenyls, absorbed on microplastics, have a carcinogenic, teratogenic, and mutagenic effect on organisms

(Wang et al., 2018). Besides microplastic, additives and plasticizers like bisphenol A, boric acid, brominated flame retardant, nonylphenol, octyl phenol, etc., are also found in natural waters (Hu et al., 2020). Microplastics enter the food chain when are ingested by the fish, and this leads to trophic exchange. Omnivore fishes consume more plastics than carnivores, as plastic have an inferior impact on excretion (Zhang et al., 2021). Many pesticides, like dichlorodiphenyltrichloroethane and polychlorinated biphenyls, are present in the water bodies. These get sorbed by microplastic, and when ingested by aquatic organisms, the toxicity increases. The contaminants work synergistically, causing increased oxidative stress and compromising the immune system of the organism (Xu et al., 2021). Along with microplastics, the marine environment is polluted with human pathogens, heavy metals, antibiotics, etc. These have evolved as a global, major threat to the ecosystem as these microplastics have passed through the food chain and have affected various strata immensely (Imran et al., 2019; Xu et al., 2020).

13.4.5 IMPACT ON HUMANS: EPIDEMIOLOGICAL AND EXPERIMENTAL EVIDENCE

Microplastics have infiltrated the human food chain in several ways, including consumption by animals in their natural environment, contamination during production of food, and/or leaching from packaging materials of food and beverages (Li et al., 2018). Micro- and nanoparticles have been detected in honey, beer, salt, sugar, fish, prawns, and bivalves (Chenxi et al., 2018). The human body can eliminate these pollutants via the biliary, urinary, faecal, and other physiological excretion routes.

However, these particles may accumulate, infiltrate the lymphatic and blood systems via the airway or gastrointestinal tract, and then accumulate in the organs. Upon ingestion, these microplastics can cause the release of the associated chemicals and inflammation (Mock, 2020). As it has a larger ratio of surface area to volume, microplastics were shown to alter the immune mechanisms during pregnancy. It also affects the after-implantation stages of growth factor signalling. It also has been shown to disrupt embryo-utero communication (Ilekis et al., 2016). The microplastics and its plasticizers have shown to have transgenerational effects on reproduction and metabolism (Lee, 2018). Even at modest concentrations, microplastics can increase genome variability (Zhang et al., 2021a, 2021b; Rochman, Kross et al., 2015; Rochman, Tahir et al., 2015). Studies have shown that consuming fish containing microplastics adversely affect humans and resulted in inflammation and cell necrosis.

Humans can be exposed to microplastics via inhalation, ingestion, and skin contact (Prata et al., 2020; Rahman et al., 2021). The airborne microplastics present in the urban dust can be inhaled, ingestion can be caused by the methods explained earlier, and it cannot breach the skin barrier but can pass through wounds and cuts, hair follicles, or sweat glands (Schneider et al., 2009). However, the most significant route is through ingestion. Subject to exposure and susceptibility, microplastics may lead to oxidative stress, cytotoxicity, and translocation to other tissues.

As microplastics are persistent in nature, their removal from the organisms results in various chronic conditions like inflammation and also an elevated risk of cancer. As a particulate matter, it may also risk the immune system and also lead to neurodegenerative diseases. Microplastics may also release various chemicals adsorbed on their matrices into the environment where it is present (Crawford & Quinn, 2017), acting as a vector for various disruptive microbial communities (Kirstein et al., 2016). The potential health complications related with microplastics are as follows:

13.4.5.1 Oxidative Stress and Cytotoxicity

As previously discussed, microplastic may contains various additives and chemicals on its surface, which may be oxidative in nature, like metals. These oxidizing compounds can cause oxidative stress and inflammation in the host due to the production of reactive oxygen species (ROS) (Kelly & Fussel, 2012; Valavanidis et al., 2013). The weathering of MPs also result in free radical production, which oxidizes the target tissues (Gewert et al., 2015; Sternschuss et al., 2012).

The limb and joint prostheses contain microplastics, which release free radicals and acute toxins, which cause acute inflammation. These free radicals may lead to rejecting prosthesis from the human body, as the oxidants induce hydrolysis, resulting in polymer degradation and leaching and cracking (Sternschuss et al., 2012)

13.4.5.1.1 Altering Metabolism and Energy Balance

The digestive enzymes are directly or indirectly influenced by microplastics, thus troubling the energy balance. Studies have revealed that exposure to microplastics in fishes and rodents has increased lactose dehydrogenase. This causes an increased rate of anaerobic metabolism (Wen et al., 2018; Deng et al., 2017). A significant decrease in ATP production has been seen in the case of mice livers, causing reduced lipid metabolism (Deng et al., 2017). Microplastics have been shown to disturb the body's homeostasis in various studies, thus disrupting the energy balance. It has been found to reduce food uptake in various animals like marine worms, clams, and crabs, thus decreasing the energy intake. It also reduces predatory and digestive activities in fish (Wright et al., 2013; Watts et al., 2015; Wen et al., 2018). Similarly, microplastics may show similar effects in the human system, as seen in other organisms. It may alter the metabolic systems by decreasing or increasing the expenditure of energy, modulating enzymatic pathways, lowering nutrient intake, etc.

13.4.5.1.2 Disruption of Immune Function

Depending upon the dissemination and host response, microplastic induces a local or systemic immune response. In the case of genetically susceptible individuals, microplastic exposure causes autoimmune diseases of immunosuppression due to disruptive immune function (Prata, 2018; Prata et al., 2020). These types of immune diseases may be caused due to oxidative stress in the cells, the release of various modulators of immune system, and aberrant activation of the immune cells (Farhat et al., 2011). Particulate matter exposure is associated with systemic lupus erythematosus and systemic rheumatic diseases (Bernatsky et al., 2016; Fernandes et al., 2015). Studies have also shown that microplastic exposure causes enhanced production of anti-inflammatory cytokines, decreased T-effector production, and suppression of T-helper cells (Saravia et al., 2014).

13.4.5.1.3   *Translocation to Distant Tissues*

On exposure to microplastic, they may enter the circulatory system and get translocated to various tissues, interfering with their normal function. The first major effect which microplastics may cause is systemic and vascular inflammation, along with cytotoxicity to the blood cells and occlusions (Canesi et al., 2015; Wright & Kelly, 2017; Campanale et al., 2020). The microplastics, after inhalation and ingestion in rats, were found to be translocated to the spleen and liver via the circulatory system (Campanale et al., 2020). It has also been found to be accumulated in the gut, liver, and kidney of mice using tissue accumulation kinetics (Deng et al., 2017). Microplastics have been found even in the bone, which may cause activation of osteoclasts, leading to the loss of bone (Liu et al., 2015; Ormsby et al., 2016). Several studies revealed that micro/nanoplastics can pass through the placental barrier without modifying the explant viability (Wick et al., 2010; Grafmueller et al., 2015).

13.4.5.1.4   *Neurotoxicity*

Various *in vivo* studies were carried out by different models to study the neurotoxic effect of microplastics. They could have an impact on neural function and behaviour. Chronic exposure to microplastics leads to the activation of immune cell and oxidative stress in the brain. This can cause damage to the neuron due to pro-inflammatory cytokines (MohanKumar et al., 2008; Barboza et al., 2018).

13.5   RISK ASSESSMENT AND ITS MITIGATION

Microplastic pollution has been environmentally persistent and complex and has emerged as a global threat. It is necessary to examine its environmental, economic, and societal implications due to its persistent nature and create an outline for future activities. Policymakers have taken various initiatives and regulations to combat the increasing plastic pollution. In the case of single-use plastics and the use of micro-beads by industries, a restricted usage has been imposed, especially in the case of cosmetics and personal care products. The USA has a policy of the Micro-Bead-Free Water Act 2015, which requires plastic manufacturers to ensure zero pellet-loss from production to transport and also prohibits the use of microbeads in rinse-off cosmetics. Similar actions have been taken by Canada (2017) and United Kingdom (2018) (Mitrano & Wohlleben, 2020; Miranda et al., 2020). The UNEP Honolulu Strategy 2011 encourages a circular economy initiative to reduce plastic pollution and also intends to solve the marine debris problems using holistic approaches by reducing land-based plastic litter, shoreline litters, and sea contaminants, which are caused due to usage of various fishing gears and cargos (Pettipas et al., 2016). The G7 nations, in 2015, formulated a policy to lower marine litter passing into the inland and coastal waters to improve the waste management system of these countries (Brennholt et al., 2018). Apart from the policies, various clean-up drives and initiatives, like Sol-gel-based agglomeration and air purifiers, have been introduced to remove microplastics from water and air, respectively. Various fungi have also been used to degrade microplastics in soil (Ali et al., 2014; Herbort et al., 2018).

13.6 REMEDIATION OF MICROPLASTICS

Plastic particulates continue to transform and migrate from one ecosystem to the other, contaminating all human interactions. Ultimately, microplastics (MPs) impact the health of animals and humans by entering our food systems and contaminating ecosystems. It becomes essential to understand the remediation of these microplastics in various ecosystems. Methods like bioremediation, photocatalysis, oxidation process of microplastics, and microwave have been used to breakdown microplastics in the soil and water. In this section, we shall briefly discuss the methods employed for the remediation process of microplastics.

13.6.1 COAGULATION/AGGLOMERATION/FLOCCULATION

This process has been used to eliminate large particles of microplastics from water, and the efficiency of removal depends majorly on the size of the microplastics. It is one of the most conventional and frequently used processes (Park et al., 2020; Ma et al., 2020). In this process, a coagulating agent (aluminum-based or iron-based) is used, which destabilizes dissolved and suspended particles, enabling sedimentation (Shirasaki et al., 2016).

13.6.2 MEMBRANE BIOREACTOR TECHNOLOGY

It has been used as one of the most reliable methods for the treatment of wastewater with high contaminants and has been recently used for the removal of microplastics. The membrane technology has good removal efficiency, high rejection potency of the contaminants, and gives high effluent quality.

13.6.3 ADVANCED PHOTOOXIDATION

Phototdegradation has been considered as a highly promising and effective method for the treatment of wastewater containing organic pollutants and microplastics. It basically consists of a visible or UV-light-absorbing semiconductor, which, in turn, generates free radicals. Microplastics are degraded by the generated free radicals like reactive oxygen species, superoxide, and hydroxyl radicals (Zhu et al., 2019), causing volatile organics, CO_2 and H_2O. The microwave-assisted catalysis process uses an iron-based catalyst and requires only 30–90 s. This process converts into carbon nanotubes and hydrogen. On the other hand, photocatalysis utilizes Nb_2O_5 and forms CO_2 and CH_3COOH without using any sacrificial agent.

13.6.4 BIOREMEDIATION

Microorganisms played an important role in the degradation of micro/nanoplastics in the soil and water (Dussud & Ghiglione, 2014; Lwanga et al., 2018; Bhatt et al., 2021; Mohanan et al., 2020). Bioremediation is carried out at various stages—i.e., biofragmentation, biodeterioration, assimilation and mineralization of polymer. Microorganisms initiate physicochemical degradation, then fragment polymers into

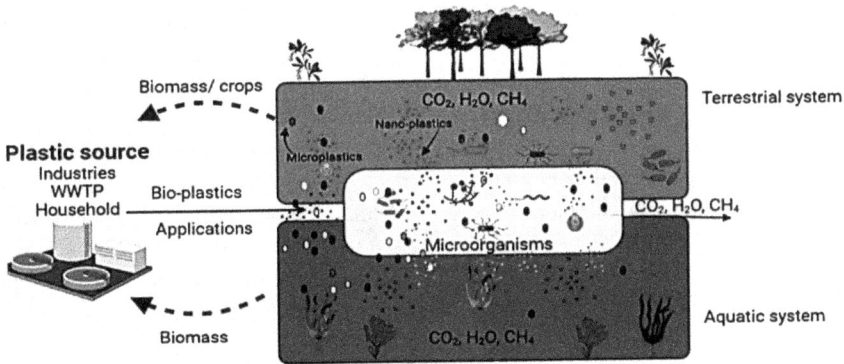

FIGURE 13.4 Bioremediation routes of microplastics in the environment.

monomers and oligomers by the production of exoenzymes, integrate molecules into metabolism, and finally, expel oxidized metabolites (Dussud & Ghiglione, 2014). Exoenzymes (oxygenases) destabilize polymers' long chains of carbon and hydrogen and can add oxygen to produce alcohol and carboxylic and peroxyl compounds. The microbial metabolic process then assimilates and mineralizes these compounds (Conkle et al., 2018). Studies have also shown that algae and earthworms can degrade microplastics (Chia et al., 2020; Nakanishi et al., 2020). Biodegradable plastics are more prone to microbial attack, and hence, are considered a safer option than synthetic plastics (Liao & Chen, 2021). Biodegradable plastics are a better replacement for synthetic plastics. This has been seen even in cosmetic industries, where chitosan beads can completely degrade in soil without toxic effects. The development of synthetic plastic and engineered microbes can act as key to microplastic remediation (Ju et al., 2021). It is essential to generate a circular economy using biobased models without hampering the benefits of the plastics. The tentative route of microplastic bioremediation in the environment if explained in Figure 13.4. The aquatic and terrestrial plants have great potential to absorb micro/nanoplastics. The microbial community can convert these pollutants into water, carbon dioxide, and methane, and plants in the environment utilize these products. Bioremediation can be done using non-edible plants and algae, along with the usage of bioplastics.

13.7 CONCLUSION

Microplastic pollution is currently recognized as an alarming issue. Economic, environmental, and societal concerns regarding the impact of microplastic pollution on the environment have attracted a great deal of interest from all sectors (the general public, environmental activists, policymakers, and scientific communities) for research on potential replacements and remediation pathways. Even though massive efforts are being made to substitute synthetic plastics with biodegradable plastics in order to reduce plastic pollution in the environment, additional attention must be paid to avoid the negative impact of microplastics from biodegradable plastics in order to prevent any unintended hazard to the human health and the environment. Along with technological advancements, economic and social implications, and microplastic pollution remediation, bioremediation probably is a potential solution for microplastics-related problems. However, their identification and characterization methods have not yet been standardized, and a consensus must be reached on them in order to simplify the development of impact indicators, and thus, the evaluation of their environmental impacts. The production of biodegradable polymers from non-edible biomass like algae could be a major approach for eradicating microplastic pollution in sustainable ecosystems. A broader sustainability analysis must support any innovative effort to reduce or eliminate micro/nanoplastic pollution to avoid investment and environmental risk. Alternatives to conventional plastics should be used, along with an integrated strategy to eradicate these pollutants. Government agencies, policymakers, and environmental activists should work with the end-users to formulate alternatives to conventional plastics.

REFERENCES

Act of 2015 [WWW Document]. www.congress.gov/bill/114th-congress/h.

Ali, M. I., Ahmed, S., Robson, G., Javed, I., Ali, N., Atiq, N., & Hameed, A. (2014). Isolation and molecular characterization of polyvinyl chloride (PVC) plastic degrading fungal isolates. *Journal of Basic Microbiology*, 54(1), 18–27.

Alimi, O. S., Farner Budarz, J., Hernandez, L. M., & Tufenkji, N. (2018). Microplastics and nanoplastics in aquatic environments: Aggregation, deposition, and enhanced contaminant transport. *Environmental Science & Technology*, 52(4), 1704–1724.

Andrady, A. L. (2011). Microplastics in the marine environment. *Marine Pollution Bulletin*, 62(8), 1596–1605.

Andrady, A. L. (2017). The plastic in microplastics: A review. *Marine Pollution Bulletin*, 119(1), 12–22.

Anger, P. M., von der Esch, E., Baumann, T., Elsner, M., Niessner, R., & Ivleva, N. P. (2018). Raman microspectroscopy as a tool for microplastic particle analysis. *TrAC Trends in Analytical Chemistry*, 109, 214–226.

Araujo, C. F., Nolasco, M. M., Ribeiro, A. M., & Ribeiro-Claro, P. J. (2018). Identification of microplastics using Raman spectroscopy: Latest developments and future prospects. *Water Research*, 142, 426–440.

Arthur, C., Baker, J. E., & Bamford, H. A. (2009). Proceedings of the international research workshop on the occurrence, effects, and fate of microplastic marine debris, September 9–11, 2008, University of Washington Tacoma, Tacoma, WA, USA.

Barboza, L. G. A., Vieira, L. R., Branco, V., Figueiredo, N., Carvalho, F., Carvalho, C., & Guilhermino, L. (2018). Microplastics cause neurotoxicity, oxidative damage and energy-related changes and interact with the bioaccumulation of mercury in the European seabass, Dicentrarchus labrax (Linnaeus, 1758). *Aquatic Toxicology*, 195, 49–57.

Barnes, D. K., Galgani, F., Thompson, R. C., & Barlaz, M. (2009). Accumulation and fragmentation of plastic debris in global environments. *Philosophical Transactions of the Royal Society B: Biological Sciences*, 364(1526), 1985–1998.

Benítez, A., Sánchez, J. J., Arnal, M. L., Müller, A. J., Rodríguez, O., & Morales, G. (2013). Abiotic degradation of LDPE and LLDPE formulated with a pro-oxidant additive. *Polymer Degradation and Stability*, 98(2), 490–501.

Bernatsky, S., Smargiassi, A., Barnabe, C., Svenson, L. W., Brand, A., Martin, R. V., … Joseph, L. (2016). Fine particulate air pollution and systemic autoimmune rheumatic disease in two Canadian provinces. *Environmental Research*, 146, 85–91.

Bhatt, P., Pathak, V. M., Bagheri, A. R., & Bilal, M. (2021). Microplastic contaminants in the aqueous environment, fate, toxicity consequences, and remediation strategies. *Environmental Research, 200*, 111762.

Bombelli, P., Howe, C. J., & Bertocchini, F. (2017). Polyethylene bio-degradation by caterpillars of the wax moth Galleria mellonella. *Current Biology, 27*(8), R292–R293.

Boots, B., Russell, C. W., & Green, D. S. (2019). Effects of microplastics in soil ecosystems: Above and below ground. *Environmental Science & Technology, 53*(19), 11496–11506.

Bouwmeester, H., Hollman, P. C., & Peters, R. J. (2015). Potential health impact of environmentally released micro-and nanoplastics in the human food production chain: Experiences from nanotoxicology. *Environmental Science & Technology, 49*(15), 8932–8947.

Bradney, L., Wijesekara, H., Palansooriya, K. N., Obadamudalige, N., Bolan, N. S., Ok, Y. S., . . . Kirkham, M. B. (2019). Particulate plastics as a vector for toxic trace-element uptake by aquatic and terrestrial organisms and human health risk. *Environment International, 131*, 104937.

Brennholt, N., Heß, M., & Reifferscheid, G. (2018). Freshwater microplastics: Challenges for regulation and management. *Freshwater Microplastics: Emerging Environmental Contaminants?* 239–272.

Cabernard, L., Roscher, L., Lorenz, C., Gerdts, G., & Primpke, S. (2018). Comparison of Raman and Fourier transform infrared spectroscopy for the quantification of microplastics in the aquatic environment. *Environmental Science & Technology, 52*(22), 13279–13288.

Campanale, C., Massarelli, C., Savino, I., Locaputo, V., & Uricchio, V. F. (2020). A detailed review study on potential effects of microplastics and additives of concern on human health. *International Journal of Environmental Research and Public Health, 17*(4), 1212.

Canesi, L., Ciacci, C., Bergami, E., Monopoli, M. P., Dawson, K. A., Papa, S., . . . Corsi, I. (2015). Evidence for immunomodulation and apoptotic processes induced by cationic polystyrene nanoparticles in the hemocytes of the marine bivalve Mytilus. *Marine Environmental Research, 111*, 34–40.

Carr, S. A., Liu, J., & Tesoro, A. G. (2016). Transport and fate of microplastic particles in wastewater treatment plants. *Water Research, 91*, 174–182.

Chang-Bum, J., Eun-Ji, W., Hye-Min, K., Min-Chul, L., Dae-Sik, H., Un-Ki, H., . . . Jae-Seong, L. (2016). Microplastic size-dependent toxicity, oxidative stress induction, and p-JNK and p-p38 Activation in the Monogonont Rotifer (Brachionus koreanus).

Chenxi, W. U., Xiangliang, P. A. N., Huahong, S. H. I., & Jinping, P. E. N. G. (2018). Microplastic pollution in freshwater environment in China and watershed management strategy. *Bulletin of Chinese Academy of Sciences (Chinese Version), 33*(10), 1012–1020.

Chia, W. Y., Tang, D. Y. Y., Khoo, K. S., Lup, A. N. K., & Chew, K. W. (2020). Nature's fight against plastic pollution: Algae for plastic biodegradation and bioplastics production. *Environmental Science and Ecotechnology, 4*, 100065.

Cole, M., Lindeque, P., Fileman, E., Halsband, C., Goodhead, R., Moger, J., & Galloway, T. S. (2013). Microplastic ingestion by zooplankton. *Environmental Science & Technology, 47*(12), 6646–6655.

Conkle, J. L., Báez Del Valle, C. D., & Turner, J. W. (2018). Are we underestimating microplastic contamination in aquatic environments? *Environmental Management, 61*(1), 1–8.

Costigan, E., Collins, A., Hatinoglu, M. D., Bhagat, K., Macrae, J., Perreault, F., & Apul, O. (2022). Adsorption of organic pollutants by microplastics: Overview of a dissonant literature. *Journal of Hazardous Materials Advances*, 100091.

Crawford, C. B., & Quinn, B. (2017). The interactions of microplastics and chemical pollutants. *Microplastic Pollutants*, 131–157.

de Souza Machado, A. A., Lau, C. W., Till, J., Kloas, W., Lehmann, A., Becker, R., & Rillig, M. C. (2018). Impacts of microplastics on the soil biophysical environment. *Environmental Science & Technology, 52*(17), 9656–9665.

Dehaut, A., Cassone, A. L., Frère, L., Hermabessiere, L., Himber, C., Rinnert, E., . . . Paul-Pont, I. (2016). Microplastics in seafood: Benchmark protocol for their extraction and characterization. *Environmental Pollution, 215,* 223–233.

Deng, H., He, J., Feng, D., Zhao, Y., Sun, W., Yu, H., & Ge, C. (2021). Microplastics pollution in mangrove ecosystems: A critical review of current knowledge and future directions. *Science of the Total Environment, 753,* 142041.

Deng, Y., Zhang, Y., Lemos, B., & Ren, H. (2017). Tissue accumulation of microplastics in mice and biomarker responses suggest widespread health risks of exposure. *Scientific Reports, 7*(1), 46687.

Du, S., Zhu, R., Cai, Y., Xu, N., Yap, P. S., Zhang, Y., . . . Zhang, Y. (2021). Environmental fate and impacts of microplastics in aquatic ecosystems: A review. *RSC Advances, 11*(26), 15762–15784.

Dussud, C., & Ghiglione, J. F. (2014). Bacterial degradation of synthetic plastics. In *CIESM Workshop Monographs* (Vol. 46, pp. 49–54). health-effects-on-humans-are-still.

Eerkes-Medrano, D., Thompson, R. C., & Aldridge, D. C. (2015). Microplastics in freshwater systems: A review of the emerging threats, identification of knowledge gaps and prioritisation of research needs. *Water Research, 75,* 63–82.

Eriksen, M., Lebreton, L. C., Carson, H. S., Thiel, M., Moore, C. J., Borerro, J. C., . . . Reisser, J. (2014). Plastic pollution in the world's oceans: More than 5 trillion plastic pieces weighing over 250,000 tons afloat at sea. *PLoS One, 9*(12), e111913.

Farhat, S. C., Silva, C. A., Orione, M. A. M., Campos, L. M., Sallum, A. M., & Braga, A. L. (2011). Air pollution in autoimmune rheumatic diseases: A review. *Autoimmunity Reviews, 11*(1), 14–21.

Fernandes, E. C., Silva, C. A., Braga, A. L., Sallum, A. M., Campos, L. M., & Farhat, S. C. (2015). Exposure to air pollutants and disease activity in juvenile-onset systemic lupus erythematosus patients. *Arthritis Care & Research, 67*(11), 1609–1614.

Fok, L., & Cheung, P. K. (2015). Hong Kong at the pearl river estuary: A hotspot of microplastic pollution. *Marine Pollution Bulletin, 99*(1–2), 112–118.

Forte, M., Iachetta, G., Tussellino, M., Carotenuto, R., Prisco, M., De Falco, M., . . . Valiante, S. (2016). Polystyrene nanoparticles internalization in human gastric adenocarcinoma cells. *Toxicology in Vitro, 31,* 126–136.

Frias, J. P., & Nash, R. (2019). Microplastics: Finding a consensus on the definition. *Marine Pollution Bulletin, 138,* 145–147.

Galafassi, S., Nizzetto, L., & Volta, P. (2019). Plastic sources: A survey across scientific and grey literature for their inventory and relative contribution to microplastics pollution in natural environments, with an emphasis on surface water. *Science of the Total Environment, 693,* 133499.

Gewert, B., Plassmann, M. M., & MacLeod, M. (2015). Pathways for degradation of plastic polymers floating in the marine environment. *Environmental Science: Processes & Impacts, 17*(9), 1513–1521.

Geyer, R., Jambeck, J. R., & Law, K. L. (2017). Production, use, and fate of all plastics ever made. *Science Advances, 3*(7), e1700782.

Gigault, J., Ter Halle, A., Baudrimont, M., Pascal, P. Y., Gauffre, F., Phi, T. L., . . . Reynaud, S. (2018). Current opinion: What is a nanoplastic? *Environmental Pollution, 235,* 1030–1034.

Grafmueller, S., Manser, P., Diener, L., Diener, P. A., Maeder-Althaus, X., Maurizi, L., . . . Wick, P. (2015). Bidirectional transfer study of polystyrene nanoparticles across the placental barrier in an ex vivo human placental perfusion model. *Environmental Health Perspectives, 123*(12), 1280–1286.

Gray, A. D., & Weinstein, J. E. (2017). Size-and shape-dependent effects of microplastic particles on adult daggerblade grass shrimp (Palaemonetes pugio). *Environmental Toxicology and Chemistry, 36*(11), 3074–3080.

Guo, J. J., Huang, X. P., Xiang, L., Wang, Y. Z., Li, Y. W., Li, H., . . . Wong, M. H. (2020). Source, migration and toxicology of microplastics in soil. *Environment International, 137*, 105263.

Guzzetti, E., Sureda, A., Tejada, S., & Faggio, C. (2018). Microplastic in marine organism: Environmental and toxicological effects. *Environmental Toxicology and Pharmacology, 64*, 164–171.

Hartline, N. L., Bruce, N. J., Karba, S. N., Ruff, E. O., Sonar, S. U., & Holden, P. A. (2016). Microfiber masses recovered from conventional machine washing of new or aged garments. *Environmental Science & Technology, 50*(21), 11532–11538.

Heo, N. W., Hong, S. H., Han, G. M., Hong, S., Lee, J., Song, Y. K., . . . Shim, W. J. (2013). Distribution of small plastic debris in cross-section and high strandline on Heungnam beach, South Korea. *Ocean Science Journal, 48*, 225–233.

Herbort, A. F., Sturm, M. T., & Schuhen, K. (2018). A new approach for the agglomeration and subsequent removal of polyethylene, polypropylene, and mixtures of both from freshwater systems–a case study. *Environmental Science and Pollution Research, 25*, 15226–15234.

Hermabessiere, L., Dehaut, A., Paul-Pont, I., Lacroix, C., Jezequel, R., Soudant, P., & Duflos, G. (2017). Occurrence and effects of plastic additives on marine environments and organisms: A review. *Chemosphere, 182*, 781–793.

Ho, N. H. E., & Not, C. (2019). Selective accumulation of plastic debris at the breaking wave area of coastal waters. *Environmental Pollution, 245*, 702–710.

Ho, W. K., Law, J. C. F., Lo, J. C. W., Chng, I. K. X., Hor, C. H. H., & Leung, K. S. Y. (2022). Sorption Behavior, Speciation, and Toxicity of Microplastic-Bound Chromium in Multisolute Systems. *Environmental Science & Technology Letters*. http s://www.discover magazine.com/health/microplastics-are-everywhere-but-their-h.

Hu, B., Li, Y., Jiang, L., Chen, X., Wang, L., An, S., & Zhang, F. (2020). Influence of microplastics occurrence on the adsorption of 17β-estradiol in soil. *Journal of Hazardous Materials, 400*, 123325.

Huang, Y., Zhao, Y., Wang, J., Zhang, M., Jia, W., & Qin, X. (2019). LDPE microplastic films alter microbial community composition and enzymatic activities in soil. *Environmental Pollution, 254*, 112983.

Huang, Z., Hu, B., & Wang, H. (2023). Analytical methods for microplastics in the environment: A review. *Environmental Chemistry Letters, 21*(1), 383–401.

Hurley, R. R., & Nizzetto, L. (2018). Fate and occurrence of micro (nano) plastics in soils: Knowledge gaps and possible risks. *Current Opinion in Environmental Science & Health, 1*, 6–11.

Hurley, R. R., Woodward, J., & Rothwell, J. J. (2018). Microplastic contamination of river beds significantly reduced by catchment-wide flooding. *Nature Geoscience, 11*(4), 251–257.

Ilekis, J. V., Tsilou, E., Fisher, S., Abrahams, V. M., Soares, M. J., Cross, J. C., . . . Bidwell, G. (2016). Placental origins of adverse pregnancy outcomes: Potential molecular targets: An executive workshop summary of the Eunice Kennedy Shriver national institute of child health and human development. *American Journal of Obstetrics and Gynecology, 215*(1), S1–S46.

Imran, M., Das, K. R., & Naik, M. M. (2019). Co-selection of multi-antibiotic resistance in bacterial pathogens in metal and microplastic contaminated environments: An emerging health threat. *Chemosphere, 215*, 846–857.

Jambeck, J. R., Geyer, R., Wilcox, C., Siegler, T. R., Perryman, M., Andrady, A., . . . Law, K. L. (2015). Plastic waste inputs from land into the ocean. *Science, 347*(6223), 768–771.

Ju, S., Shin, G., Lee, M., Koo, J. M., Jeon, H., Ok, Y. S., . . . Park, J. (2021). Biodegradable chito-beads replacing non-biodegradable microplastics for cosmetics. *Green Chemistry, 23*(18), 6953–6965.

Judy, J. D., Williams, M., Gregg, A., Oliver, D., Kumar, A., Kookana, R., & Kirby, J. K. (2019). Microplastics in municipal mixed-waste organic outputs induce minimal short to long-term toxicity in key terrestrial biota. *Environmental Pollution*, *252*, 522–531.

Käppler, A., Windrich, F., Löder, M. G., Malanin, M., Fischer, D., Labrenz, M., . . . Voit, B. (2015). Identification of microplastics by FTIR and Raman microscopy: A novel silicon filter substrate opens the important spectral range below 1300 cm– 1 for FTIR transmission measurements. *Analytical and Bioanalytical Chemistry*, *407*, 6791–6801.

Karbalaei, S., Hanachi, P., Walker, T. R., & Cole, M. (2018). Occurrence, sources, human health impacts and mitigation of microplastic pollution. *Environmental Science and Pollution Research*, *25*, 36046–36063.

Kay, P., Hiscoe, R., Moberley, I., Bajic, L., & McKenna, N. (2018). Wastewater treatment plants as a source of microplastics in river catchments. *Environmental Science and Pollution Research*, *25*, 20264–20267.

Kelly, F. J., & Fussell, J. C. (2012). Size, source and chemical composition as determinants of toxicity attributable to ambient particulate matter. *Atmospheric Environment*, *60*, 504–526.

Kershaw, P. J., Turra, A., & Galgani, F. (2019). Guidelines for the monitoring and assessment of plastic litter and microplastics in the ocean. IMO/FAO/UNESCO-IOC/UNIDO/WMO/IAEA/UN/UNEP/UNDP/ISA Joint Group of Experts on the Scientific Aspects of Marine Environmental Protection). Rep. Stud. GESAMP No. 99, 130p. United Kingdom. http://www.gesamp.org/publications/guidelines-for-the-monitoring-and-assessment-of-plastic-litter-in-the-ocean

Kirstein, I. V., Kirmizi, S., Wichels, A., Garin-Fernandez, A., Erler, R., Löder, M., & Gerdts, G. (2016). Dangerous hitchhikers? Evidence for potentially pathogenic Vibrio spp. on microplastic particles. *Marine Environmental Research*, *120*, 1–8.

Koelmans, A. A., Besseling, E., & Shim, W. J. (2015). Nanoplastics in the aquatic environment. Critical review. *Marine Anthropogenic Litter*, 325–340.

Krueger, M. C., Harms, H., & Schlosser, D. (2015). Prospects for microbiological solutions to environmental pollution with plastics. *Applied Microbiology and Biotechnology*, *99*, 8857–8874.

Lebreton, L. C., Van Der Zwet, J., Damsteeg, J. W., Slat, B., Andrady, A., & Reisser, J. (2017). River plastic emissions to the world's oceans. *Nature Communications*, *8*(1), 15611.

Lee, D. H. (2018). Evidence of the possible harm of endocrine-disrupting chemicals in humans: Ongoing debates and key issues. *Endocrinology and Metabolism*, *33*(1), 44–52.

Lehmann, A., Leifheit, E. F., Gerdawischke, M., & Rillig, M. C. (2021). Microplastics have shape-and polymer-dependent effects on soil aggregation and organic matter loss–an experimental and meta-analytical approach. *Microplastics and Nanoplastics*, *1*(1), 1–14.

Li, J., Liu, H., & Chen, J. P. (2018). Microplastics in freshwater systems: A review on occurrence, environmental effects, and methods for microplastics detection. *Water Research*, *137*, 362–374.

Li, W. C., Tse, H. F., & Fok, L. (2016). Plastic waste in the marine environment: A review of sources, occurrence and effects. *Science of the Total Environment*, *566*, 333–349.

Liao, J., & Chen, Q. (2021). Biodegradable plastics in the air and soil environment: Low degradation rate and high microplastics formation. *Journal of Hazardous Materials*, *418*, 126329.

Lim, L. T., Auras, R., & Rubino, M. (2008). Processing technologies for poly (lactic acid). *Progress in Polymer Science*, *33*(8), 820–852.

Liu, A., Richards, L., Bladen, C. L., Ingham, E., Fisher, J., & Tipper, J. L. (2015). The biological response to nanometre-sized polymer particles. *Acta Biomaterialia*, *23*, 38–51.

Liu, P., Shi, Y., Wu, X., Wang, H., Huang, H., Guo, X., & Gao, S. (2021). Review of the artificially-accelerated aging technology and ecological risk of microplastics. *Science of the Total Environment*, *768*, 144969.

Löder, M. G., & Gerdts, G. (2015). Methodology used for the detection and identification of microplastics—a critical appraisal. *Marine Anthropogenic Litter*, 201–227.

Luo, Y. Y., Not, C., & Cannicci, S. (2021). Mangroves as unique but understudied traps for anthropogenic marine debris: A review of present information and the way forward. *Environmental Pollution*, *271*, 116291.

Lusher, A. L., Hernandez-Milian, G., O'Brien, J., Berrow, S., O'Connor, I., & Officer, R. (2015). Microplastic and macroplastic ingestion by a deep diving, oceanic cetacean: The True's beaked whale Mesoplodon mirus. *Environmental Pollution*, *199*, 185–191.

Lwanga, E. H., Gertsen, H., Gooren, H., Peters, P., Salánki, T., van der Ploeg, M., . . . Geissen, V. (2017). Incorporation of microplastics from litter into burrows of Lumbricus terrestris. *Environmental Pollution*, *220*, 523–531.

Lwanga, E. H., Thapa, B., Yang, X., Gertsen, H., Salánki, T., Geissen, V., & Garbeva, P. (2018). Decay of low-density polyethylene by bacteria extracted from earthworm's guts: A potential for soil restoration. *Science of the Total Environment*, *624*, 753–757.

Ma, H., Pu, S., Liu, S., Bai, Y., Mandal, S., & Xing, B. (2020). Microplastics in aquatic environments: Toxicity to trigger ecological consequences. *Environmental Pollution*, *261*, 114089.

Mariano, S., Tacconi, S., Fidaleo, M., Rossi, M., & Dini, L. (2021). Micro and nanoplastics identification: Classic methods and innovative detection techniques. *Frontiers in Toxicology*, *3*, 636640.

Martin, C., Baalkhuyur, F., Valluzzi, L., Saderne, V., Cusack, M., Almahasheer, H., . . . Duarte, C. M. (2020). Exponential increase of plastic burial in mangrove sediments as a major plastic sink. *Science Advances*, *6*(44), eaaz5593.

Mattsson, K., Jocic, S., Doverbratt, I., & Hansson, L. A. (2018). Nanoplastics in the aquatic environment. *Microplastic Contamination in Aquatic Environments*, 379–399.

Miranda, M. N., Silva, A. M., & Pereira, M. F. R. (2020). Microplastics in the environment: A DPSIR analysis with focus on the responses. *Science of the Total Environment*, *718*, 134968.

Mitrano, D. M., & Wohlleben, W. (2020). Microplastic regulation should be more precise to incentivize both innovation and environmental safety. *Nature Communications*, *11*, 1–12. doi:10.1038/s41467-020-19069-1.

Mohanan, N., Montazer, Z., Sharma, P. K., & Levin, D. B. (2020). Microbial and enzymatic degradation of synthetic plastics. *Frontiers in Microbiology*, *11*, 580709.

MohanKumar, S. M., Campbell, A., Block, M., & Veronesi, B. (2008). Particulate matter, oxidative stress and neurotoxicity. *Neurotoxicology*, *29*(3), 479–488.

Nakanishi, A., Iritani, K., & Sakihama, Y. (2020). Developing neo-bioplastics for the realization of carbon sustainable society. *Journal of Nanotechnology and Nanomaterials*, *1*(2), 72–85.

Ng, E. L., Lwanga, E. H., Eldridge, S. M., Johnston, P., Hu, H. W., Geissen, V., & Chen, D. (2018). An overview of microplastic and nanoplastic pollution in agroecosystems. *Science of the Total Environment*, *627*, 1377–1388.

Nizzetto, L., Bussi, G., Futter, M. N., Butterfield, D., & Whitehead, P. G. (2016). A theoretical assessment of microplastic transport in river catchments and their retention by soils and river sediments. *Environmental Science: Processes & Impacts*, *18*(8), 1050–1059.

Not, C., Lui, C. Y. I., & Cannicci, S. (2020). Feeding behavior is the main driver for microparticle intake in mangrove crabs. *Limnology and Oceanography Letters*, *5*(1), 84–91.

Nowack, B., & Bucheli, T. D. (2007). Occurrence, behavior and effects of nanoparticles in the environment. *Environmental Pollution*, *150*(1), 5–22.

Ormsby, R. T., Cantley, M., Kogawa, M., Solomon, L. B., Haynes, D. R., Findlay, D. M., & Atkins, G. J. (2016). Evidence that osteocyte perilacunar remodelling contributes to polyethylene wear particle induced osteolysis. *Acta Biomaterialia, 33*, 242–251.

Park, T. J., Lee, S. H., Lee, M. S., Lee, J. K., Park, J. H., & Zoh, K. D. (2020). Distributions of microplastics in surface water, fish, and sediment in the vicinity of a sewage treatment plant. *Water, 12*(12), 3333.

Peng, L., Mehmood, T., Bao, R., Wang, Z., & Fu, D. (2022). An overview of micro (nano) plastics in the environment: Sampling, identification, risk assessment and control. *Sustainability, 14*(21), 14338.

Pettipas, S., Bernier, M., & Walker, T. R. (2016). A Canadian policy framework to mitigate plastic marine pollution. *Marine Policy, 68*, 117–122.

PlasticsEurope. (2019). *Plastics the Facts 2019: An Analysis of European Plastics.* PlasticsEurope, Wemmel – Belgium.

Prajapati, A., Narayan Vaidya, A., & Kumar, A. R. (2022). Microplastic properties and their interaction with hydrophobic organic contaminants: A review. *Environmental Science and Pollution Research, 29*(33), 49490–49512.

Prata, J. C. (2018). Airborne microplastics: Consequences to human health? *Environmental Pollution, 234*, 115–126.

Prata, J. C., da Costa, J. P., Lopes, I., Duarte, A. C., & Rocha-Santos, T. (2020). Environmental exposure to microplastics: An overview on possible human health effects. *Science of the Total Environment, 702*, 134455.

Qi, Y., Ossowicki, A., Yang, X., Lwanga, E. H., Dini-Andreote, F., Geissen, V., & Garbeva, P. (2020). Effects of plastic mulch film residues on wheat rhizosphere and soil properties. *Journal of Hazardous Materials, 387*, 121711.

Rahman, A., Sarkar, A., Yadav, O. P., Achari, G., & Slobodnik, J. (2021). Potential human health risks due to environmental exposure to nano-and microplastics and knowledge gaps: A scoping review. *Science of the Total Environment, 757*, 143872.

Ren, Z., Gui, X., Xu, X., Zhao, L., Qiu, H., & Cao, X. (2021). Microplastics in the soil-groundwater environment: Aging, migration, and co-transport of contaminants–a critical review. *Journal of Hazardous Materials, 419*, 126455.

Ribeiro, F., O'Brien, J. W., Galloway, T., & Thomas, K. V. (2019). Accumulation and fate of nano-and micro-plastics and associated contaminants in organisms. *TrAC Trends in Analytical Chemistry, 111*, 139–147.

Rillig, M. C., Lehmann, A., de Souza Machado, A. A., & Yang, G. (2019). Microplastic effects on plants. *New Phytologist, 223*(3), 1066–1070.

Rillig, M. C., Ziersch, L., & Hempel, S. (2017). Microplastic transport in soil by earthworms. *Scientific Reports, 7*(1), 1362.

Rocha-Santos, T., & Duarte, A. C. (2015). A critical overview of the analytical approaches to the occurrence, the fate and the behavior of microplastics in the environment. *TrAC Trends in Analytical Chemistry, 65*, 47–53.

Rochman, C. M., Tahir, A., Williams, S. L., Baxa, D. V., Lam, R., Miller, J. T., . . . Teh, S. J. (2015). Anthropogenic debris in seafood: Plastic debris and fibers from textiles in fish and bivalves sold for human consumption. *Scientific Reports, 5*(1), 1–10.

Rodriguez-Seijo, A., Lourenço, J., Rocha-Santos, T. A. P., Da Costa, J., Duarte, A. C., Vala, H., & Pereira, R. (2017). Histopathological and molecular effects of microplastics in Eisenia Andrei Bouché. *Environmental Pollution, 220*, 495–503.

Rychter, P., Pospíšil, L., Černý, M., & Šlouf, M. (2012). Recent progress in understanding and modeling the radiation and thermo-oxidative degradation of polymers. *Polymer Degradation and Stability, 97*(11), 2023–2044.

Saravia, J., You, D., Thevenot, P., Lee, G. I., Shrestha, B., Lomnicki, S., & Cormier, S. A. (2014). Early-life exposure to combustion-derived particulate matter causes pulmonary immunosuppression. *Mucosal Immunology*, 7(3), 694–704.

Schneider, M., Stracke, F., Hansen, S., & Schaefer, U. F. (2009). Nanoparticles and their interactions with the dermal barrier. *Dermato-Endocrinology*, 1(4), 197–206.

Sen, S. K., & Raut, S. (2015). Microbial degradation of low density polyethylene (LDPE): A review. *Journal of Environmental Chemical Engineering*, 3(1), 462–473.

Shah, A. A., Hasan, F., Hameed, A., & Ahmed, S. (2008). Biological degradation of plastics: A comprehensive review. *Biotechnology Advances*, 26(3), 246–265.

Shirasaki, N., Matsushita, T., Matsui, Y., & Marubayashi, T. (2016). Effect of aluminum hydrolyte species on human enterovirus removal from water during the coagulation process. *Chemical Engineering Journal*, 284, 786–793.

Sivan, A. (2011). New perspectives in plastic biodegradation. *Current Opinion in Biotechnology*, 22(3), 422–426.

Song, Y., Qiu, R., Hu, J., Li, X., Zhang, X., Chen, Y., . . . He, D. (2020). Biodegradation and disintegration of expanded polystyrene by land snails Achatina Fulica. *Science of the Total Environment*, 746, 141289.

Sternschuss, G., Ostergard, D. R., & Patel, H. (2012). Post-implantation alterations of polypropylene in the human. *The Journal of Urology*, 188(1), 27–32.

Sussarellu, R., Suquet, M., Thomas, Y., Lambert, C., Fabioux, C., Pernet, M. E. J., . . . Huvet, A. (2016). Oyster reproduction is affected by exposure to polystyrene microplastics. *Proceedings of the National Academy of Sciences*, 113(9), 2430–2435.

Talvitie, J., Mikola, A., Setälä, O., Heinonen, M., & Koistinen, A. (2017). How well is microlitter purified from wastewater?–A detailed study on the stepwise removal of microlitter in a tertiary level wastewater treatment plant. *Water Research*, 109, 164–172.

Tang, K. H. D. (2021). Interactions of microplastics with persistent organic pollutants and the ecotoxicological effects: A review. *Tropical Aquatic and Soil Pollution*, 1(1), 24–34.

Thompson, R. C., Moore, C. J., Vom Saal, F. S., & Swan, S. H. (2009). Plastics, the environment and human health: Current consensus and future trends. *Philosophical Transactions of the Royal Society B: Biological Sciences*, 364(1526), 2153–2166.

Thompson, R. C., Olsen, Y., Mitchell, R. P., Davis, A., Rowland, S. J., John, A. W. G., . . . Russell, A. E. (2004). Lost at sea: Where is all the plastic? *Science*, 304.

Valavanidis, A., Vlachogianni, T., Fiotakis, K., & Loridas, S. (2013). Pulmonary oxidative stress, inflammation and cancer: Respirable particulate matter, fibrous dusts and ozone as major causes of lung carcinogenesis through reactive oxygen species mechanisms. *International Journal of Environmental Research and Public Health*, 10(9), 3886–3907.

Vo, H. C., & Pham, M. H. (2021). Ecotoxicological effects of microplastics on aquatic organisms: A review. *Environmental Science and Pollution Research*, 28, 44716–44725.

Wagner, S., Klöckner, P., Stier, B., Römer, M., Seiwert, B., Reemtsma, T., & Schmidt, C. (2019). Relationship between discharge and river plastic concentrations in a rural and an urban catchment. *Environmental Science & Technology*, 53(17), 10082–10091.

Waldman, W. R., & Rillig, M. C. (2020). Microplastic research should embrace the complexity of secondary particles. *Environmental Science & Technology*, 54(13), 7751–7753. doi:10.1021/acs.est.0c02194.

Wang, F., Wang, Q., Adams, C. A., Sun, Y., & Zhang, S. (2022). Effects of microplastics on soil properties: Current knowledge and future perspectives. *Journal of Hazardous Materials*, 424, 127531.

Wang, F., Wong, C. S., Chen, D., Lu, X., Wang, F., & Zeng, E. Y. (2018). Interaction of toxic chemicals with microplastics: A critical review. *Water Research*, 139, 208–219.

Wang, F., Zhang, M., Sha, W., Wang, Y., Hao, H., Dou, Y., & Li, Y. (2020). Sorption behavior and mechanisms of organic contaminants to nano and microplastics. *Molecules, 25*(8), 1827.

Wang, J., Coffin, S., Sun, C., Schlenk, D., & Gan, J. (2019). Negligible effects of microplastics on animal fitness and HOC bioaccumulation in earthworm Eisenia Fetida in soil. *Environmental Pollution, 249*, 776–784.

Wang, J., Jiang, J., Sun, Y., Wang, X., Li, M., Pang, S., . . . Tsang, D. C. (2021). Catalytic degradation of waste rubbers and plastics over zeolites to produce aromatic hydrocarbons. *Journal of Cleaner Production, 309*, 127469.

Watts, A. J., Urbina, M. A., Corr, S., Lewis, C., & Galloway, T. S. (2015). Ingestion of plastic microfibers by the crab Carcinus maenas and its effect on food consumption and energy balance. *Environmental Science & Technology, 49*(24), 14597–14604.

Wei, R., Oeser, T., Then, J., Kühn, N., Barth, M., Schmidt, J., & Zimmermann, W. (2014). Functional characterization and structural modeling of synthetic polyester-degrading hydrolases from Thermomonospora Curvata. *AMB Express, 4*, 1–10.

Wei, R., & Zimmermann, W. (2017). Microbial enzymes for the recycling of recalcitrant petroleum-based plastics: How far are we? *Microbial Biotechnology, 10*(6), 1308–1322.

Wen, B., Zhang, N., Jin, S. R., Chen, Z. Z., Gao, J. Z., Liu, Y., . . . Xu, Z. (2018). Microplastics have a more profound impact than elevated temperatures on the predatory performance, digestion and energy metabolism of an Amazonian cichlid. *Aquatic Toxicology, 195*, 67–76.

Wick, P., Malek, A., Manser, P., Meili, D., Maeder-Althaus, X., Diener, L., . . . von Mandach, U. (2010). Barrier capacity of human placenta for nanosized materials. *Environmental Health Perspectives, 118*(3), 432–436.

Wright, S. L., & Kelly, F. J. (2017). Plastic and human health: A micro issue? *Environmental Science & Technology, 51*(12), 6634–6647.

Wright, S. L., Rowe, D., Thompson, R. C., & Galloway, T. S. (2013). Microplastic ingestion decreases energy reserves in marine worms. *Current Biology, 23*(23), R1031–R1033.

Xu, K., Zhang, Y., Huang, Y., & Wang, J. (2021). Toxicological effects of microplastics and phenanthrene to zebrafish (Danio rerio). *Science of the Total Environment, 757*, 143730.

Xu, S., Ma, J., Ji, R., Pan, K., & Miao, A. J. (2020). Microplastics in aquatic environments: Occurrence, accumulation, and biological effects. *Science of the Total Environment, 703*, 134699.

Yang, C. Z., Yaniger, S. I., Jordan, V. C., Klein, D. J., & Bittner, G. D. (2011). Most plastic products release estrogenic chemicals: A potential health problem that can be solved. *Environmental Health Perspectives, 119*(7), 989–996.

Yang, J., Yang, Y., Wu, W. M., Zhao, J., & Jiang, L. (2014). Evidence of polyethylene biodegradation by bacterial strains from the guts of plastic-eating waxworms. *Environmental Science & Technology, 48*(23), 13776–13784.

Yang, Y., Yang, J., Wu, W. M., Zhao, J., Song, Y., Gao, L., . . . Jiang, L. (2015). Biodegradation and mineralization of polystyrene by plastic-eating mealworms: Part 1. Chemical and physical characterization and isotopic tests. *Environmental Science & Technology, 49*(20), 12080–12086.

Yoshida, S., Hiraga, K., Takehana, T., Taniguchi, I., Yamaji, H., Maeda, Y., . . . Oda, K. (2016). A bacterium that degrades and assimilates poly (ethylene terephthalate). *Science, 351*(6278), 1196–1199.

Yu, N., Huang, B., Li, M., Cheng, P., Li, L., Huang, Z., . . . Zhou, Z. (2017). Single particle characteristics of fine particulate matter in dust. *China Environmental Science, 37*(4), 1262–1268.

Zhang, C., Wang, J., Zhou, A., Ye, Q., Feng, Y., Wang, Z., . . . Zou, J. (2021a). Species-specific effect of microplastics on fish embryos and observation of toxicity kinetics in larvae. *Journal of Hazardous Materials, 403*, 123948.

Zhang, X., Luo, D., Yu, R. Q., Xie, Z., He, L., & Wu, Y. (2021b). Microplastics in the endangered Indo-Pacific humpback dolphins (Sousa chinensis) from the Pearl River Estuary, China. *Environmental Pollution, 270*, 116057.

Zhaorong, M., Yousheng, L., Qianqian, Z., & Guangguo, Y. (2020). The usage and environmental pollution of agricultural plastic film. *Asian Journal of Ecotoxicology* (4), 21–32.

Zhou, Y., Wang, J., Zou, M., Jia, Z., Zhou, S., & Li, Y. (2020). Microplastics in soils: A review of methods, occurrence, fate, transport, ecological and environmental risks. *Science of the Total Environment, 748*, 141368.

Zhu, F., Zhu, C., Wang, C., & Gu, C. (2019). Occurrence and ecological impacts of microplastics in soil systems: A review. *Bulletin of Environmental Contamination and Toxicology, 102*, 741–749.

Zubris, K. A. V., & Richards, B. K. (2005). Synthetic fibers as an indicator of land application of sludge. *Environmental Pollution, 138*(2), 201–211.

14 Green Remediation of Microplastics Using Bionanomaterials

Jyoti Bhattacharjee and Subhasis Roy

14.1 INTRODUCTION

The greatest severe environmental threat confronting our world right now may be the enormous increase in human population. Plastic usage has gotten out of hand, causing havoc on the ecosystem. Microplastics (MPs), also known as small plastic particles, are minuscule particles less than 5 mm in size that are found everywhere and in a wide range of ecosystems, including freshwater, marine, and terrestrial settings (Thompson et al., 2004) [1]. MPs are formed when people use personal care items like toothpaste and facial cleansers and when bigger plastic objects such as packaging, fishing nets, and other plastic debris degrade. MPs have a lengthy half-life and can last hundreds of years.

Green remediation technologies, such as bionanomaterials, offer a practical and environmentally acceptable alternative for microplastic remediation. Bionanomaterials are a form of material derived from biological sources with distinct properties and capabilities that make them useful in a wide range of applications, including environmental remediation [2]. This review chapter looks at the current use of green bionanomaterials, paired with artificial intelligence, for microplastic remediation and their promise as an environmentally friendly alternative to traditional remediation methods.

Cellulose is a biopolymer substrate found in plant cell walls with a high capacity for microplastic adsorption due to its high surface area and hydroxyl functional groups that allow for interactions via hydrogen bonds with microplastics. In 2006, two French ships containing toxic ashes and non-biodegradable plastic wastes headed towards India and dumped in the Indian Ocean near Alang Port in Gujarat. Several studies have looked into the use of cellulose-based materials, such as cellulose nanocrystals and cellulose aerogels, for microplastic clearance. Madani et al. (2021) [3] investigated using cellulose aerogels for microplastic adsorption in seawater. The researchers discovered cellulose aerogels have a high capacity for tiny particle adsorption, with up to 98% removal efficacy. A new Greenpeace study discovered a large vortex of old toothbrushes, beach toys, aluminum cans, and other debris in the Pacific Ocean, endangering the lives of sea species.

Magnetic nanoparticles (MNPs) are nanoscale particles that exhibit magnetic properties. They are employed in various applications, including wastewater treatment, magnetic resonance imaging (MRI), and medication administration. Several

 DOI: 10.1201/9781003449133-14

studies have investigated the use of MNPs for microplastic cleaning. For example, Zhang et al. (2020) study discovered that MNPs could remove microplastics from water with up to 94% efficiency [4]. Chitosan is a natural polymer found in crab exoskeletons formed from chitin. It has been employed in several applications, such as food packaging, medication administration, and wound healing. Chitosan nanoparticles have been found to remove microplastics from water. For example, Yang et al. discovered, in 2021 [5], that chitosan nanoparticles could remove microplastics from water with an efficiency of up to 97.3%.

Microplastic removal from ecosystems is a sophisticated and challenging task that calls for the development of new materials and techniques that may be used to accomplish it without putting the environment in danger [6]. The use of bionanomaterials and machine learning to address this issue has gained popularity recently. Since the ecosystem is harmed when used plastic garbage is discharged into rivers and streams, the present investigation focuses on ecologically friendly approaches, including bionanomaterials and artificial intelligence (AI), to reduce the risk of plastics [7]. Using bionanomaterials, AI has been utilized to improve the circumstances for microplastic breakdown. For instance, one study employed a neural network to identify the ideal conditions for the breakdown of polyethylene microplastics by TiO_2 nanoparticles. Data on the impact of several variables, such as nanoparticle concentration, pH, and temperature, on the breakdown of microplastics were used to train the neural network [8].

To examine the present status of research on this topic, a Scopus search was done, which generated 30 articles, comprising 22 research papers and eight review articles. The first article discovered was published in 2017, while the most current piece discovered was published in 2022. Most publications were published in the most recent two years, demonstrating that interest in this topic is expanding. According to the Scopus database, the most recent 2022 study, published in the journal *Water Research*, mentioned the review paper in its discussion of the use of AI for the identification of sources of microplastic pollution. In a recent review (Zhang et al., 2022), the potential of chitosan nanoparticles for green microplastic clean-up was highlighted [9].

Figure 14.1 illustrates global annual microplastic and microplastic production from the years 1950 to 2022.

This review chapter aims to (1) emphasize the gaps and limitations in the current research on microplastic pollution in the ecosystem and (2) explain and explore green techniques for microplastic contamination management using bionanomaterials and AI, as well as their future scopes.

14.2 SOURCES OF MICROPLASTICS

14.2.1 Packaging Made of Plastics

Globally, plastic goods and packaging are the main sources of microplastics. These include, among other things, things like plastic bags, bottles, and containers. These products contain plastic, which breaks down and releases microplastics into the environment over time. For instance, the plastic used in water bottles degrades into microplastics, which are then eaten by marine life. Tires are another

Annual production of microplastics worldwide from 1950 to 2022

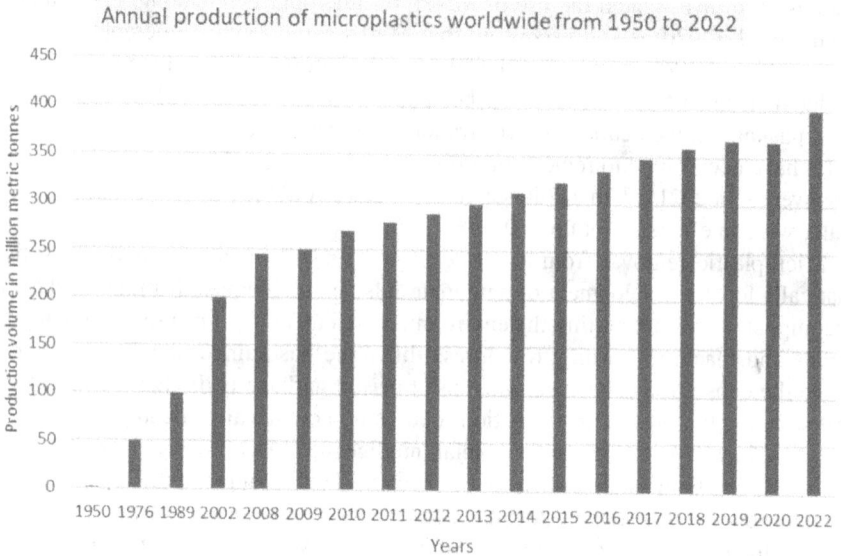

FIGURE 14.1 Global microplastic generation from the years 1950 to 2022 (data input from Web of Science).

significant source of microplastics worldwide. Tires deteriorate over time, releasing small rubber fragments containing microplastics. After being pushed into rivers and oceans, these particles may impact marine life [10]. Single-use plastic packaging contributes significantly to the worldwide plastic waste crisis and serves as a reservoir of microplastics. Packaging made from discarded plastic that winds up in the environment can disintegrate into smaller particles, including tiny plastic particles.

14.2.2 Personal Care Items

Microbeads are little bits of plastic in personal care products like toothpaste and face wash. Marine organisms can consume these items after they are used and flushed and wind up in streams and oceans. India has some of the most polluted streams in the world.

Untreated wastewater, industrial pollution, and runoff from farms all pollute waterways. These pollutants release small particles into the environment. According to one estimate, the Ganges River alone is responsible for up to 1,200 tonnes of microplastics entering the ocean. Many nations rapidly followed the United States' 2015 prohibition on producing and selling rinse-off cosmetic products containing plastic microbeads. However, some products from prior generations are still available in some regions [11].

14.2.3 Fibres of Textiles

The second-largest producer of textiles in the world is India, and this sector is vital to the Indian economy. On the other hand, the textile industry is a significant source of nanoplastics. During the washing and manufacturing procedures, textile fibres are destroyed. The industry generates a lot of plastic waste. According to the International Union for Conservation of Nature (IUCN), synthetic textiles are responsible for up to 35% of all microplastics in the ocean. According to the study, washing a single polyester fleece jacket can release up to 250,000 microfibres [12].

14.2.4 Industrial Processes

India has a sizable manufacturing economy, with industries such as plastic items, textiles, and electronics contributing significantly to the GDP of the nation. However, these businesses release microplastics into the environment, harming the well-being of both humans and animals. Microplastics have been found in various foods and beverages, including seafood, beer, and bottled water. According to a study published in the journal *Environmental Science & Technology*, microplastics were found in 83% of tap water samples sampled worldwide. According to the analysis, the microplastics were most likely produced from plastic bottles and packaging. Microplastics were found in 25% of fish sold for human consumption in California and Indonesian markets, according to another study [13].

14.2.5 Artificial Turf

Microplastics can be found in synthetic turf, extensively used in sports fields and playgrounds. The turf is made of synthetic materials that, over time, deteriorate and release small plastic particles into the environment. Synthetic turf emits approximately 2,000 tonnes of microplastics into the environment each year, according to a Dutch government study [14]. Clothing typically uses synthetic plastic materials such as polyester, nylon, and acrylic. The following chart depicts the major sources of plastics and microplastics.

14.2.5.1 Microplastics in Food Products, Atmosphere, and the Marine Environment

Microplastics have been discovered in various foods, including seafood, drinking water, and salt. Intake of polluted water or creatures is the primary source of microplastics in dietary supplements, which can result in microplastics being transported up the food chain.

Seafood is one of the most common sources of microplastics in people. According to a study, plastic debris has been detected in a range of seafood, including shellfish, fish, and crabs. According to one investigation, microplastics were found in 73% of the fish analyzed from Indonesian markets (Setiawan et al., 2020). Similarly, microfibres were found in 18% of shellfish samples collected in the United Kingdom (Van Cauwenberghe & Janssen, 2014) [15].

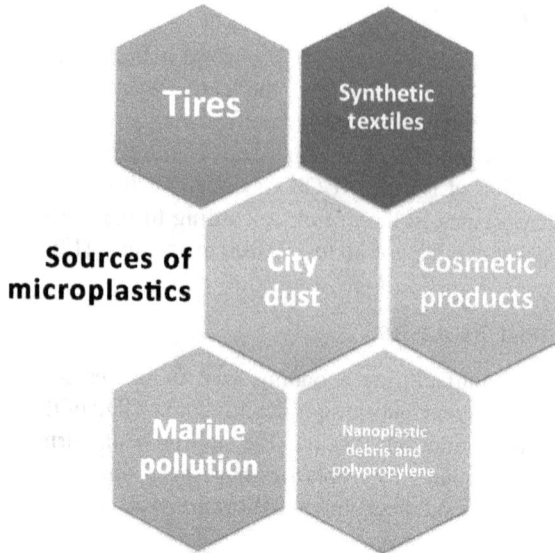

FIGURE 14.2 Sources of microplastics and emerging plastic pollution.

The health effects of eating microplastics through food products are unknown. Microplastics, however, have been linked to organ damage, inflammation, and even cancer in some studies (Gambardella et al., 2020). More research is needed to fully understand the potential adverse effects of microplastic ingestion [16]. Furthermore, microplastics have been discovered in honey. Plastic particles were found in 50% of honey samples gathered from around the world, according to a study published in the *Environmental Science & Technology* journal. Although their origin is uncertain, bees are suspected to collect these tiny plastics from nature, including polluted flowers and drinking water sources [17]. Food items, the seas, and the atmosphere all contain plastic debris. Numerous pathways, including tire wear, road dust, and the breakdown of bigger plastic particles, can release microplastics into the atmosphere. Microplastics can travel a long distance once in the atmosphere; according to some studies, they can be transported more than 1,000 kilometers from their initial source.

A study in the journal *Environmental Science & Technology Letters* predicts that 3 to 10 tonnes of microplastics will be dumped in Paris annually [18]. Another investigation reported in the journal *Environmental Pollution* estimated that Dongguan, air of China had 330 million microplastic particles. Chemical exposure: Chemicals like persistent organic pollutants (POPs) can build up on micro- and nanoplastics and adsorb there before being released into the intestinal tract.

Drainage from land: Additionally, drainage from the land can carry microplastics and nanoplastics into the water. Several methods exist for this to happen, including stormwater runoff, sewage discharge, and agricultural runoff. Urban and suburban regions, where plastic is widely used in consumer goods, can produce stormwater

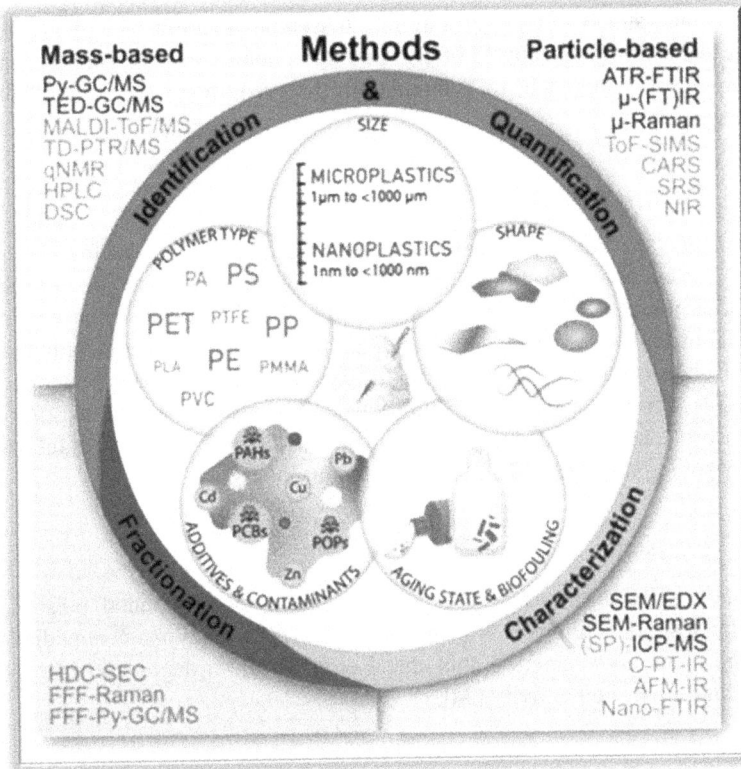

FIGURE 14.3 Chemical Evaluation of Microplastics and Nanoplastics [20]. (Reproduced with permission.)

runoff that can transport microplastics and nanoplastics to surrounding waterways and the ocean. Additionally, sewage may discharge microplastics and miniplastics from consumer goods and other sources into the ocean. Microplastics and nanoplastics from plastic-based agricultural products, such as mulch films and irrigation tubing, can be transported by agricultural runoff into surrounding streams, and eventually, into the ocean [19]. The emerging particle anthropogenic contaminants, microplastics, and nanoplastics are shown in the figure that follows.

14.3 GREEN REMEDIATION USING BIO-NANOTECHNOLOGY

Green remediation is a type of environmental rehabilitation emphasizing sustainability and environmental care. Green remediation solutions seek to lessen the environmental impact of remediation while providing effective clean-up. Sustainable restoration methods include using renewable energy, reducing waste, and using natural and sustainable materials [21].

Bio-nanotechnology is an environmentally friendly method of cleaning up contaminated soil and water. It is a long-term strategy that employs natural biological processes to create nanostructures capable of eliminating contaminants. Bio-nanotechnology is a low-impact approach to remediation that requires fewer chemicals and less energy [22].

14.3.1 Phytoremediation

Phytoremediation, a green remediation procedure that eliminates toxins from both water and soil, employs plants. Plant roots can collect pollutants from water and dirt and store them in their tissues. Toxins can be eliminated from the environment by harvesting [23].

Bio-nanotechnology has the potential to boost botanical remediation performance. To boost the ability of plant organs to remove contaminants, nanoparticles, for example, can be synthesized. Nanoparticles can also be applied to soil to help plants absorb contaminants. Furthermore, nanoparticles can stimulate plant development, improving the capacity to eliminate contaminants.

14.3.2 Enzymatic Remediation

Enzymatic remediation uses enzymes to eliminate environmental poisons. Bio-nanotechnology can improve enzyme performance in enzymatic remediation by producing nano-sized materials that shield enzymes from degradation, boost their stability, and raise their activity [24].

14.3.3 Nano-Adsorbents

Adsorbents are nano-sized materials that may selectively adsorb contaminants from the environment. Bio-nanotechnology can generate selective, efficient, and eco-friendly nano-adsorbents [25]. Bio-nanotechnology can be used to change existing nanoparticles to make them more effective at removing pollutants and using biological processes to create nanoparticles. Researchers, for example, have devised a way to coat titanium dioxide nanoparticles with a protein called cysteine, which boosts their ability to degrade organic contaminants in water. The cysteine-coated nanoparticles can break down a wide range of contaminants, including colours, pharmaceuticals, and insecticides, and can be easily removed from water using a magnetic field [26].

14.3.4 Chitosan-Coated Sand

One study investigated the efficacy of chitosan-coated sand for microplastic disposal in a simulated aquatic environment. The findings revealed that chitosan-coated sand was effective at removing microplastics, with up to 95% removal efficiency. The study also discovered that chitosan-coated sand may be reused multiple times without losing efficiency.

Graphene oxide is a two-dimensional graphite compound with a vast surface area and adsorption capabilities. According to Zhang et al. (2021), graphene oxide nanoparticles can degrade up to 85% of polyethylene microplastics in water in 48 hours [27].

14.3.5 Machine Intelligence

Another cutting-edge technique for purifying microplastics is artificial intelligence (AI). Artificial intelligence (AI) systems may be trained to recognize and categorize microplastics in images, making it simpler to locate and remove them from the environment [28]. AI can be used to forecast the occurrence of microplastics in the environment and pinpoint potential pollution sources. One example of how artificial intelligence (AI) is being used for microplastic remediation is the creation of autonomous underwater vehicles (AUVs) [29]. Cameras and AI algorithms on AUVs enable them to find and locate microplastics on the ocean floor. The AUV can gather the tiny plastic particles for disposal once they have been acknowledged. A different instance of how AI is being used to remove microplastics is the development of predictive models. These models use weather patterns, ocean currents, and other variables to forecast the dispersion of microplastics in the environment. This data can be used to direct remediation operations and identify areas of high toxicity. Another application method of AI in the remediation of microplastics is using robotics [30]. Robots can be used to gather and remove microplastics from the environment. Researchers have created a robot that can navigate across the water and collect plastic waste, and a brief outlook of the degradation of plastics is shown in Figure 14.4.

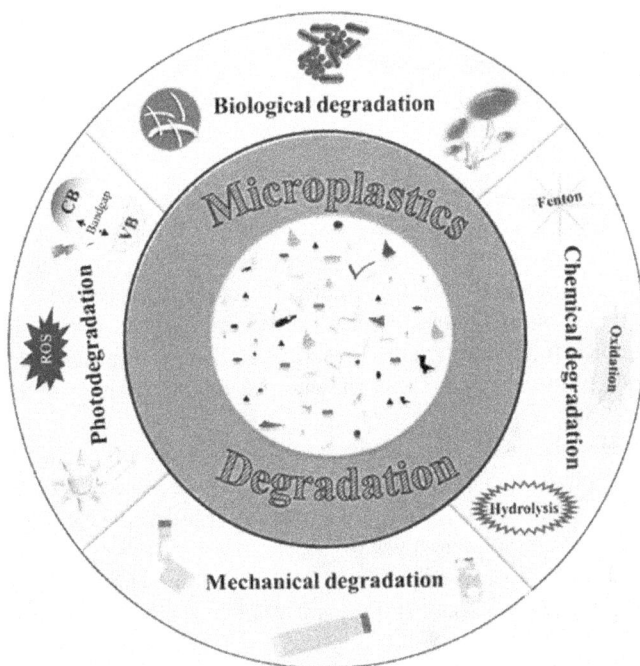

FIGURE 14.4 Mechanisms for green degradation of microplastics [31]. (Reproduced with permission.)

14.3.5.1 Advantages and Drawbacks of Bio-nanomaterials in Microplastic Remediation

Biodegradability, biocompatibility, and low toxicity are only a few of the benefits of bio-nanomaterials over standard microplastic remediation materials. They also have high surface area-to-volume ratios, allowing them to absorb tiny plastic particles from water easily. Furthermore, bio-nanomaterials can be created from renewable resources, making them less damaging to the environment than conventional materials [32].

14.4 MITIGATION USING BIONANOMATERIALS

Green bio-nanomaterials are biologically generated nanomaterials. They are a new family of materials that have piqued the curiosity of scientists due to their remarkable properties such as biocompatibility, biodegradability, and non-toxicity. Bio-nanostructures can be applied in a wide range of fields, including medicine, energy, and the environment [33]. Recently, bio-nanoparticles have been investigated for removing microplastics from the environment. Bio-nanomaterials, which are recyclable and made from renewable resources, offer a green alternative to microplastic treatment. To increase interactions with microplastics, bio-nanomaterials can be functionalized with different groups such as carboxyl, amino, and hydroxyl groups. Amino groups, for example, can electrostatically interact with the surface of microplastics, whereas carboxyl groups can create hydrogen bonds with the ester groups in the microplastics.

Cyclodextrin-based nano-sponges: Cyclodextrins are cyclic oligosaccharides derived from starch. They can be made into functional nano-sponges by crosslinking polymers with them; these nano-sponges have the potential to encapsulate microplastics in water [34].

Enzymes are another type of bio-nanomaterial that has been studied for microplastic remediation. Enzymes are biological catalysts that may selectively destroy specific polymers, including plastics. They can be designed or changed to improve their activity and selectivity towards microplastics. For instance, Lu et al. (2021) created an enzyme-based system to break down microplastics. The nano-zyme was created by immobilizing horseradish peroxidase (HRP) on a magnetic iron oxide nanoparticle. Within three hours, the enzyme was able to destroy up to 98% of the microplastics [35].

Bacterial cellulose nanofibers: Microorganisms produce microbial cellulose and have distinctive physical and mechanical attributes that enable it to be employed for various purposes. Using bacterial cellulose nanofibers, microplastics can be successfully removed from contaminated water.

Algal-based nano-sorbents offer exceptional adsorption characteristics generated from natural and renewable sources. They are efficient at eliminating plastic particles from water [36].

Liposomes are composite nanoparticles that contain phospholipids. They have been used for miniature plastic treatment due to their ability to encapsulate microplastics. The resulting complex is easily removed by sedimentation or screening.

14.4.1   CORAL REEF DAMAGE REVERSAL USING BIO-NANO PARTICLES

Coral reefs are among the most diversified ecosystems in the world, sometimes called the ocean's rainforests. Many coastal communities rely on them for food, money, and habitat for a wide variety of plant and animal species. Coral reefs, on the other hand, face several threats, including pollution, overfishing, coastal development, and climate change. Little plastic particles are one of the most serious threats to coral reefs [37]. Corals are filter feeders, which means they get their nutrition by filtering minute water particles. Corals may consume microplastics with their food if they are in the water. Corals may die or experience stress as a result [38]. Microplastics can also physically separate corals from their surroundings, preventing them from absorbing the sunlight and nutrients they require to survive.

Microplastics significantly affect coral reefs. According to a study in the journal *Marine Pollution Bulletin*, every sample of coral tissue collected from the Great Barrier Reef contained microplastics.

Several research projects have been conducted to investigate the usage of bio-nanomaterials for removing microplastics from water. For example, researchers utilized chitosan nanoparticles to remove microplastics from water in a laboratory environment. The chitosan nanoparticles absorbed up to 90% of the microplastics in the water.

Other researchers have also utilized cellulose nanocrystals (CNCs) to remove microplastics from water. CNCs are made from cellulose, a naturally occurring polymer found in plants. They have a large surface area and are quite strong, making them ideal for removing microplastics. In one case, researchers used CNCs to remove microplastics from water with up to 95% effectiveness [39].

Traditional microplastic monitoring methods entail manually collecting and counting particles from water or sediment. This procedure can be time-consuming and tedious and may not provide a complete picture of microplastic distribution and abundance in the ocean. AI and machine learning approaches can aid in automating the detection and tracking of microplastics in the ocean. Drones and underwater robots are one technique for employing AI for microplastic identification [40]. These devices may be outfitted with cameras and sensors capable of detecting and identifying microplastics in the water column and on the seafloor. These devices' photos and data can then be evaluated using machine learning techniques to classify and quantify the microplastics in the sample. Using remote-sensing techniques instead of AI for microplastic identification is an alternative approach. Remote sensing is collecting data from the Earth's surface using satellites and other flying equipment. This data can be used to detect and categorize microplastics in water using machine learning algorithms. This method can provide a large-scale, high-resolution perspective of microplastic dispersion in the ocean, facilitating the discovery of areas of major accumulation and informing management decisions [41]. The detection and removal of microplastics from the ocean using autonomous underwater vehicles (AUVs): Unmanned submerged vehicles (AUVs) are unmanned vehicles that may be programmed to traverse the ocean and collect data on microplastic concentrations. Scientists may employ AUVs to build highly accurate maps of microplastic pollution, which can subsequently be used to develop

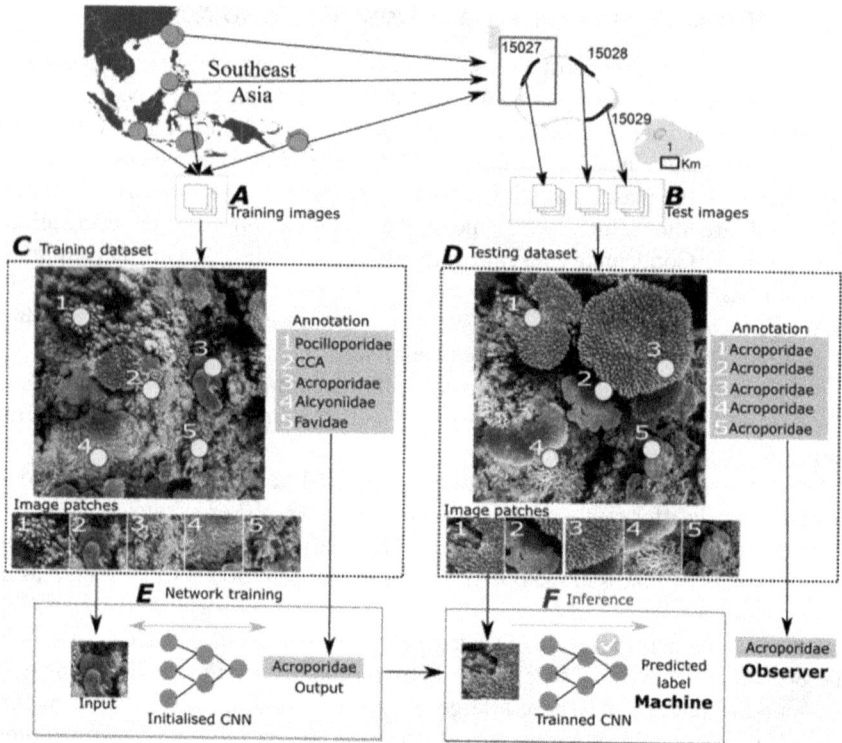

FIGURE 14.5 Using Remote Sensing as an Efficient and Economical Strategy [42]. (Reproduced with permission.)

targeted initiatives to clean up. Figure 14.5A depicts the training images of remote sensing near Southeast Asia; Figure 14.5B shows the test images; Figures 14.5C and 14.5D depict the training and testing dataset, respectively. Figure 14.5E shows the network training to produce Acroporidae output, and Figure 14.5F portrays the predicted label machine.

14.5 BIOSENSORS FOR MICROPLASTIC DETECTION

14.5.1 ANALYTICAL EQUIPMENT THAT DETECTS AND QUANTIFIES SPECIFIC ANALYTES IN A SAMPLE IS KNOWN AS BIOSENSORS

Biosensors were used to detect microplastics in environmental samples. Microplastics have been detected using the following biosensors:

14.5.2 ENZYME-BASED BIOSENSORS

Enzymes are used as the recognition element in enzyme-based biosensors. The enzymes catalyze a reaction that generates a signal related to the analyte concentration.

Enzyme-based biosensors have been utilized to detect microplastics by employing enzymes capable of degrading the microplastics [43].

14.5.3 APTAMER-BASED BIOSENSORS

Aptamers are employed in aptamer-based biosensors as the recognition factor. The aptamers are synthetic molecules that can bind to a specific analyte. The resulting complex generates a signal proportional to the concentration of the analyte. Aptamer-based biosensors have been used to detect small plastic particles by using aptamers that can accurately bind to microplastics.

14.5.4 ANTIBODY-BASED BIOSENSORS

Antibodies serve as the recognition element in antibody-based biological sensors. The antibodies preferentially attach to the analyte, and the resulting complex produces an indication proportional to the analyte concentration. Microplastics have been detected using antibody-based biosensors that use antibodies that recognize specific nanoplastic pieces. Bacteriophages—viruses that may infect bacteria—are another example of a bio-nano-sensor for microplastic detection [44]. Researchers developed phages that attach particularly to microplastics, allowing them to be detected in ambient samples. These bacteriophages can be used with a biosensor platform, such as a paper-based assay, to give a cheap and simple technique for detecting tiny plastics [45].

14.6 NOVEL ACTIVITIES

Traditional plastic reduction strategies, such as mechanical recycling and landfilling, are insufficient to address this issue. This review chapter suggests a unique, in-situ microplastic and plastic mitigation method based on bio-nano-particles and artificial intelligence [46].

Our procedure uses bio-nano-particles to break down plastics and microplastics in a controlled environment. The bio-nano-granules utilized in the experiment were created by the bacterial strain discovered at a site with plastic garbage. The bacterial strain was cultivated in a nutrient broth medium, and then the bio-nano-particles were separated by centrifugation. The bio-nano-particles have been investigated using TEM and Dynamic Light Scattering (DLS) to ascertain their size, shape, and charge on the exterior [47]. The experiment's microplastics and plastic fragments were gathered from a nearby beach after the tides had carried them there. The recyclable debris was separated and meticulously cleaned to remove any contaminants, like grit and biological compounds. The experiment was carried out in a batch reactor system, which included a glass vessel with a magnetic stirrer and a temperature controller. The reactor was filled with bio-nano-particle suspension and plastic trash, and the system was swirled at a constant speed of 200 rpm. The temperature was kept at 30°C since it was ideal for the bacterial strain employed in the experiment. Using a buffer solution, the pH of the system was adjusted to 7.0. The AI system of the experiment comprised a computer program to monitor the process and optimize the circumstances for optimal efficiency. The program was created to analyze sensor

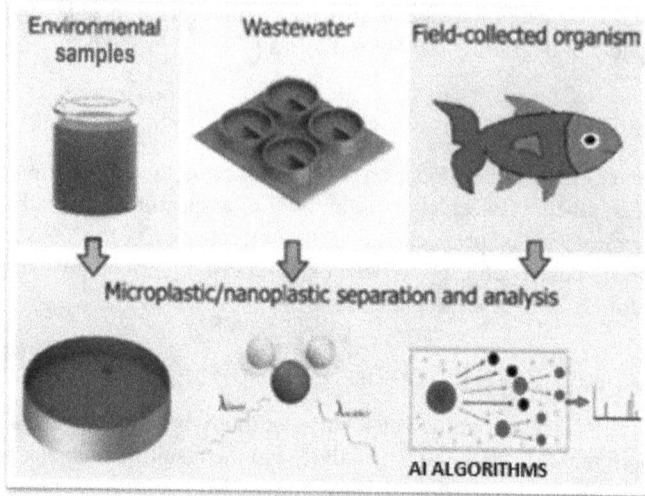

FIGURE 14.6 Analysis, separation, treatment, and purification of microplastics using AI algorithms [48]. (Reproduced with permission.)

data, such as temperature and pH sensors, and to change the parameters. We expect our proposed experiment to remove plastics from the contaminated site successfully. Bio-nanomaterials have been shown, in laboratory settings, to be successful at breaking down plastics, and we anticipate that this effectiveness will translate to in-situ applications. Using AI to monitor the treatment process will also allow for treatment optimization and identification of any potential environmental hazards associated with the therapy.

14.6.1 MICROPLASTIC ALTERNATIVES

The possibilities to nanoplastics are substances that can be used in place of in a variety of products and processes, such as personal care items, cleaning supplies, and industrial procedures. Green bio-nanomaterials can create microplastic replacements, which have several advantages over microplastics, such as biodegradability, low toxicity, and minimal carbon [49].

One type of substitute for microplastics is made of wood-pulp-derived cellulose microbeads. In personal care products like scrubs and exfoliants, cellulose microbeads can be used in place of conventional plastic microbeads. Microbeads made of cellulose are organic and less toxic than plastic ones, making them safer. Polylactic acid (PLA), generated from corn starch, is another example of a microplastic alternative [50]. PLA has a wide range of applications, including packaging materials and disposable utensils. PLA is biodegradable and low in toxicity, making it an excellent substitute for conventional plastic [51].

Microplastics are made from petroleum, a limited resource contributing to greenhouse gas emissions and climate change. Plastics as an energy source have been proposed as a solution to this problem. This problem can be handled by developing new, ecologically friendly bio-nanomaterials capable of converting microplastics into useable energy sources such as hydrogen or methane. Furthermore, the creation of these materials can benefit from artificial intelligence, which allows researchers to quickly screen a huge number of prospective candidates and select the most promising ones for further investigation. The process of pyrolysis is one potential method for converting microplastics into energy. Pyrolysis is a thermal process that decomposes organic material at high temperatures without oxygen. During pyrolysis, plastic trash is broken down into smaller molecules, which can then be converted into fuels or other valuable compounds [52]. Using bio-nanomaterials as catalysts in the pyrolysis process can improve conversion efficiency and reduce energy consumption [53]. Catalysts for microplastic pyrolysis can be made from a variety of bio-nanomaterials. These are examples of metal nanoparticles, metal oxide nanoparticles, and carbon-based compounds. Metal nanoparticles, such as platinum, are effective plastic pyrolysis catalysts [54].

Microplastics to convert energy presents many challenges that must be overcome to be both sustainable and cost-effective. One of the biggest challenges is the variety of microplastics. Because microplastics vary in size, shape, and chemical composition, developing a universal conversion process capable of managing all types of plastic debris is difficult. Another issue is the likelihood of contamination of the environment [55–56]. The conversion of microplastics to energy may produce greenhouse gases, toxic vapors, and other pollutants that are detrimental to the environment and the well-being of humans. As a result, it is critical to develop conversion procedures that minimize environmental contamination [57]. Sustainable bio-nanomaterials are environmentally friendly and made from sustainable and renewable resources. To synthesize them, green chemistry methods are employed, removing harmful chemicals and solvents and making the process safer and more sustainable. Microplastics are being converted into biodegradable packaging materials using bio-nanoparticles. It may create more environmentally friendly packaging materials than regular plastics by assembling tiny particles into new architectures using bio-nanoparticles. In surveys, using bio-nanoparticles to break down microplastics has shown promise. For instance, a recent study published in the journal *Environmental Science and Technology* revealed the use of protein nanotubes to decompose polystyrene microplastics. The nanoparticles were made from milk protein casein and were extremely effective at degrading microfibers [58].

14.6.2   BIO-NANOPARTICLE TYPES

Many bio nanoparticles can be employed to convert microplastics into value-added goods. Enzymes are biological catalysts that can degrade a wide range of molecules, including plastics. Several enzymes have been demonstrated to be useful in breaking down various plastics. PETase, for example, has been demonstrated to degrade polyethylene terephthalate (PET), which is extensively used in plastic bottles [59].

Bacteria can degrade a wide range of things, including plastics. Several bacteria species have been demonstrated to be effective at degrading various plastics.

The bacterium *Ideonella sakaiensis*, for example, can degrade PET. Oxidation is the process of adding oxygen to a chemical molecule. Plastics such as polyvinyl chloride (PVC) and polystyrene (PS) can oxidize when exposed to illnesses such as *Pseudomonas* and *Comamonas* [60]. AI can be used to improve microplastic manufacturing parameters. Machine learning algorithms can be taught using data from prior processing runs to determine the ideal processing settings for a specific application. This can help to reduce recycling time and increase the efficiency of microplastic processing. Green, new bio-nanomaterials have developed as a potential solution to the microplastic problem [61]. Because they are generated from renewable resources such as plant extracts and microorganisms, these products can remove small plastic particles from the environment. The use of ecologically friendly innovative bio-nanomaterials is part of the circular economy, highlighting the importance of reducing waste and recycling materials to create a sustainable economy [62].

14.7 CHALLENGES

Despite the potential benefits of using bio-nanomaterials to remediate microplastics, several challenges must be overcome before this technique can be extensively used. Among these limitations are:

1. Lack of understanding of the interactions between bio-microstructures and microplastics: Bio-nanomaterial interactions with microplastics are difficult and poorly understood. Researchers must understand these interactions' mechanisms to determine the ideal conditions for using bio-nanomaterials to remediate microplastics.
2. Limited utilization: Using environmentally friendly based bio to remove microplastics also has this issue. Although these materials have produced encouraging results in lab tests, their usefulness in practical uses has not yet been established. Numerous variables can affect its effectiveness, including the kind of pollutants in the environment used and the intensity of the and the surrounding conditions [63].
3. Cost: Using artificial intelligence to remove microscopic plastics from the surroundings can be expensive. AI system testing and deployment can be costly, limiting its applicability in countries with limited financial resources. Keeping the system up-to-date might also be expensive [64].
4. Scaling up production can be difficult: While bio-nanomaterials can be produced in various ways, scaling up manufacturing to meet the need for large-scale applications can be challenging. Researchers must figure out how to increase output while maintaining bio-nanomaterial quality.
5. Environmental effects of bio-nanomaterials: Bio-nanomaterials are typically from organic materials such as fungi, bacteria, and plants. As a result, applying them for washing up microfibers may have unforeseen environmental consequences. For example, the production of bio-nanomaterials may entail the usage of hazardous chemicals and other resources. In addition, the accumulation of bio-nanomaterials in nature may harm species and produce ecological imbalances [65].

6. Selectivity of bio-nanomaterials: The selectivity of bio-nanomaterials for microplastic remediation is another key barrier. The size, shape, and chemical composition of microplastics vary. As a result, bio-nanomaterials used in their repair must be selective and unique to the microplastic type in question. On the other hand, the bulk of bio-nanomaterials are non-selective and can adsorb a wide range of particles, including important nutrients and other organic compounds, resulting in unintended consequences [66].

7. Despite providing immediate benefits, it is impossible to predict whether breakthrough artificial intelligence sensors will give long-term benefits. Fouling, corrosion, and maintenance requirements may all impact the ability of these sensors to detect or remove microplastics effectively over time, which is why continuous monitoring and upkeep are required for long-term efficacy. Furthermore, there is currently a dearth of legal frameworks governing bio-nanomaterials paired with AI in environmental restoration, further complicating problems [67].

14.8 CONCLUSION

Systems based on artificial intelligence have also been implemented to increase the efficiency and effectiveness of microplastic remediation methods. In this chapter, we provide an overview of green microplastic remediation using bio-nanomaterials and AI and some critical remarks on this current state and future possibilities of this technology. Chitosan is a great substitute for microplastic remediation because it degrades naturally and has no negative environmental effects. A distinctive micro called cellulose nanocrystals (CNCs) has shown promise in eliminating microplastics. Since CNCs are cellulose-based nanoscale crystalline structures, which are a key component of plant cell walls, they are desirable choices for microplastic solubilization. Chitosan and cellulose-based bio-nano-particles have demonstrated excellent capabilities in binding and removing microplastics from various situations, including soil and water sources. Their use has several advantages over previous remediation methods: They are less expensive, non-toxic, biodegradable, and easily synthesized from renewable resources, giving an efficient long-term solution to plastic pollution. Using Artificial intelligence approaches improves design precision for bio-nano-particle formulations by predicting binding capabilities and discovering optimal utilization conditions that lead to the optimal output. Furthermore, AI enables real-time monitoring during the repair process, ensuring that relevant adjustments are made to ensure efficient outcomes. One of the primary benefits of employing magnetic bio-nanoparticles for microplastic disposal is their ease of recovery from the environment. A magnetic field can be used to recover magnetic nanoparticles, making the procedure very efficient and cost-effective. Furthermore, magnetic nanoparticles can be reused numerous times, reducing the need for the synthesis of new nanomaterials and lowering the environmental effect further.

Although the use of machine learning and bio-nano-granules in microplastic bioremediation is a promising strategy, there are still considerable challenges to be solved. The scalability of the technology is one important issue. Although it has been demonstrated that bio-nano-particles can effectively remove plastic particles in lab settings, it is unclear whether this procedure can be scaled up to treat larger volumes of water or soil. In some cases, the cost of producing bio-nano-particles may also be prohibitive. The potential for unanticipated consequences is another problem. More research is needed to determine the long-term effects of bio-nano-particles on the environmental impacts, even though they are typically considered safe and sustainable. A promising solution that could assist India in achieving a circular economy uses bio-nano-particles in conjunction with AI to clean up green microplastic. Over traditional microplastic remediation techniques, bio-nano-particles have several benefits, including biodegradability, cost-effectiveness, and scalability. Artificial intelligence can be incorporated into the process to increase the remediation accuracy and efficiency of the process while also increasing its economic viability and sustainability. Green microplastic clean-up using bio-nano-particles combined with AI is a promising solution that could help India achieve a circular economy. Bio-nano-particles have various advantages over standard microplastic remediation technologies, including biodegradability, cost-effectiveness, and the ability to scale up. Incorporating artificial intelligence into the process can improve the efficiency and accuracy of the remediation process, boosting its economic feasibility and sustainability. In conclusion, green nanoplastics remediation using bio-nano-particles and AI is a promising solution to plastic pollution globally and in India. While AI can boost the effectiveness and precision of the repair process, this technology provides a long-term and ecologically sound solution. To make sure that this technology is secure, efficient, and scalable, it is crucial to conduct more research in this area. With the correct assistance and collaboration, the approach mentioned in the review book chapter can alleviate India's waste-management problem while creating new economic opportunities and minimizing the adverse environmental effects generated by plastic trash.

REFERENCES

[1] Issac, M.N., Kandasubramanian, B., 2021. Effect of microplastics in water and aquatic systems. Environmental Science and Pollution Research. https://doi.org/10.1007/s11356-021-13184-2.

[2] Khan, I., Saeed, K., Khan, I., 2019. Nanoparticles: Properties, applications, and toxicities. Arabian Journal of Chemistry. https://doi.org/10.1016/j.arabjc.2017.05.011.

[3] Yadav, M., Chiu, F.-C., 2019. Cellulose nanocrystals reinforced κ-carrageenan based UV resistant transparent bio-nanocomposite films for sustainable packaging applications. Carbohydrate Polymers. https://doi.org/10.1016/j.carbpol.2019.01.114.

[4] Kundu, A., Shetti, N.P., Basu, S., Raghava Reddy, K., Nadagouda, M.N., Aminabhavi, T.M., 2021. Identification and removal of micro- and nano-plastics: Efficient and cost-effective methods. Chemical Engineering Journal. https://doi.org/10.1016/j.cej.2021.129816.

[5] Aramesh, N., Bagheri, A.R., Bilal, M., 2021. Chitosan-based hybrid materials for adsorptive removal of dyes and underlying interaction mechanisms. International Journal of Biological Macromolecules. https://doi.org/10.1016/j.ijbiomac.2021.04.158.

[6] Lamichhane, G., Acharya, A., Marahatha, R., Modi, B., Paudel, R., Adhikari, A., Raut, B.K., Aryal, S., Parajuli, N., 2022. Microplastics in the environment: Global concern, challenges, and control measures. International Journal of Environmental Science and Technology. https://doi.org/10.1007/s13762-022-04261-1.

[7] Maraveas, C., Kotzabasaki, M.I., Bartzanas, T., 2023. Intelligent technologies, enzyme-embedded and microbial degradation of agricultural plastics. AgriEngineering. https://doi.org/10.3390/agriengineering5010006.

[8] Tayal, S., Singla, P., Nandi, A., Davim, J.P., 2021. Computational Technologies in Materials Science. CRC Press. https://doi.org/10.1201/9781003121954.

[9] Cole, M., Lindeque, P.K., Fileman, E., Clark, J., Lewis, C., Halsband, C., Galloway, T.S., 2016. Microplastics alter the properties and sinking rates of zooplankton faecal pellets. Environmental Science & Technology. https://doi.org/10.1021/acs.est.5b05905.

[10] Worm, B., Lotze, H.K., Jubinville, I., Wilcox, C., Jambeck, J., 2017. Plastic as a Persistent Marine Pollutant. Annual Review of Environment and Resources. https://doi.org/10.1146/annurev-environ-102016-060700.

[11] Periyasamy, A.P., Tehrani-Bagha, A., 2022. A review on microplastic emission from textile materials and its reduction techniques. Polymer Degradation and Stability. https://doi.org/10.1016/j.polymdegradstab.2022.109901.

[12] Zhang, Y.-Q., Lykaki, M., Markiewicz, M., Alrajoula, M.T., Kraas, C., Stolte, S., 2022. Environmental contamination by microplastics originating from textiles: Emission, transport, fate, and toxicity. Journal of Hazardous Materials. https://doi.org/10.1016/j.jhazmat.2022.128453.

[13] Mercogliano, R., Avio, C.G., Regoli, F., Anastasio, A., Colavita, G., Santonicola, S., 2020. Occurrence of microplastics in commercial seafood under the perspective of the human food chain. A review. Journal of Agricultural and Food Chemistry. https://doi.org/10.1021/acs.jafc.0c01209.

[14] Zuccaro, P., Thompson, D.C., de Boer, J., Watterson, A., Wang, Q., Tang, S., Shi, X., Llompart, M., Ratola, N., Vasiliou, V., 2022. Artificial turf and crumb rubber infill: An international policy review concerning the current state of regulations. Environmental Challenges. https://doi.org/10.1016/j.envc.2022.100620.

[15] Van Cauwenberghe, L., Janssen, C.R., 2014. Microplastics in bivalves cultured for human consumption. Environmental Pollution. https://doi.org/10.1016/j.envpol.2014.06.010.

[16] Yang, W., Jannatun, N., Zeng, Y., Liu, T., Zhang, G., Chen, C., Li, Y., 2022. Impacts of microplastics on immunity. Frontiers in Toxicology. https://doi.org/10.3389/ftox.2022.956885.

[17] Al Naggar, Y., Brinkmann, M., Sayes, C.M., AL-Kahtani, S.N., Dar, S.A., El-Seedi, H.R., Grünewald, B., Giesy, J.P., 2021. Are honey bees at risk from microplastics? Toxics. https://doi.org/10.3390/toxics9050109.

[18] Mihai, F.-C., Gündoğdu, S., Markley, L.A., Olivelli, A., Khan, F.R., Gwinnett, C., Gutberlet, J., Reyna-Bensusan, N., Llanquileo-Melgarejo, P., Meidiana, C., Elagroudy, S., Ishchenko, V., Penney, S., Lenkiewicz, Z., Molinos-Senante, M., 2021. Plastic pollution, waste management issues, and circular economy opportunities in rural communities. Sustainability. https://doi.org/10.3390/su14010020.

[19] Lebreton, L.C.M., van der Zwet, J., Damsteeg, J.-W., Slat, B., Andrady, A., Reisser, J., 2017. River plastic emissions to the world's oceans. Nature Communications. https://doi.org/10.1038/ncomms15611.

[20] Ivleva, N.P., 2021. Chemical analysis of microplastics and nanoplastics: Challenges, advanced methods, and perspectives. Chemical Reviews. https://doi.org/10.1021/acs.chemrev.1c00178.

[21] Cárdenas-Alcaide, M.F., Godínez-Alemán, J.A., González-González, R.B., Iqbal, H.M.N., Parra-Saldívar, R., 2022. Environmental impact and mitigation of micro(nano) plastics pollution using green catalytic tools and green analytical methods. Green Analytical Chemistry. https://doi.org/10.1016/j.greeac.2022.100031.

[22] Lumio, R.T., Tan, M.A., Magpantay, H.D., 2021. Biotechnology-based microbial degradation of plastic additives. 3 Biotech. https://doi.org/10.1007/s13205-021-02884-8.

[23] Zou, J., Wang, C., Li, J., Wei, J., Liu, Y., Hu, L., Liu, H., Bian, H., Sun, D., 2022. Effect of polyethylene (LDPE) microplastic on remediation of cadmium contaminated soil by Solanum nigrum L. GEP. https://doi.org/10.4236/gep.2022.101004.

[24] Löder, M.G.J., Imhof, H.K., Ladehoff, M., Löschel, L.A., Lorenz, C., Mintenig, S., Piehl, S., Primpke, S., Schrank, I., Laforsch, C., Gerdts, G., 2017. Enzymatic purification of microplastics in environmental samples. Environmental Science & Technology. https://doi.org/10.1021/acs.est.7b03055.

[25] Lee, M., Kim, H., 2022. COVID-19 pandemic and microplastic pollution. Nanomaterials. https://doi.org/10.3390/nano12050851.

[26] Knott, B.C., Erickson, E., Allen, M.D., Gado, J.E., Graham, R., Kearns, F.L., Pardo, I., Topuzlu, E., Anderson, J.J., Austin, H.P., Dominick, G., Johnson, C.W., Rorrer, N.A., Szostkiewicz, C.J., Copić, V., Payne, C.M., Woodcock, H.L., Donohoe, B.S., Beckham, G.T., McGeehan, J.E., 2020. Characterization and engineering of a two-enzyme system for plastics depolymerization. Proceedings of the National Academy of Sciences United States of America. https://doi.org/10.1073/pnas.2006753117.

[27] Mehmood, T., Mustafa, B., Mackenzie, K., Ali, W., Sabir, R.I., Anum, W., Gaurav, G.K., Riaz, U., Liu, X., Peng, L., 2023. Recent developments in microplastic contaminated water treatment: Progress and prospects of carbon-based two-dimensional materials for membranes separation. Chemosphere. https://doi.org/10.1016/j.chemosphere.2022.137704.

[28] Zhang, Y., Zhang, D., Zhang, Z., 2023. A critical review on artificial intelligence—based microplastics imaging technology: Recent advances, hot-spots, and challenges. IJERPH. https://doi.org/10.3390/ijerph20021150.

[29] Di Ciaccio, F., Troisi, S., 2021. Monitoring marine environments with autonomous underwater vehicles: A bibliometric analysis. Results in Engineering. https://doi.org/10.1016/j.rineng.2021.100205.

[30] Zhang, Y., Zhang, D., Zhang, Z., 2023. A critical review on artificial intelligence—based microplastics imaging technology: Recent advances, hot-spots, and challenges. IJERPH. https://doi.org/10.3390/ijerph20021150.

[31] Bacha, A.-U.-R., Nabi, I., Zhang, L., 2021. Mechanisms and the engineering approaches for the degradation of microplastics. ACS EST Engineering. https://doi.org/10.1021/acsestengg.1c00216.

[32] Rahman, M.M., Ahmed, L., Anika, F., Riya, A.A., Kali, S.K., Rauf, A., Sharma, R., 2023. Bioinorganic nanoparticles for the remediation of environmental pollution: Critical appraisal and potential avenues. Bioinorganic Chemistry and Applications. https://doi.org/10.1155/2023/2409642.

[33] Salata, O., 2004. Journal of Nanobiotechnology. https://doi.org/10.1186/1477-3155-2-3.

[34] Iravani, S., Varma, R.S., 2022. Nanosponges for water treatment: Progress and challenges. Applied Sciences. https://doi.org/10.3390/app12094182.

[35] Heo, Y., Lee, E.-H., Lee, S.-W., 2022. Adsorptive removal of micron-sized polystyrene particles using magnetic iron oxide nanoparticles. Chemosphere. https://doi.org/10.1016/j.chemosphere.2022.135672.

[36] Mondal, S., Bera, S., Mishra, R., Roy, S., 2022. Redefining the role of microalgae in industrial wastewater remediation. Energy Nexus. https://doi.org/10.1016/j.nexus.2022.100088.

[37] Béraud, E., Bednarz, V., Otto, I., Golbuu, Y., Ferrier-Pagès, C., 2022. Plastics are a new threat to Palau's coral reefs. PLoS One. https://doi.org/10.1371/journal.pone.0270237.

[38] Deleja, M., Paula, J.R., Repolho, T., Franzitta, M., Baptista, M., Lopes, V., Simão, S., Fonseca, V.F., Duarte, B., Rosa, R., 2022. Effects of hypoxia on coral photobiology and oxidative stress. Biology. https://doi.org/10.3390/biology11071068.

[39] Leppänen, I., Lappalainen, T., Lohtander, T., Jonkergouw, C., Arola, S., Tammelin, T., 2022. Capturing colloidal nano- and microplastics with plant-based nano-cellulose networks. Nature Communications. https://doi.org/10.1038/s41467-022-29446-7.

[40] Sharafi, R., Anam, N., Kamal, T., Tazwar, S.T., 2022. Effective microplastic extraction using drone submarine. Figshare. https://doi.org/10.6084/M9.FIGSHARE.1885 8623.V1.

[41] Kapoor, R.T., Rafatullah, M., Qamar, M., Qutob, M., Alosaimi, A.M., Alorfi, H.S., Hussein, M.A., 2022. Review of recent developments in bioinspired materials for sustainable energy and environmental applications. Sustainability. https://doi.org/10.3390/su142416931.

[42] González-Rivero, M., Beijbom, O., Rodriguez-Ramirez, A., Bryant, D.E.P., Ganase, A., Gonzalez-Marrero, Y., Herrera-Reveles, A., Kennedy, E.V., Kim, C.J.S., Lopez-Marcano, S., Markey, K., Neal, B.P., Osborne, K., Reyes-Nivia, C., Sampayo, E.M., Stolberg, K., Taylor, A., Vercelloni, J., Wyatt, M., Hoegh-Guldberg, O., 2020. Monitoring of coral reefs using artificial intelligence: A feasible and cost-effective approach. Remote Sensing. https://doi.org/10.3390/rs12030489.

[43] Fan, Y.-F., Guo, Z.-B., Ge, G.-B., 2023. Enzyme-based biosensors and their applications. Biosensors. https://doi.org/10.3390/bios13040476.

[44] Niu, L., Zhao, S., Chen, Y., Li, Y., Zou, G., Tao, Y., Zhang, W., Wang, L., Zhang, H., 2023. Diversity and potential functional characteristics of phage communities colonizing microplastic biofilms. Environmental Research. https://doi.org/10.1016/j.envres.2022.115103.

[45] Nishat, S., Jafry, A.T., Martinez, A.W., Awan, F.R., 2021. Paper-based microfluidics: Simplified fabrication and assay methods. Sensors and Actuators B: Chemical. https://doi.org/10.1016/j.snb.2021.129681.

[46] Mssr, T., Pathak, P., Singh, L., Raj, D., Gupta, D.K., 2023. A novel circular approach to analyze the challenges associated with micro-nano plastics and their sustainable remediation techniques. Journal of Environmental Science and Health, Part A. https://doi.org/1 0.1080/10934529.2023.2208507.

[47] Stetefeld, J., McKenna, S.A., Patel, T.R., 2016. Dynamic light scattering: A practical guide and applications in biomedical sciences. Biophysical Reviews. https://doi.org/10.1007/s12551-016-0218-6.

[48] Nguyen, B., Claveau-Mallet, D., Hernandez, L.M., Xu, E.G., Farner, J.M., Tufenkji, N., 2019. Separation and analysis of microplastics and nano-plastics in complex environmental samples. Accounts of Chemical Research. https://doi.org/10.1021/acs.accounts.8b00602.

[49] Patra, J.K., Baek, K.-H., 2014. Green nanobiotechnology: Factors affecting synthesis and characterization techniques. Journal of Nanomaterials. https://doi.org/10.1155/2014/417305.

[50] Muller, J., González-Martínez, C., Chiralt, A., 2017. Combination of poly(lactic) acid and starch for biodegradable food packaging. Materials. https://doi.org/10.3390/ma10080952.

[51] Sun, C., Wei, S., Tan, H., Huang, Y., Zhang, Y., 2022. Progress in upcycling polylactic acid waste as an alternative carbon source: A review. Chemical Engineering Journal. https://doi.org/10.1016/j.cej.2022.136881.

[52] Miandad, R., Rehan, M., Barakat, M.A., Aburiazaiza, A.S., Khan, H., Ismail, I.M.I., Dhavamani, J., Gardy, J., Hassanpour, A., Nizami, A.-S., 2019. Catalytic pyrolysis of plastic waste: Moving toward pyrolysis based biorefineries. Frontiers in Energy Research. https://doi.org/10.3389/fenrg.2019.00027.

[53] Muhammad, I., Manos, G., 2021. Improving the conversion of biomass in catalytic pyrolysis via intensification of biomass—catalyst contact by co-pressing. Catalysts. https://doi.org/10.3390/catal11070805.

[54] Yoshioka, T., Handa, T., Grause, G., Lei, Z., Inomata, H., Mizoguchi, T., 2005. Effects of metal oxides on the pyrolysis of poly(ethylene terephthalate). Journal of Analytical and Applied Pyrolysis. https://doi.org/10.1016/j.jaap.2005.01.004.

[55] Li, N., Liu, H., Cheng, Z., Yan, B., Chen, G., Wang, S., 2022. Conversion of plastic waste into fuels: A critical review. Journal of Hazardous Materials. https://doi.org/10.1016/j.jhazmat.2021.127460.

[56] Bhuyan, M.S., 2022. Effects of microplastics on fish and in human health. Frontiers in Environmental Science. https://doi.org/10.3389/fenvs.2022.827289.

[57] Campanale, M.S., Locaputo, U., 2020. A detailed review study on potential effects of microplastics and additives of concern on human health. IJERPH. https://doi.org/10.3390/ijerph17041212.

[58] Rashidi, M., Bijari, S., Khazaei, A.H., Shojaei-Ghahrizjani, F., Rezakhani, L., 2022. The role of milk-derived exosomes in the treatment of diseases. Frontiers in Genetics. https://doi.org/10.3389/fgene.2022.1009338.

[59] Benyathiar, P., Kumar, P., Carpenter, G., Brace, J., Mishra, D.K., 2022. Polyethylene terephthalate (PET) bottle-to-bottle recycling for the beverage industry: A review. Polymers. https://doi.org/10.3390/polym14122366.

[60] Mouafo Tamnou, E.B., Tamsa Arfao, A., Nougang, M.E., Metsopkeng, C.S., Noah Ewoti, O.V., Moungang, L.M., Nana, P.A., Atem Takang-Etta, L.-R., Perrière, F., Sime-Ngando, T., Nola, M., 2021. Biodegradation of polyethylene by the bacterium Pseudomonas aeruginosa in acidic aquatic microcosm and effect of the environmental temperature. Environmental Challenges. https://doi.org/10.1016/j.envc.2021.100056.

[61] Yee, M.S.-L., Hii, L.-W., Looi, C.K., Lim, W.-M., Wong, S.-F., Kok, Y.-Y., Tan, B.-K., Wong, C.-Y., Leong, C.-O., 2021. Impact of microplastics and nanoplastics on human health. Nanomaterials. https://doi.org/10.3390/nano11020496.

[62] Syberg, K., Nielsen, M.B., Oturai, N.B., Clausen, L.P.W., Ramos, T.M., Hansen, S.F., 2022. Circular economy and reduction of micro(nano)plastics contamination. Journal of Hazardous Materials Advances. https://doi.org/10.1016/j.hazadv.2022.100044.

[63] Dey, T.K., Uddin, Md.E., Jamal, M., 2021. Detection and removal of microplastics in wastewater: Evolution and impact. Environmental Science and Pollution Research. https://doi.org/10.1007/s11356-021-12943-5.

[64] Remya, R.R., Julius, A., Suman, T.Y., Mohanavel, V., Karthick, A., Pazhanimuthu, C., Samrot, A.V., Muhibbullah, M., 2022. Role of nanoparticles in biodegradation and their importance in environmental and biomedical applications. Journal of Nanomaterials. https://doi.org/10.1155/2022/6090846.

[65] Hou, W.-C., Westerhoff, P., Posner, J.D., 2013. Biological accumulation of engineered nanomaterials: A review of current knowledge. Environmental Science: Processes Impacts. https://doi.org/10.1039/c2em30686g.

[66] Ray, P.C., Yu, H., Fu, P.P., 2009. Toxicity and environmental risks of nanomaterials: Challenges and future needs. Journal of Environmental Science and Health, Part C. https://doi.org/10.1080/10590500802708267.

[67] Ligozat, A.-L., Lefevre, J., Bugeau, A., Combaz, J., 2022. Unraveling the hidden environmental impacts of AI solutions for environment life cycle assessment of AI solutions. Sustainability. https://doi.org/10.3390/su14095172.

15 Emerging Applications of Magnetic Nanomaterials in the Remediation of Microplastics from the Aquatic Environment

Uma Sankar Mondal, Anisha Karmakar, Aritri Paul and Subhankar Paul

15.1 INTRODUCTION

The sustainability of aquatic ecosystems, human populations, and animals is seriously threatened by the inorganic and organic pollutants that have contaminated water resources (Das et al., 2022, 2023) One of these pollutants in contaminated water is microplastics, which are regarded as emerging contaminants of concern (Goh et al., 2022). Microplastics, which can exist in the form of fragments, granules, fibres, and small beads and have a diameter of less than 5 mm, is a general term that refers to the waste products of all different kinds of plastic sources, such as polyethylene terephthalate, polyester, and polyvinyl chloride. Microplastics can be found in the oceans, freshwater, and soil. (Bhatt et al., 2021).

The presence of microplastics in the ecosystem and their frequent interactions with other compounds, such as dyes, heavy metals, and other organic and inorganic contaminants, pose serious threats to human health, marine life, and the ecosystem. This is because microplastics are composed of very small pieces of plastic. It is now well-recognized that ingesting microplastics might have detrimental effects on a variety of essential metabolic functions. This notion has gained widespread acceptance. It has been shown that the interaction of microplastics with other pollutants, such as heavy metals, pesticides, and pharmaceuticals, may lead to secondary pollution and poisoning of biota when the microplastics are ingested. The hazardous effects of microplastics might be exhibited as physical harm and chemical toxicity, especially when substantial numbers of microplastics are discharged into water bodies (Elgarahy et al., 2021). Numerous creatures, including humans, are at risk from microplastics because microplastics may enter and bioaccumulate in the food chain. Microplastic contamination is a worldwide concern; thus, we must find solutions that can be implemented. Material reduction, greater recycling capacity, and the development of nontoxic feedstocks are only a few examples of the enormous efforts that have been executed. Several agreements and policies have also

DOI: 10.1201/9781003449133-15

been put into place as measures to deal with the problems caused by microplastics. The identification of all aspects of microplastics in terms of their varieties, forms, morphologies, and surface chemistry is the first obstacle that must be overcome in order to successfully eliminate microplastic contaminants. Methods for the detection, identification, and characterization of microplastics in aquatic environments have been developed in order to establish reliable information on the characteristics, quantities, and toxicity of this pollutant. This will allow for the development of effective strategies for mitigating its adverse impacts. So far, a variety of removal and treatment approaches and engineering tools have been developed to mitigate the adverse effects that are caused by microplastics. The removal of microplastics has been attempted, using a great number of the traditional methods that are used to treat wastewater. Adsorption, oxidation, membrane separation, and chemical flocculation are all examples of these processes (Cherniak et al., 2022). However, some of the processes are not cost-effective because of their complicated nature, which involves number of materials and associated with high expenses in their maintenance. There has been a significant increase in the number of technical advancements, with new materials and methods being continually investigated to treat and removal of microplastics from water bodies.

The use of nanotechnology, a major enabling technology, may solve a broad variety of problems facing society. Techniques for the elimination of pollutants that are enabled by nanotechnology have emerged as feasible solutions for addressing the challenges posed by the status quo. In recent years, nanomaterials-based adsorbents and photocatalysts have been explored for efficient treatments of organic and inorganic water-pollutants (Misra et al., 2020; Das et al., 2021; Mondal et al., 2023). These techniques may either improve performance or fill niches that have not previously been addressed. Several investigations have been performed on the possibility of employing magnetic nanoparticles in a variety of removal strategies for microplastics in aquatic environments (Zandieh and Liu, 2022; Heo et al., 2022; Tang et al., 2021). This chapter provides a concise overview of the problem of microplastic pollution in aquatic environments and the methods that have been developed toward a solution, including the use of magnetic nanomaterials through adsorption, photocatalysis, and filtration processes.

15.2 SOURCES AND PHYSICOCHEMICAL PROPERTIES OF MICROPLASTICS

Primary and secondary microplastics are two distinct categories that may be found among microplastics that pollute aquatic environments. Primary microplastics are made directly in the micron range (for example, the pellets used in cosmetics and personal care items), and secondary microplastics break down from larger plastic materials when they are exposed to environmental stress (Wang et al., 2020).

There have been observations of microplastic in a variety of surface waterways. The number of microplastics in environments with freshwater may vary anywhere from a few tons to millions of tons, on average. The locations, the natural environments, and the activities carried out by humans are the primary contributors to these vast disparities. Microplastics may make their way into the surface water environment via the discharge of wastewater that already contains microplastics. Microplastics at high concentrations were found to originate mostly from the industrial sector, wastewater treatment plant effluent, fishing activities, and direct throwing out of household products. Furthermore, microplastic emissions are also affected by the features of the residing place. Compared to rural regions, densely populated places with greater living conditions may have more sources and emissions of microplastics. However, wastewater treatment systems are better in metropolitan areas, which means that microplastic contamination is likely to be less in these locations. The higher levels of microplastic contamination were seen in rural locations because environmental protection measures are less efficient and not implemented properly. As a result, human activities play a substantial role in the pollution of microplastics in the aquatic environment (Bui et al., 2020; Shen et al., 2020).

15.3 IMPACT OF MICROPLASTICS ON THE AQUATIC ENVIRONMENT AND HUMAN HEALTH

The aquatic ecosystem is affected by microplastics in many ways. Different aquatic species, including fishes, get affected by decreased food intake and slow growth, thereby resulting in oxidative damage and abnormal behavioural patterns. Furthermore, microplastics tend to penetrate through the physiological barriers and cause accumulation in tissues, thus generating reactive oxygen species, which hinders lipid metabolism within their body. Due to their microscopic size, they are susceptible to being consumed by marine species and affect their reproductive systems as well. Likewise, polystyrene microplastics have a negative effect on the reproductive ability of *Crassostrea gigas* by yielding very few egg cells and a smaller number of sperm, along with decreased sperm motility. As microplastics come in contact with biological tissues and organs, an immunological response is observed (Li et al., 2021b). Human exposure to microplastics can occur through the ingestion of seafood. In aquatic ecosystem, prawns, crabs, and other small fishes present consume microplastics, which accumulate within their intestinal tracts. During consumption of these fishes by humans, the intestinal tract is not thrown away, thereby causing their entry into the human food chain. A similar kind of case is not observed for big fishes, as their intestinal tract is removed before human consumption. Therefore, humans are susceptible to being affected by both chemical and physical means, causing toxicity within the lungs, liver, and even brain cells (Smith et al., 2018).

15.4 CONVENTIONAL STRATEGIES

The widely used traditional techniques for removing microplastics fall under the following categories: chemical, physical, or biological separation. Density floatation, filtration, and sedimentation fall under the category of physical method of separation. Though these techniques provide high removal efficiencies, the cost of the material is high, owing to membrane fouling and the inability to exude smaller particles (less than 100 μm) for flotation and sedimentation. The subsequent energy requirement is also high. Coagulation and flocculation fall under chemical separation techniques,

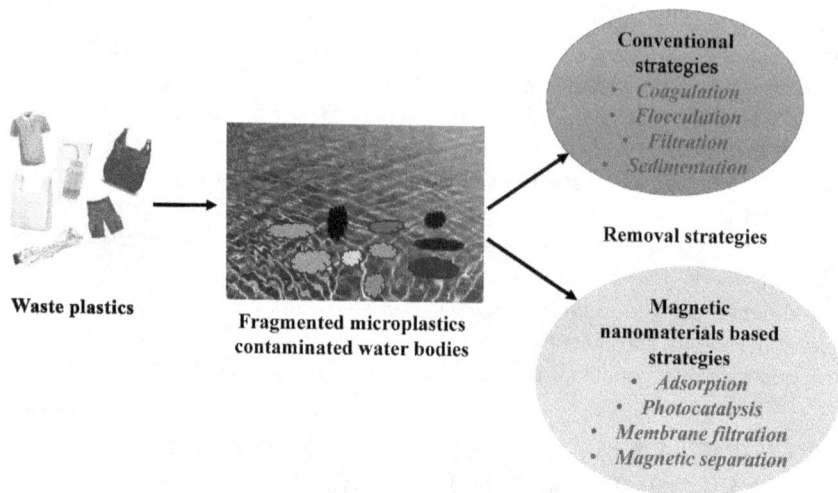

FIGURE 15.1 Schematic representation of microplastic-mediated pollution and the conventional as well as magnetic nanomaterials-based removal strategies.

TABLE 15.1

Various Conventional Methods for the Removal of Microplastics from the Aquatic Environment

Methods	Materials	Microplastic Type	Removal Efficiency	Ref.
Coagulation	$Fe_2(SO_4)_3$, $Al_2(SO_4)_3$	polyvinyl chloride (< 50 µm)	80%	(Prokopova et al., 2021)
coagulation	$FeCl_3$	polystyrene and polyethylene (< 0.5 mm)	62%	(Zhou et al., 2021)
Coagulation-flocculation	$Al_2(SO_4)_3$ (Alum)	polystyrene (100 µm)	98.9%	(Li et al., 2021a)
Coagulation/ ultrafiltration	$AlCl_3$, PAM	Polyethelyne	45.34%	(Ma et al., 2019)
Filtration	Biochar	polystyrene (10 µm)	90%	(Wang et al., 2020)
coagulation-flocculation-sedimentation	Alum	Polystyrene (6 µm)	85%	(Xue et al., 2021)

where high removal efficiency is also achieved and can be used in wastewater treatment facilities with little modification. Certain drawbacks faced by these techniques include inefficient removal of the tiny particles of a specific shape. For biological separation, biodegradation and bioactive sludges are the two main components. These methods are feasible because they are inexpensive, and upscaling can be achieved. But the efficiency of these methods is much less. Certain alternative strategies for the effective removal of microplastics are of current interest due to the natural constraints of existing procedures (Pasanen et al., 2023).

15.5   MICROPLASTICS REMOVAL STRATEGIES USING MAGNETIC NANOMATERIALS

It is very difficult to collect and remove microplastics from aquatic environment in a reliable manner. The manual treatment of small microplastics is challenging, and it is unpredictable how successful the various extraction strategies currently in use are for the recovery of these particles from water bodies. Researchers have developed and shown successful approaches for microplastics removal from aquatic environments through various strategies.

15.5.1   ADSORPTION AND MAGNETIC SEPARATION

The process of purifying water using adsorption is simple, environmentally friendly, and efficient. It is controlled by surface phenomena, which include the attachment of pollutant molecules from the fluid bulk to the solid surface. Nano-adsorbents exhibit a high specific surface area and porosity; thus, they are capable of significantly enhancing the amount of pollutant removal from wastewater. Several investigations that focused on the development and implementation of magnetic nanomaterials-based adsorbents have been investigated to extract and separate microplastics from a variety of aquatic environments, such as freshwater, river water, and the ocean. This method is intended to complement current methods and assist in the recovery of smaller microplastic particles (Ouda et al., 2023).

In order to magnetize and enable magnetic extraction and the separation of microplastics, Grbic et al. developed a technique using hydrophobic-coated iron (Fe) nanoparticles. They have investigated on large (> 1 mm), medium (200 µm–1 mm), and small (< 20 µm) particles of polyethylene, polyethylene terephthalate, polystyrene, polyurethane, polyvinyl chloride, and polypropylene from seawater. It was shown that 93%, 92%, and 81% of large, medium, and small microplastics, respectively, were recovered from seawater samples. The efficiency of this technique depends on the surface area to volume ratio; for smaller particles, more Fe nanoparticles may bind per unit mass of plastic. Therefore, this technique is very helpful for tiny microparticles (Grbic et al., 2019).

Micron-sized polystyrene (PS) was used as a model material for microplastics remediation from wastewater in recent research (Heo et al., 2022). Polystyrene is a major contributor to the plastic pollution that is seen in aquatic environments and wastewater. Iron oxide nanoparticles (Fe_3O_4) was used for the adsorptive removal of PS from water. An investigation of the kinetics and isothermal adsorption characteristics was carried out in order to assess the adsorption of PS particles onto Fe_3O_4. Experiments involving adsorption were also carried out in order to study the adsorption efficiencies of Fe_3O_4 for sets of PS particles that varied in terms of their average diameter. In this study, the electrostatic and hydrophobic interactions that may play

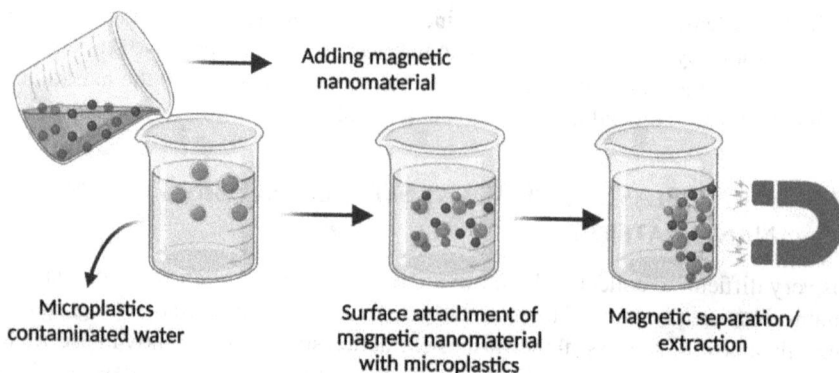

FIGURE 15.2 Schematic representation of magnetic nanomaterials-based adsorption and separation of microplastics from water.

a role in the adsorption of PS particles onto Fe_3O_4 were dissected and analyzed. Depending on the size of the PS particles, Fe_3O_4 exhibited varying adsorption efficiencies towards the PS particles. Fe_3O_4 was able to obtain an adsorption efficiency of 94% for the removal of PS particles with a diameter of 1 μm, and it was able to reach adsorption efficiencies of 58% for the removal of PS particles with a diameter of 0.7 μm.

In another study, the optimal magnetization of microplastics was done via surface absorption using nano-Fe_3O_4 for 150 min. Then, magnetized microplastics were removed from the water conveniently by suction of the magnet. The average removal rate of four common types of microplastics, including polystyrene, polyethylene, polyethylene terephthalate, and polypropylene in the size range of 200–900 μm, was 86%, 87%, 63%, and 85%, respectively. The density of nano-Fe_3O_4 played an important role in the clearance rate of the microparticles which were absorbed onto microparticle surfaces. Additionally, the removal rate of microparticles in pure water was less compared to artificial seawater. Furthermore, this approach was effective in removal of microplastics greater than 80% from environmental water bodies like domestic sewage, river water, and natural seawater. In a nutshell, this current study provided a unique approach to removing microplastics in water, with a broad application prospect (Shi et al., 2022).

Fe-C-NH_2 (FNP) magnetic nanoparticles use the principle of gradient magnetic-field, which is an efficient method for removing microparticles of polyethylene (PE) and polyethylene terephthalate (PET) from water (Bakhteeva et al., 2023). These plastics are widely used across the globe and are among the most popular manufactured solid polymers. The plastic particles that were utilized in this investigation were in the form of fragmented flakes. Magnetic core-shell-structured FNP composite nanomaterials were added to an aqueous medium to begin and speed up the sedimentation of plastic microparticles in water. In this investigation, 0.005–0.05 g/L range of concentrations of the FNP magnetic nanoseeds were added. This was a very low concentration, in comparison to the conventional coagulants that are often utilized. The kinetics of the generation of heteroaggregates are directly related to the rate at which magnetic sedimentation occurs. Within 15 min, the complete elimination of MPET was achieved under the magnetic field system with the highest gradient. However, the MPE removal rate was comparatively low. The MPE concentration was reduced by 88% after 15 min of magnetic sedimentation and decreased by 94% after 30 min of magnetic sedimentation in the magnetic field with the greatest gradient. Heteroaggregation of magnetic and plastic particles fuels the process and is reliant on magnetic and electro-steric interactions between the particles. Iron is biocompatible; thus, it does not contaminate in the level of secondary pollution in the water that has been treated. As it exhibits electrostatic attraction, it has the potential to form heteroaggregates with plastic particles in water. Therefore, this particle shell of carbon with connected amino groups was selected for this research.

Certain microplastics—i.e., polystyrene of 1 μm, polysulphone of 100–500 μm, and PET particles—are being removed by adsorptive magnetic extraction from potable and saltwater using magnetic organo-polyoxometalate nanocomposites (Fe_3O_4-PWA/amine) in another study (Bhore and Kamble, 2022). They synthesized magnetic organo-polyoxometalate nanocomposites with amine group n-Butyl, n-Hexyl, and n-Octylamines. It was found that the Fe_3O_4-PWA/amine nanocomposite was quite useful for the separation of microplastics. The Fe_3O_4-PWA/nOct composite showed the maximum removal capability of microplastics, which was around polyethylene terephthalate (PET), 99% for polystyrene and polysulphone.

Martin et al. studied the removal of plastic particulate trash from sea and river water by magnetizing them using hydrophobic coatings on iron oxide nanoparticles (IONPs). They prepared and evaluated IONPs, which were coated with various polydimethylsiloxane (PDMS)-based hydrophobic coatings and were synthesized in vacuum circumstances as well as in atmospheric air. They have fabricated amphiphilic co-block polymers together with other PDMS-functionalized hydrophobic IONPs with different PDMS-functionalization. Using a magnet, 100% removal of particles in a size range of 2 to 5 mm and approximately 90% removal of nanoplastic particles in a size range of 100 to 1,000 nm was achieved. A practical technique for separating microplastics from water samples for measurement, characterization, purification, and water remediation is magnetization utilizing IONPs (Martin et al., 2022).

A magnetic, porous nanocomposite with a zeolitic imidazolate framework (ZIF-8) modified with n-butylamine (nano-Fe@ZIF-8) was fabricated in water at room temperature for the removal of microplastics from aquatic environment (Pasanen et al., 2023). The nanocomposite framework was thoroughly characterized before being applied in microplastics removal experiments. The elimination of endocrine-disrupting phenols (bisphenol A and 4-tert-butylphenol) and polystyrene beads (1.1 μm) was made possible by nano-Fe@ZIF-8. In comparison to unmodified Fe@ZIF-8, nano-Fe@ZIF-8 demonstrated higher removal efficiency, and under optimum circumstances, 20 mg of nano-Fe@ZIF-8 was able to remove 98% of polystyrene microbeads within 5 min. It was also able to remove 94% of bisphenol A and 4-tert-butylphenol.

TABLE 15.2

Various Strategies for the Removal of Microplastics from the Aquatic Environment Using Magnetic Nanomaterials

Methods	Materials	Microplastic Type	Removal Efficiency	Reference
Magnetic seperation	Fe nanoparticles	Polyethylene, Polystyrene, Polyurethane, Polyvinyl chloride, Polypropylene	81–93%	(Grbic et al., 2019)
Adsorption	Nano-Fe$_3$O$_4$	Polypropylene, Polystyrene, Polyethylene terephthalate (200–900 µm)	87%, 85%, 86%	(Shi et al., 2022)
Adsorption	Fe$_3$O$_4$ nanoparticles	Polystyrene (1 µm)	93.9%	(Heo et al., 2022)
Adsorption	M–CNTs	Polyethylene, Polyethylene terephthalate, Polyamide	100%	(Tang et al., 2021)
Magnetic seperation	Magnetic Fe$_2$O$_3$/SiO$_2$ core–shell nanoparticles	Polystyrene (1–10 µm)	100%	(Misra et al., 2020)
Magnetic seperation	IONPs (Fe$_3$O$_4$, α-Fe$_2$O$_3$)	Polyethelene (2–5 mm), Polystyrene beads (100 nm–1 µm)	100%, 90%	(Martin et al., 2022)
Magnetic sedimentation	Fe–C–NH$_2$ nanoparticles	Polyethylene (10–200 µm), Polyethylene terephthalate (5–30 µm)	94%	(Bakhteeva et al., 2023)
Adsorption	Fe$_3$O$_4$–PWA/nOct	Polystyrene, Polyethylene terephthalate, Polysulphone	99%	(Bhore and Kamble, 2022)
Adsorption-Photocatalysis	BiVO$_4$/Fe$_3$O$_4$	PLA, PCL	70%	(Beladi-Mousavi et al., 2021)
Membrane filtration	Nano-ferrofluid (magnetite and cobalt ferrite)	Polyethylene, Polyvinyl chloride, Polyester (75, 150, and 300 µm)	85%, 82%, 69%	(Pramanik et al., 2021)

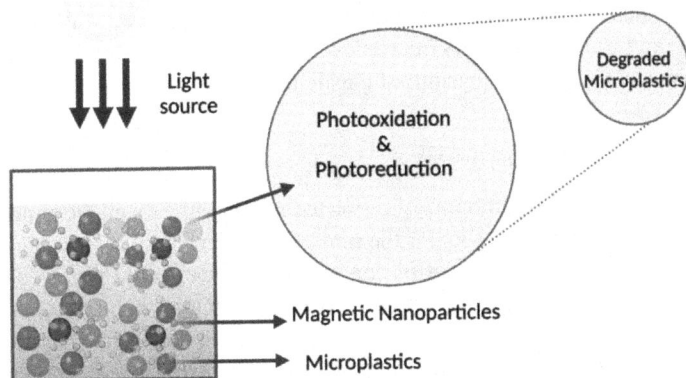

FIGURE 15.3 Schematic illustration of photocatalytic degradation of microplastics.

15.5.2   PHOTOCATALYSIS

Under the influence of visible light or UV light as a source of energy, photocatalysis is an environmentally safe, low-cost, and effective process that can mineralize a broad range of organic pollutants into H_2O, CO_2, and other mineralized products. This method has several benefits, some of which include the usage of sunlight as a source of clean energy, high efficiency in the process of degradation, and the production of harmless by-products. This process can be implemented for the removal of microplastics from the aquatic environment.

Smart, magnetic microrobots were used in a significant proof-of-concept research to collect microplastics and their photocatalytic degradation in an isolated region (Beladi-Mousavi et al., 2021). They were large-scale hydrothermally produced star-shaped $BiVO_4$ microparticles (4–8 m in size) embedded with noble metal-free Fe_3O_4 nanoparticles. A prominent, visible light-activated photocatalyst is $BiVO_4$. Microrobots were able to move autonomously in water while being exposed to visible light with the presence of a little amount of H_2O_2 because of their inherently asymmetrical design. Comparing this outcome to the traditional light-driven microrobots, which need pricey noble metals to enable the self-propulsion capability, shows a significant advancement. Due to the included magnetic nano-particles, $BiVO_4/Fe_3O_4$ microrobots may also be gathered by permanent magnets and readily navigated utilizing external magnetic fields. Such dual-movement was used to target several microplastics in macroscale channels, such as polylactic acid (PLA), polycaprolactone (PCL), polyethylene terephthalate (PET), and polypropylene (PP). It was shown, specifically, that these microrobots' light-induced active mobility caused them to reach the microplastic bits when placed at one end of the channel. The adsorption/precipitation process was used to explain how microrobots attached to the surface of microplastics. Even after two hours of vigorous shaking, the majority of the microrobots remained attached to the plastic pieces, owing to their high stickiness. They entered to the channel and cleaned it by setting a magnet on the other end of it. The removal effectiveness declined with channel length and was greater for hydrophilic microplastics (70% for PLA and PCL) than hydrophobic ones (40% for PET and 20% for PP). A system of five linked channels with varying diameters was used to verify the capture and transport of PLA microplastics, demonstrating their remarkable performance in a complicated labyrinth of plastic trash.

15.5.3　MEMBRANE FILTRATION

The implementation of a membrane filtration technique using magnetic nanomaterials could prove to be an efficient tool for the removal of microplastics from aquatic environments. As part of a recent proof-of-concept study, the effect of water's shear force on the breaking down of larger-size microplastics into smaller microplastic fragments and their removal by nano-ferrofluid (i.e., magnetite and cobalt ferrite particles as a coagulant) through membrane filtration processes was investigated (Pramanik et al., 2021). It was observed that a mechanical impeller with two blades could fragment microplastics with sizes ranging from 75 μm to 300 μm into mean sizes of 0.7, 1.1, and 1.8 μm, respectively. The findings indicated that the air flotation technique was capable of achieving a maximum removal efficiency of 89% for polyethylene, 82% for polyvinyl chloride, and 69% for polyester, respectively. The effectiveness of the air flotation procedure for the removal of microplastics was significantly increased by increasing the dosage of surfactant from 2 to 10 mg/L. Additionally, it has been shown that the removal effectiveness of microplastics by the air flotation system is dependent on the pH of the solution, the size, and the different types of microplastics. Two types of ferrofluid employed in this research were magnetite and cobalt ferrite. The average removal efficiency of magnetite was 43%, while the average removal efficiency of cobalt ferrite was 55%. This study also demonstrated a less significant removal efficiency of microplastics by both types of ferrofluid. The nano-ferrofluid particles were effective at removing all three types of microplastics from water that were evaluated, and with showing similar removal efficiency, it can be concluded that the effectiveness of this removal method is not contingent on the plastic component type. The membrane procedures of ultrafiltration and microfiltration were found to be very successful, eliminating more than 90% of the microplastic fragment particles. These techniques were among those that were examined. This research has shown that it is possible to effectively remove microplastics from wastewater by combining the processes of air flotation with membrane filtration.

15.6　REGENERATION AND REUSE OF MAGNETIC NANOMATERIALS

The ability of magnetic nanomaterials regeneration in water purification to manage the economics of various water treatment technologies is among the most essential components of this process. pH-dependent solvents are critical for nanoparticle regeneration. In the treatment system, this may be performed by the use of a separation device or through the immobilization of nanomaterials. The currently available techniques of immobilization have not proven to be significantly efficient. Simple and inexpensive techniques must be developed in order to immobilize nanoparticles without affecting their functionality. In recent years, a simple desorption procedure was used to remove Fe_3O_4 particles from Fe_3O_4-polystyrene complexes (Heo et al., 2022). Iron oxide particles were forced out of the Fe_3O_4-PS complexes by using an ultrasonic bath, and this process was made more effective by treating the complexes with a 5% NaCl solution rather than DI water. Additionally, when the quantity of desorption cycles grew, so did the desorption efficiency. The effectiveness of desorption rose to 85% after three cycles. By releasing them from the Fe_3O_4-PS complexes, the separated PS microplastics may be concentrated, and the gathered PS particles can be disposed of using other processes, such as heat treatment. Reusing the separated iron oxide particles to adsorb PS particles resulted in an 83% adsorption effectiveness. The result suggests that iron oxide particles may be utilized frequently as an adsorbent to remove PS microparticles.

Thus, magnetic nanomaterials are an effective material for microplastics remediation, since they can be recycled and used again. Thus, recyclable properties are an attractive feature for the wastewater treatment industry.

15.7   LIMITATIONS, CHALLENGES, AND FUTURE PROSPECTS

Microplastic removal has become highly essential, as reports of its prevalence in water systems throughout the world are constantly increasing. While legislation and regulations are developed to improve microplastic waste management, the rapid impact of water and wastewater treatments is lowering the amount of microplastics from different water sources. The nanotechnology-based water treatments for the removal of microplastics are gaining massive interest worldwide. The efficacy of nanomaterials in removing microplastics is reported to be far more efficient than other conventional bulk materials. Moreover, nanomaterials are found to lower the operational cost at large scale according to many research studies.

For instance, nanomaterial-based adsorbents can be recycled better than their bulk forms. This means that the material's functional lifespan can be prolonged. The superparamagnetic nanomaterials are also known for their easy recovery after the treatment of microplastics in water. Despite the advantages, a nanomaterial-based approach for microplastic removal is still in the preliminary stage of research. In addition, industries are also figuring out the best way to use nanomaterials in their current systems, making the replacement of traditional materials or processes at a broader scale.

Most of the investigations reporting the removal of microplastic have been carried out using artificial wastewater. Even though these experiments showed excellent microplastics removal efficiency, the understanding of the interaction between the macroplastic particles and nanomaterials and other important aspects related to actual working have been skipped. The stability of the nanomaterials, their behaviour, and interactions with microplastics is expected to drastically change in salinity and pH of real water systems. For example, adsorption of microplastic using nano-adsorbent might get affected by competitive adsorption of several kinds of plastics present in the water body. Nano-adsorbents might be inhibited from adsorbing microplastics if small, suspended particles present in waterbodies occupy their pores. Natural organic matters and bacteria present as pollutants can foul the membrane and can severely impair the membrane's filtration performance. Additionally, the separation process is primarily dependent on the interaction between magnetic nanomaterials and microplastics, so the presence of multiple types of microplastics in water bodies that differ in the polymer composition and size of the microplastics may result in discrepancies in how efficiently the nanomaterials are removed by them. Consequently, further experiments in complex, realistic water environments are needed to establish an accurate comparison of microplastic removal efficiency by nanomaterials.

Better methods for analyzing and minimizing the source of microplastics should be the focus of future research into microplastics mitigation. Accurately identifying microplastics requires improvements to infrastructure, product design, organization, and sequencing. Regulating plastics and methods for microplastics removal from water need constant and persistent study. Large-scale unit and individual-level education on waste management and health risks associated with plastics are required to eradicate the issues and establish a sustainable environment.

15.8 CONCLUSIONS

This study provides a summary of the risks posed by microplastics and the methods using magnetic nanomaterials in use to mitigate those risks. Microplastics may have adverse impacts on living things when inhaled or ingested, and their full removal is challenging due to their tiny sizes. Magnetic nanoparticles are a key class of materials employed in the effective removal of microplastics from aquatic environment. The novel adsorbents with photocatalysis may have more advantages than the conventional process in the removal of microplastics, but the influence of the size of microplastics on the process is still unclear, and most of the studies have focused on microplastics with size > 20 µm, but smaller microplastics are also frequently found in the aquatic environment. In addition, the investigation of the regeneration and reuse of magnetic nanomaterials needs to be performed efficiently.

REFERENCES

Bakhteeva, I. A. *et al.* (2023) 'Removal of microplastics from water by using magnetic sedimentation', *International Journal of Environmental Science and Technology*. doi: 10.1007/s13762-023-04776-1.

Beladi-Mousavi, S. M. *et al.* (2021) 'A maze in plastic wastes: Autonomous motile photocatalytic microrobots against microplastics', *ACS Applied Materials & Interfaces*. American Chemical Society, 13(21), pp. 25102–25110. doi: 10.1021/acsami.1c04559.

Bhatt, P. *et al.* (2021) 'Microplastic contaminants in the aqueous environment, fate, toxicity consequences, and remediation strategies', *Environmental Research*. Elsevier, 200, p. 111762.

Bhore, R. K., and Kamble, S. B. (2022) 'Nano adsorptive extraction of diverse microplastics from the potable and seawater using organo-polyoxometalate magnetic nanotricomposites', *Journal of Environmental Chemical Engineering*, 10(6), p. 108720. https://doi.org/10.1016/j.jece.2022.108720.

Bui, X.-T. *et al.* (2020) 'Microplastics pollution in wastewater: Characteristics, occurrence and removal technologies', *Environmental Technology & Innovation*, 19, p. 101013. https://doi.org/10.1016/j.eti.2020.101013.

Cherniak, S. L. *et al.* (2022) 'Conventional and biological treatment for the removal of microplastics from drinking water', *Chemosphere*, 288, p. 132587. https://doi.org/10.1016/j.chemosphere.2021.132587.

Das, S., Mondal, U. S., and Paul, S. (2022) '21—Nanophytoremediation technology: A better approach for environmental remediation of toxic metals and dyes from water', in Kumar, V., Shah, M. P., Shahi, S. K. B. T.-P. T. for the R. of H. M., and O. C. from S. W. (eds). Elsevier, pp. 459–481. https://doi.org/10.1016/B978-0-323-85763-5.00002-7.

Das, S., Mondal, U. S., and Paul, S. (2023) '1 Emerging nanotechnologies for detection and removal', in *Emerging Technologies in Wastewater Treatment*. CRC Press, p. 1.

Das, S., Somu, P., and Paul, S. (2021) 'Visible light induced efficient photocatalytic degradation of azo dye into nontoxic byproducts by CdSe quantum dot conjugated nano graphene oxide', *Journal of Molecular Liquids*, 340, p. 117055. https://doi.org/10.1016/j.molliq.2021.117055.

Elgarahy, A. M., Akhdhar, A., and Elwakeel, K. Z. (2021) 'Microplastics prevalence, interactions, and remediation in the aquatic environment: A critical review', *Journal of Environmental Chemical Engineering*, 9(5), p. 106224. https://doi.org/10.1016/j.jece.2021.106224.

Goh, P. S. *et al.* (2022) 'Nanomaterials for microplastic remediation from aquatic environment: Why nano matters?' *Chemosphere*, 299, p. 134418. https://doi.org/10.1016/j.chemosphere.2022.134418.

Grbic, J. *et al.* (2019) 'Magnetic extraction of microplastics from environmental samples', *Environmental Science & Technology Letters*. American Chemical Society, 6(2), pp. 68–72. doi: 10.1021/acs.estlett.8b00671.

Heo, Y., Lee, E.-H., and Lee, S.-W. (2022) 'Adsorptive removal of micron-sized polystyrene particles using magnetic iron oxide nanoparticles', *Chemosphere*, 307, p. 135672. https://doi.org/10.1016/j.chemosphere.2022.135672.

Li, C. *et al.* (2021a) 'Preliminary study on low-density polystyrene microplastics bead removal from drinking water by coagulation-flocculation and sedimentation', *Journal of Water Process Engineering*, 44, p. 102346. https://doi.org/10.1016/j.jwpe.2021.102346.

Li, Y. *et al.* (2021b) 'Research on the influence of microplastics on marine life', *IOP Conference Series: Earth and Environmental Science*. IOP Publishing, 631(1), p. 12006. doi: 10.1088/1755-1315/631/1/012006.

Ma, B. *et al.* (2019) 'Characteristics of microplastic removal via coagulation and ultrafiltration during drinking water treatment', *Chemical Engineering Journal*, 359, pp. 159–167. https://doi.org/10.1016/j.cej.2018.11.155.

Martin, L. M. A. *et al.* (2022) 'Testing an iron oxide nanoparticle-based method for magnetic separation of nanoplastics and microplastics from water', *Nanomaterials*. doi: 10.3390/nano12142348.

Misra, A. *et al.* (2020) 'Water purification and microplastics removal using magnetic polyoxometalate-supported ionic liquid phases (magPOM-SILPs)', *Angewandte Chemie International Edition*. John Wiley & Sons, Ltd, 59(4), pp. 1601–1605. https://doi.org/10.1002/anie.201912111.

Mondal, U. S. *et al.* (2023) 'Silica sand–supported nano zinc oxide–graphene oxide composite induced rapid photocatalytic decolorization of azo dyes under sunlight and improved antimicrobial activity', *Environmental Science and Pollution Research*, 30(7), pp. 17226–17244. doi: 10.1007/s11356-022-23248-6.

Ouda, M. *et al.* (2023) 'Recent advances on nanotechnology-driven strategies for remediation of microplastics and nanoplastics from aqueous environments', *Journal of Water Process Engineering*, 52, p. 103543. https://doi.org/10.1016/j.jwpe.2023.103543.

Pasanen, F., Fuller, R. O., and Maya, F. (2023) 'Fast and simultaneous removal of microplastics and plastic-derived endocrine disruptors using a magnetic ZIF-8 nanocomposite', *Chemical Engineering Journal*, 455, p. 140405. https://doi.org/10.1016/j.cej.2022.140405.

Pramanik, B. K., Pramanik, S. K., and Monira, S. (2021) 'Understanding the fragmentation of microplastics into nano-plastics and removal of nano/microplastics from wastewater using membrane, air flotation and nano-ferrofluid processes', *Chemosphere*, 282, p. 131053. https://doi.org/10.1016/j.chemosphere.2021.131053.

Prokopova, M. *et al.* (2021) 'Coagulation of polyvinyl chloride microplastics by ferric and aluminium sulphate: Optimisation of reaction conditions and removal mechanisms', *Journal of Environmental Chemical Engineering*, 9(6), p. 106465. https://doi.org/10.1016/j.jece.2021.106465.

Shen, M. *et al.* (2020) 'Removal of microplastics via drinking water treatment: Current knowledge and future directions', *Chemosphere*. Elsevier, 251, p. 126612.

Shi, X. *et al.* (2022) 'Removal of microplastics from water by magnetic nano-Fe_3O_4', *Science of the Total Environment*, 802, p. 149838. https://doi.org/10.1016/j.scitotenv.2021.149838.

Smith, M. *et al.* (2018) 'Microplastics in seafood and the implications for human health', *Current Environmental Health Reports*, 5(3), pp. 375–386. doi: 10.1007/s40572-018-0206-z.

Tang, Y. *et al.* (2021) 'Removal of microplastics from aqueous solutions by magnetic carbon nanotubes', *Chemical Engineering Journal*, 406, p. 126804. https://doi.org/10.1016/j.cej.2020.126804.

Wang, F. *et al.* (2020) 'Occurrence and distribution of microplastics in domestic, industrial, agricultural and aquacultural wastewater sources: A case study in Changzhou, China', *Water Research*, 182, p. 115956. https://doi.org/10.1016/j.watres.2020.115956.

Wang, Z., Sedighi, M., and Lea-Langton, A. (2020) 'Filtration of microplastic spheres by biochar: Removal efficiency and immobilisation mechanisms', *Water Research*, 184, p. 116165. https://doi.org/10.1016/j.watres.2020.116165.

Xue, J. *et al.* (2021) 'Removal of polystyrene microplastic spheres by alum-based coagulation-flocculation-sedimentation (CFS) treatment of surface waters', *Chemical Engineering Journal*, 422, p. 130023. https://doi.org/10.1016/j.cej.2021.130023.

Zandieh, M., and Liu, J. (2022) 'Removal and degradation of microplastics using the magnetic and nanozyme activities of bare iron oxide nanoaggregates', *Angewandte Chemie International Edition*. John Wiley & Sons, Ltd, 61(47), p. e202212013. https://doi.org/10.1002/anie.202212013.

Zhou, G. *et al.* (2021) 'Removal of polystyrene and polyethylene microplastics using PAC and $FeCl_3$ coagulation: Performance and mechanism', *Science of the Total Environment*, 752, p. 141837. https://doi.org/10.1016/j.scitotenv.2020.141837.

Index

For Product Safety Concerns and Information please contact our EU
representative GPSR@taylorandfrancis.com
Taylor & Francis Verlag GmbH, Kaufingerstraße 24, 80331 München, Germany